INTRODUCTION TO
COMPUTATIONAL SCIENCE

INTRODUCTION TO COMPUTATIONAL SCIENCE

MODELING AND SIMULATION FOR THE SCIENCES

Angela B. Shiflet and George W. Shiflet

PRINCETON UNIVERSITY PRESS

PRINCETON AND OXFORD

© 2006 by Princeton University Press
Published by Princeton University Press, 41 William Street, Princeton,
New Jersey 08540
In the United Kingdom: Princeton University Press, 3 Market Place, Woodstock,
Oxfordshire OX20 1SY

Library of Congress Cataloging-in-Publication Data

Shiflet, Angela B.
Introduction to computational science : modeling and simulation for the sciences /
Angela B. Shiflet and George W. Shiflet.
    p.   cm.
Includes bibliographical references and index.
ISBN-13: 978-0-691-12565-7 (cloth : alk. paper)
ISBN-10: 0-691-12565-1 (cloth : alk. paper)
1. Computer science.   2. Computer simulation.   3. Mathematical models.
4. Computational complexity.   I. Shiflet, George W., 1947–   II. Title.
QA76.6.S54143 2006
004—dc22   2005029492

British Library Cataloging-in-Publication Data is available

This book has been composed in Times Roman
Printed on acid-free paper. ∞
pup.princeton.edu

Printed in the United States of America
1 3 5 7 9 10 8 6 4 2

Dedicated to

Robert K. Cralle, Theodore H. Einwohner, and George A. Michael
Lawrence Livermore National Laboratory
and
Robert M. Panoff
The Shodor Education Foundation,
whose friendship and guidance we have treasured

CONTENTS

Preface   xix

1 OVERVIEW

Module 1.1 Overview of Computational Science   3
    Projects   5
    References   5

Module 1.2 The Modeling Process   6
    Introduction   6
    Model Classifications   7
    Steps of the Modeling Process   8
    Exercises   11
    References   11

2 FUNDAMENTAL CONSIDERATIONS

Module 2.1 Computational Toolbox—Tools of the Trade:
Tutorial 1   15
    Download   15
    Introduction   16

Module 2.2 Errors   17
    Introduction   17
    Data Errors   17
    Modeling Errors   17
    Implementation Errors   18
    Precision   18
    Absolute and Relative Errors   19
    Round-off Error   21
    Overflow and Underflow   22
    Arithmetic Errors   23
    Error Propagation   24
    Violation of Numeric Properties   27
    Comparison of Floating Point Numbers   27
    Truncation Error   29
    Exercises   31
    Projects   32
    Answers to Quick Review Questions   34
    References   35

Module 2.3 Rate of Change   36
    Introduction   36
    Velocity   36
    Derivative   41
    Slope of Tangent Line   42

Differential Equations 47
Second Derivative 48
Exercises 49
Project 51
Answers to Quick Review Questions 51
Reference 52

Module 2.4 Fundamental Concepts of Integral Calculus 53
Introduction 53
Total Distance Traveled and Area 53
Definite Integral 60
Total Change 61
Fundamental Theorem of Calculus 62
Differential Equations Revisited 64
Exercises 64
Project 66
Answers to Quick Review Questions 66
References 67

**3 SYSTEM DYNAMICS PROBLEMS WITH
RATE PROPORTIONAL TO AMOUNT**

Module 3.1 System Dynamics Tool: Tutorial 1 71
Download 71
Introduction 71

Module 3.2 Unconstrained Growth and Decay 73
Introduction 73
Differential Equation 73
Difference Equation 74
Simulation Program 78
Analytical Solution Introduction 79
Analytical Solution: Explanation with Indefinite Integrals 79
Analytical Solution: Explanation without Indefinite Integrals 80
Completion of Analytical Solution 80
Further Refinement 82
Unconstrained Decay 82
Exercises 84
Projects 85
Answers to Quick Review Questions 86
Reference 86

Module 3.3 Constrained Growth 87
Introduction 87
Carrying Capacity 87
Revised Model 89
Equilibrium and Stability 91
Exercises 92
Projects 93
Answers to Quick Review Questions 95
References 96

Contents

Module 3.4 System Dynamics Tool: Tutorial 2   97
   Download   97
   Introduction   97

Module 3.5 Drug Dosage   98
   Downloads   98
   Introduction   98
   One-Compartment Model of Single Dose   99
   One-Compartment Model of Repeated Doses   101
   Mathematics of Repeated Doses   103
   Sum of Finite Geometric Series   106
   Two-Compartment Model   106
   Exercises   107
   Projects   108
   Answers to Quick Review Questions   109
   References   110

4 FORCE AND MOTION

Module 4.1 Modeling Falling and Skydiving   113
   Downloads   113
   Introduction   113
   Acceleration, Velocity, and Position   114
   Physics Background   117
   Friction During Fall   120
   Modeling a Skydive   122
   Assessment of the Skydive Model   124
   Exercises   125
   Projects   125
   Answers to Quick Review Questions   127
   References   128

Module 4.2 Modeling Bungee Jumping   129
   Downloads   129
   Introduction   129
   Physics Background   130
   Vertical Springs   132
   Modeling a Bungee Jump   135
   Exercises   137
   Projects   137
   Answers to Quick Review Questions   138
   References   139

Module 4.3 Tick Tock—The Pendulum Clock   140
   Download   140
   Introduction   140
   Simple Pendulum   141
   Linear Damping   144
   Pendulum Clock   144
   Exercises   145
   Projects   146

Answers to Quick Review Questions   147
References   147

Module 4.4 Up, Up, and Away—Rocket Motion   149
Download   149
Introduction   149
Physics Background   150
System Dynamics Model   152
Exercises   154
Projects   155
Answers to Quick Review Questions   157
References   157

5 SIMULATION TECHNIQUES

Module 5.1 Computational Toolbox—Tools of the Trade: Tutorial 2   161
Download   161
Introduction   161

Module 5.2 Euler's Method   162
Download   162
Introduction   162
Reasoning behind Euler's Method   162
Algorithm for Euler's Method   164
Error   165
Exercises   167
Projects   167
Answers to Quick Review Questions   168
References   169

Module 5.3 Runge-Kutta 2 Method   170
Introduction   170
Euler's Estimate as a Predictor   170
Corrector   170
Runge-Kutta 2 Algorithm   173
Error   174
Exercises   175
Projects   175
Answers to Quick Review Questions   175
References   175

Module 5.4 Runge-Kutta 4 Method   176
Introduction   176
First Estimate ∂_1 Using Euler's Method   176
Second Estimate ∂_2 177
Third Estimate ∂_3 179
Fourth Estimate ∂_4 181
Using the Four Estimates   183
Runge-Kutta 4 Algorithm   184
Error   185
Exercises   186
Projects   186

Answers to Quick Review Questions   186
References   187

6 SYSTEM DYNAMICS MODELS WITH INTERACTIONS

Module 6.1 Competition   191
Download   191
Community Relations   191
Competition Introduction   191
Modeling Competition   192
Exercises   195
Projects   195
Answers to Quick Review Questions   197
References   197

Module 6.2 Spread of SARS   198
Downloads   198
Introduction   198
SIR Model   199
SARS Model   202
Reproductive Number   207
Exercises   208
Projects   208
Answers to Quick Review Questions   210
References   211

Module 6.3 Enzyme Kinetics   213
Download   213
Introduction   213
Michaelis-Menten Equation   214
Differential Equations   217
Model   218
Exercises   219
Projects   221
Answers to Quick Review Questions   222
References   223

Module 6.4 Predator-Prey Model   224
Download   224
Introduction   224
Lotka-Volterra Model   225
Particular Situations   227
Exercises   230
Projects   231
Answers to Quick Review Questions   235
References   235

Module 6.5 Modeling Malaria   237
Download   237
Introduction   237
Background Information   238
Analysis of Problem   238

Formulating a Model: Gather Data   239
Formulating a Model: Make Simplifying Assumptions   240
Formulating a Model: Determine Variables and Units   241
Formulating a Model: Establish Relationships   242
Formulating a Model: Determine Equations and Functions   243
Solving the Model   244
Verifying and Interpreting the Model's Solution   247
Exercises   249
Projects   249
Answers to Quick Review Questions   251
References   251

7 ADDITIONAL DYNAMIC SYSTEMS PROJECTS
Overview   253

Module 7.1 Radioactive Chains—Never the Same Again   255
Introduction   255
Modeling the Radioactive Chain   255
Projects   257
Answers to Quick Review Question   258
Reference   258

Module 7.2 Turnover and Turmoil—Blood Cell Populations   259
Introduction   259
Formation and Destruction of Blood Cells   259
Basic Model   260
Model Parameters   260
Projects   262
Answers to Quick Review Questions   263
References   264

Module 7.3 Deep Trouble—Ideal Gas Laws and Scuba Diving   265
Pressure   265
Ideal Gas   266
Dalton's Law   266
Boyle's Law   267
Charles' Law   268
Henry's Law   269
Rate of Absorption   270
Decompression Sickness   271
Projects   271
Answers to Quick Review Questions   272
References   273

Module 7.4 What Goes Around Comes Around—
The Carbon Cycle   274
Introduction   274
Flow between Subsystems   274
Fossil Fuels   275
Projects   276
References   276

Contents

Module 7.5 A Heated Debate—Global Warming | 278
Greenhouse Effect | 278
Global Warming | 279
Greenhouse Gases | 279
Consequences | 279
Projects | 280
References | 281

Module 7.6 Cardiovascular System—A Pressure-Filled Model | 283
Circulation | 283
Blood Pressure | 284
Heart Rate | 284
Stroke Volume | 285
Venous Return | 285
Systemic Vascular Resistance | 285
Blood Flow | 285
Projects | 286
References | 287

Module 7.7 Electrical Circuits—A Complete Story | 288
Defibrillators | 288
Current and Potential | 288
Resistance | 290
Capacitance | 291
Inductance | 292
Circuit for Defibrillator | 292
Kirchhoff 's Voltage Law | 293
Kirchhoff 's Current Law | 295
Projects | 296
Answers to Quick Review Questions | 297
References | 297

Module 7.8 Fueling Our Cells—Carbohydrate Metabolism | 299
Glycolysis | 299
Recycling NAD^+'s | 300
Aerobic Respiration | 301
Projects | 301
References | 302

Module 7.9 Mercury Pollution—Getting on Our Nerves | 303
Introduction | 303
Projects | 304
References | 307

Module 7.10 Managing to Eat—What's the Catch? | 308
Introduction | 308
Economics Background | 309
Gordon-Schaefer Fishery Production Function | 314
Projects | 314
Answers to Quick Review Questions | 316
References | 316

8 DATA-DRIVEN MODELS

Module 8.1 Computational Toolbox—Tools of the
Trade: Tutorial 3   321
    Download   321
    Introduction   321

Module 8.2 Function Tutorial   322
    Download   322
    Introduction   322
    Linear Function   323
    Quadratic Function   324
    Polynomial Function   325
    Square Root Function   326
    Exponential Function   327
    Logarithmic Functions   328
    Logistic Function   330
    Trigonometric Functions   331

Module 8.3 Empirical Models   335
    Downloads   335
    Introduction   336
    Linear Empirical Model   336
    Predictions   338
    Linear Regression   339
    Nonlinear One-Term Model   340
    Solving for y in a One-Term Model   346
    Multiterm Models   349
    Exercises   351
    Projects   351
    Answers to Quick Review Questions   351
    References   352

9 MONTE CARLO SIMULATIONS

Module 9.1 Computational Toolbox—Tools of the
Trade: Tutorial 4   357
    Download   357
    Introduction   357

Module 9.2 Simulations   358
    Introduction   358
    Element of Chance   359
    Disadvantages   359
    Genesis of Monte Carlo Simulations   359
    Multiplicative Linear Congruential Method   360
    Different Ranges of Random Numbers   361
    Exercises   364
    Projects   365
    Answers to Quick Review Questions   366
    References   366

Module 9.3 Area Through Monte Carlo Simulation   367
   Download   367
   Introduction   367
   Throwing Darts for Area   368
   Measure of Quality   370
   Algorithm   371
   Implementation   371
   Exercises   371
   Projects   372
   Answers to Quick Review Questions   373
   Reference   373

Module 9.4 Random Numbers from Various Distributions   374
   Downloads   374
   Introduction   374
   Statistical Distributions   374
   Discrete Distributions   377
   Normal Distributions   380
   Exponential Distributions   382
   Rejection Method   384
   Exercises   385
   Projects   387
   Answers to Quick Review Questions   387
   References   388

10 RANDOM WALK SIMULATIONS

Module 10.1 Computational Toolbox—Tools of the Trade: Tutorial 5   391
   Download   391
   Introduction   391

Module 10.2 Random Walk   392
   Downloads   392
   Introduction   392
   Algorithm for Random Walk   393
   Animate Path   395
   Average Distance Covered   398
   Relationship between Number of Steps and Distance Covered   400
   Exercises   400
   Projects   401
   Answers to Quick Review Questions   402
   References   402

11 DIFFUSION

Module 11.1 Computational Toolbox—Tools of the Trade: Tutorial 6   405
   Download   405
   Introduction   405

Module 11.2 Spreading of Fire   406
   Downloads   406
   Introduction   406

Initializing the System   407
Updating Rules   408
Periodic Boundary Conditions   411
Applying a Function to Each Grid Point   414
Simulation Program   416
Display Simulation   417
Exercises   417
Projects   419
Answers to Quick Review Questions   420
References   420

Module 11.3 Movement of Ants   422
Downloads   422
Introduction   422
Analysis of Problem   423
Formulating a Model: Gather Data   423
Formulating a Model: Make Simplifying Assumptions   424
Formulating a Model: Determine Variables   424
Formulating a Model: Establish Relationships and Submodels   424
Formulating a Model: Determine Functions—Sensing   425
Formulating a Model: Determine Functions—Walking without
    Concern for Collision   425
Formulating a Model: Determine Functions—Walking with
    Concern for Collision   426
Solving the Model—A Simulation   428
Verifying and Interpreting the Model's Solution—Visualizing
    the Simulation   429
Exercises   429
Projects   431
Answers to Quick Review Questions   434
References   434

12 HIGH PERFORMANCE COMPUTING

Module 12.1 Concurrent Processing   437
Introduction   437
Analogy   439
Types of Processing   440
Multiprocessor   441
Classification of Computer Architectures   443
Metrics   443
Exercises   446
Project   446
Answers to Quick Review Questions   446
References   447

Module 12.2 Parallel Algorithms   448
Introduction   448
Embarrassingly Parallel Algorithm: Adding Two Vectors   448
Data Partitioning: Adding Numbers   449
Divide and Conquer: Adding Numbers   452

Parallel Random Number Generator   455
Sequential Algorithm for *N*-Body Problem   457
Barnes-Hut Algorithm for *N*-Body Problem   462
Exercises   465
Projects   467
Answers to Quick Review Questions   468
References   470

13 ADDITIONAL CELLULAR AUTOMATA PROJECTS

Overview   471

Module 13.1 Polymers—Strings of Pearls   473
Introduction   473
Simulations   475
Projects   476
References   477

Module 13.2 Solidification—Let's Make It Crystal Clear!   479
Introduction   479
Projects   480
References   482

Module 13.3 Foraging—Finding a Way to Eat   483
Introduction   483
Simulations   484
Projects   485
References   488

Module 13.4 Pit Vipers—Hot Bodies, Dead Meat   489
Introduction   489
Simulations of Heat Diffusion   489
Projects   490
References   491

Module 13.5 Mushroom Fairy Rings—Just Going in Circles   492
Introduction   492
What Are Fungi?   493
What Do Fungi Look Like?   493
How Do Fungi "Feed Themselves"?   494
How Do Fungi Reproduce?   494
How Do Fungi Grow?   494
The Problem   494
How Do Fairy Rings Get Started?   495
Initializing the System   495
Updating Rules   497
Displaying the Simulation   498
Projects   498
References   499

Module 13.6 Spread of Disease—"Gesundheit!"   501
Introduction   501
Exercise   501
Projects   501

Module 13.7 HIV—The Enemy Within   504
   The Developing Epidemic   504
   Attack on the Immune System   505
   Plan of Attack   506
   Simulation of the Attack   507
   Projects   507
   References   508

Module 13.8 Predator-Prey—"Catch Me If You Can"   510
   Introduction   510
   Projects   510
   References   514

Module 13.9 Clouds—Bringing It All Together   516
   Introduction   516
   Projects   517
   References   520

Module 13.10 Fish Schooling—Hanging Together, not Separately 521
   Introduction   521
   Simulations   522
   Projects   522
   References   523

Glossary of Terms   525
Answers to Selected Exercises   543
Index   547

PREFACE

Overview

Many significant applied and basic research questions in science today are interdisciplinary in nature, involving physical and/or biological sciences, mathematics, and computer science in an area called **computational science**. Frequently, a research project has a team of professionals from a variety of fields. The ability to understand various perspectives and perform interdisciplinary work can aid communication and speed the progress of a project. Moreover, the use of computers has become an essential ingredient to many of such projects.

Much scientific investigation now involves computing as well as theory and experiment. Computing can often stimulate the insight and understanding that theory and experiment alone cannot achieve. With computers, scientists can study problems that previously would have been too difficult, time consuming, or hazardous; and, virtually instantaneously, they can share their data and results with scientists around the world.

The increasing speed and memory of computers, the emergence of distributed processing, the explosion of information available through the World Wide Web, the maturing of the area of scientific visualization, and the availability of reasonably priced computational tools all contribute to the increasing importance of computation to scientists and of computational science in education.

Introduction to Computational Science: Modeling and Simulation for the Sciences prepares the student to understand and utilize fundamental concepts of computational science, the modeling process, computer simulations, and scientific applications. The text considers **two major approaches to computational science problems: system dynamics models and cellular automaton simulations**. System dynamics models provide global views of major systems that change with time. For example, one such model considers changes over time in the numbers of predators and prey, such as hawks and squirrels. To model such dynamic systems, students using the text can employ any one of several tools, such as *STELLA®*, *Vensim®* Personal Learning Edition (PLE) (free for personal and educational use), and *Berkeley Madonna®*. With the tool, the student can create pictorial representations of models, develop relationships, run simulations, and generate graphs of the results.

In contrast to system dynamics, cellular automaton simulations provide local views of individuals affecting individuals. The world under consideration consists of a rectangular grid of cells, and each cell has a state that can change with time according to rules. For example, the state of one cell could represent a squirrel and the state of an adjacent cell could correspond to a hawk. One rule could be that, when adjacent, a hawk gets a squirrel with a probability of 25%. Thus, on the average at the next time step, a 25% chance exists that the particular squirrel will be no more. The text employs a generic approach for cellular automaton simulations and scientific visualizations of the results, so that students can employ any one of a variety of computational tools, such as *Maple®*, *Mathematica®*, *MATLAB®*, and *Excel®*.

Tutorials, package-specific Quick Review Questions and answers, and files to accompany the text material are **available from the text's website in various system dynamics tools**—such as *STELLA*, *Vensim* Personal Learning Edition (PLE), and *Berkeley Madonna*—**and in several computational tools**—such as *Maple*, *Mathematica*, *MATLAB*, and *Excel*. Typically, an instructor picks one system dynamics tool and one computational tool for class use during the term.

Prerequisites

Prerequisites for *Introduction to Computational Science* are minimal. While including projects for students who have had programming, the text **does *not* require computer programming experience**. The concept of rate of change, or derivative, from a first course in calculus is used throughout the text. For the student who has not had Calculus I or who would like a review of the material, the text provides **two modules on fundamental calculus concepts**. One of these modules, "Rate of Change," is important for understanding the remainder of the text, while the other module, "Fundamental Concepts of Integral Calculus," is optional. The required calculus background is minimal, and students do not need to know derivative formulas to understand the material or develop the models.

Learning Features

One of the positive aspects and challenges of computational science is its interdisciplinary nature. This challenge is particularly acute with students who have not had extensive experience in computer science, mathematics, and all areas of the sciences. Thus, the text provides the background that is necessary for the student to understand the material and confidently succeed in the course. Each module involving a scientific application **covers the prerequisite science without overwhelming the reader** with excessive detail. The **numerous application areas** for examples, exercises, and projects include astronomy, biology, chemistry, economics, engineering, finance, geology, medicine, physics, and psychology.

Introduction to Computational Science has chapters consisting of several modules. The text's website contains **two tutorials on system dynamics tools**—such as *STELLA*, *Vensim*, and *Berkeley Madonna*—**and six tutorials on computational tools**—such as *Maple*, *Mathematica*, *MATLAB*, and *Excel*. The text presents the tutorials in a just-in-time fashion, covering the features needed in the immediately subsequent material.

Module 1.2 introduces the **modeling process**, and the text consistently uses the process to guide the reader through numerous scientific examples. For instance, after covering the prerequisite scientific background, Module 6.5 develops a model of malaria by following the modeling process in a step-by-step fashion. Thus, the text helps students to learn how modelers model.

The text presents material in a clear manner with ample use of examples and figures. Most sections of a module end with **Quick Review Questions** that provide fast checks of the student's comprehension of the material. **Answers**, often with explanations, at the end of the module give immediate feedback and reinforcement to the student. In the case of system dynamics or computational tool-dependent questions, the questions and answers are on the text's website in pdf files for several tools, such as *STELLA*, *Vensim*, and *Berkeley Madonna* in the former case and *Maple*, *Mathematica*, *MATLAB*, and *Excel* in the latter case.

To further aid in understanding the material, most modules include a number of **exercises** that correlate directly to the material and that the student usually is to complete with pencil and paper. Answers to selected problems, whose exercise numbers are in color, appear in an appendix.

A subsequent "**Projects**" section provides numerous project assignments for students to develop individually or in teams. While a module, such as "Modeling Malaria," might develop one model for an application area, the projects section suggests many other refinements, approaches, and applications. The ability to work well with an interdisciplinary team is important for a computational scientist. **Chapters 7 and 13 provide modules of additional, substantial projects** from a variety of scientific areas that are particularly appropriate for teams of students. These modules indicate prerequisite text material, and earlier modules forward reference appropriate projects from Chapters 7 and 13.

Another section on "**References**," which occurs at the end of most modules, provides a list of hyperlinks, books, and articles for further study.

Appendixes include a **glossary** of scientific, modeling, and simulation terms for quick reference. The text's website provides links to **downloadable tutorials, models, pdf files, and datasets** for various tool-dependent quick review questions and answers, examples, and projects.

The Material

Because the area is emerging, a variety of departments offer introductory computational science courses; and professors approach the material in diverse ways. Thus, *Introduction to Computational Science* provides **several pathways through the material**, and the text's website suggests various alternatives. Moreover, the text provides an abundance of discipline-specific applications so that the text is suitable either for an **introductory course generally in computational science or**, with appropriate selection of applications, **specifically in computational biology**.

The text begins with an introduction to computational science and the modeling process. With computational estimates, the modeler should always be aware of sources of computational error. Thus, after a beginning tutorial on a tool we can use for computation and cellular automaton simulations, such as *Maple*, *Mathematica*, *MATLAB*, or *Excel* (tutorial versions on the text's website), Chapter 2 on "Fundamental Considerations" contains a module on "Errors." Another module in the chapter, which is on "Rate of Change," covers all the calculus required to study the material in the text.

Chapter 3 commences the discussion of system dynamics and models where the rate of change of the quantity is proportional to the quantity. Two tutorials available in a choice of several tools (such as *STELLA*, *Vensim*, and *Berkeley Madonna*) lead the student step by step through the process of implementing a model with the software. "Unconstrained Growth and Decay" discusses models that exhibit exponential growth or decay and introduces concepts of time-driven simulations. The module also develops the analytical solution to unconstrained growth and decay problems for students who have had integral calculus and for those who have not. The module "Constrained Growth" considers situations in which the quantity under change, such as a population, has a maximum value, or carrying capacity. In this context, we introduce the concepts of equilibrium and stability. The module on "Drug Dosage," which includes geometric series, provides other examples where rate is proportional to amount.

For those interested in physics models, Chapter 4 on "Force and Motion" provides modules on falling and skydiving, bungee jumping (springs), pendulum clocks, and rocket motion.

After a second computational tool tutorial (such as *Maple*, *Mathematica*, *MATLAB*, or *Excel*) from the text's website, Chapter 5 covers the simulation techniques of Euler's, Runge-Kutta 2 (Euler's Predictor-Corrector), and Runge-Kutta 4 methods. One or more of these techniques can be covered at any time after Chapter 3's module on "Unconstrained Growth and Decay." For example, the instructor may choose to discuss Euler's Method immediately after that module and delay consideration of the other two techniques until later in the term.

Numerous system models involve interactions, such as with population dynamics or chemical reactions. Chapter 6 considers such models with discussions of "Competition," "Spread of SARS," "Enzyme Kinetics," "Predator-Prey Models," and "Modeling Malaria."

Chapter 7 provides opportunities for students to learn systems dynamics modeling by completing additional extensive projects. Unlike earlier chapters, the modules of this chapter do not include examples. Instead, each module contains sufficient background in a scientific application area for students to develop their own dynamic systems models, which projects suggest. Each module lists the prerequisite material, so that students can do Chapter 7's projects at any time after covering the earlier material. These projects, and some of the more extensive projects in previous chapters, provide excellent opportunities for teamwork. The chapter includes the following topics: radioactive chains, blood cell populations, scuba diving, carbon cycle, global warming, cardiovascular system, electrical circuits, carbohydrate metabolism, mercury pollution, and economics of commercial fishing.

Chapter 8 shifts away from systems modeling. After a third tutorial on a computational tool, a tutorial covers functions that often appear in modeling. With this background, empirical models, which are based only on data and are used to predict and not to explain a system, are considered.

Monte Carlo simulations of Chapter 9 form the basis for most of the remainder of the text. The chapter considers area estimation using this technique and, after an appropriate tutorial, implementation. An instructor interested in doing so can also cover how to generate random numbers in other probability distributions for computer simulations and details of the multiplicative linear congruential method to generate uniformly distributed random numbers.

The next chapter covers the random walk method that occurs in numerous computer simulations. Chapter 11, "Diffusion," considers many applications of cellular automata. A computational tool tutorial leads into a module involving applications, such as spreading fire, as well as fundamental concepts, such as periodic boundary conditions. A module on "Movement of Ants" provides another in-depth cellular automaton simulation.

Some modeling and simulation projects require massive computational power beyond the capabilities of present-day sequential workstations. Thus, Chapter 12 provides an introduction to "High Performance Computing" (HPC). The chapter covers the basic concepts and hardware configurations of HPC as well as some parallel-processing algorithms. With this background, the student can gain an appreciation of some of HPC's potential and challenges.

As with Chapter 7, Chapter 13 provides opportunities for students, perhaps in teams, to enhance their computational science problem-solving abilities through completion of additional extensive projects that they can do at any time after covering prerequisite material. The modules do not have examples but do have sufficient scientific background for the projects. The applications of computational science empirical models, random walk and cellular automaton simulations, and high performance computing in Chapter 13 are as follows: polymers, solidification, foraging, pit vipers and heat diffusion, mushroom fairy rings, spread of disease, HIV in the body, predator-prey relationships, clouds, and fish schooling.

Supplementary Materials

Instructors and students can link to the text's website through Princeton University Press's website (http://pup.princeton.edu/) or Wofford College's Computational Science website (http://www.wofford.edu/ecs/). The following **resources** are available on the text's site:

- Two system dynamics tool tutorials in several tools, such as *STELLA*, *Vensim* PLE, and *Berkeley Madonna*
- Six computational toolbox tutorials in several tools, such as *Maple*, *Mathematica*, *MATLAB*, and *Excel*
- For a variety of tools, pdf files that contain system-dependent Quick Review Questions and answers
- In a variety of tools, files that contain models, as indicated in the "Download" sections of modules.
- Datasets
- References with links to other websites

The text's website also has an **online *Instructor's Manual***, which contains the following material:

- Solutions to all text exercises
- Solutions to the two system dynamics tool tutorials using several tools, such as *STELLA*, *Vensim* PLE, and *Berkeley Madonna*
- Solutions to the six computational toolbox tutorials and a function tutorial using several tools, such as *Maple*, *Mathematica*, *MATLAB*, and *Excel*
- Test problems with answers
- Model solutions for selected projects in various tools
- *PowerPoint* files of key figures and algorithms
- Suggested pathways through the material

Instructors who adopt the text may obtain a password from Princeton University Press to access the online *Instructor's Manual*.

Acknowledgments

This book has been a labor of love, which has encouraged us both to grow as teachers and writers. The development of the text has been fun, though complicated; and we have many to thank for their support, collaboration, and patience. First, we must acknowledge our colleagues and students from whom we learn every day. In particular, we want to single out one colleague and two very capable students. Professor David Sykes was extremely helpful in his extensive review of the manuscript, providing many useful suggestions. David Harmon was an invaluable aid in preparing the glossary and verifying answers to the exercises and tutorials, and Jonathan De-Busk worked very hard to ensure that the references conformed to the proper style for publication.

We are very appreciative of the sabbaticals that Wofford College gave us and of the consistent support and encouragement by Dan Maultsby, Senior Vice-President and Academic Dean. The financial support provided by the National Science Foundation for computational science at Wofford ("Enhancing Computation in the Sciences," NSF CCLI Proof-of-Concept Grant No. 0087979) was essential in the development of the program and supporting materials.

We are indebted to the following reviewers, who offered many valuable constructive criticisms:

Rob Cole, Evergreen State University
Richard Hull, Lenoir-Rhyne College
James Noyes, Wittenberg College
Bob Panoff, The Shodor Foundation
Sylvia Pulliam, Western Kentucky University
Joseph Sloan, Wofford College
Chuck Swanson, University of Minnesota
David Sykes, Wofford College
Peter Turner, Clarkson University
Ignatios Vakalis, Capital University

Vickie Kearn, Senior Editor at Princeton, had a clear understanding of the project and provided excellent guidance. We thank you, Vickie, for your vision, trust, and encouragement. Meera Vaidyanathan and Ellen Foos very ably orchestrated production. Dimitri Karetnikov worked his magic on our figures to transform them into art. Thanks also go to Jennifer Slater of Running Dogs Editorial Services for her accurate and insightful copyediting and to Lorraine Doneker for the attractive design.

Dr. Robert M. Panoff of the Shodor Education Foundation has been more than a source of ideas and information for this project. His passion and generosity are remarkable and have been our inspiration. Bob and his very able colleagues at Shodor have done and continue to do amazing things for computational science education.

Ultimately we thank our parents, Isabell and Carroll Buzzett and Douglas and George Shiflet, Sr., who throughout their lives have given us boundless love and support. Words are inadequate to express our appreciation. Only Isabell remains, and she continues to delight us each day with her wisdom and wit. Obviously, we are where we are today because they were there for us.

INTRODUCTION TO
COMPUTATIONAL SCIENCE

1

OVERVIEW

MODULE 1.1

Overview of Computational Science

Many significant applied and basic research questions in science today are interdisciplinary in nature, involving physical and/or biological sciences, mathematics, and computer science. For example, *Nature* reported that John Krebs, chief executive of Britain's Natural Environment Research Council, considers that the environment "requires a 'new breed' of scientist, and new ways of problem solving that cut across traditional disciplines" and that Britain expects a shortage of "environmental scientists with mathematical, computational and statistical skills." (Masood 1998)

The Human Genome Project "has created the need for new kinds of scientific specialists who can be creative at the interface of biology and other disciplines, such as computer science, engineering, mathematics, physics, chemistry, and the social sciences. As the popularity of genomic research increases, the demand for these specialists greatly exceeds the supply. . . . There is an urgent need to train more scientists in interdisciplinary areas that can contribute to genomics," according to Francis Collins in an article in *Science* (Collins et al. 1998).

Computational science is a fast-growing interdisciplinary field that is at the intersection of the sciences, computer science, and mathematics. There is a critical need for scientists who have a strong background in computational science. Much scientific investigation now involves computing as well as theory and experiment. Computing can often stimulate the insight and understanding that theory and experiment alone cannot achieve.

This field of computational science combines computer simulation, scientific visualization, mathematical modeling, computer programming and data structures, networking, database design, symbolic computation, and high performance computing with various scientific disciplines. Computer simulation and modeling offer valuable approaches to problems in many areas, as the following examples indicate.

1. Scientists at Los Alamos National Laboratory and the University of Minnesota wrote, "**mathematical modeling has impacted our understanding of HIV pathogenesis**. Before modeling was brought to bear in a serious manner, AIDS was thought to be a slow disease in which treatment could be delayed until symptoms appeared, and patients were not monitored very aggressively. In the large, multicenter AIDS cohort studies aimed at monitoring the natural history of the disease, blood typically was drawn every six months. There was a poor understanding of the biological processes that were responsible for the observed levels of virus in the blood and the rapidity at which the virus became drug resistant. Modeling, coupled with advances in technology, has changed all of this." Dynamic modeling not only has revealed important features of HIV pathogenesis but has advanced the drug treatment regime for AIDS patients (Perelson and Nelson 1999).

2. Boeing Airline engineers completely designed **The Boeing 777 jetliner** using three-dimensional computer graphics. "Preassembly" of the airplane on the computer at every stage of the design process eliminated the necessity of a costly, full-scale mock-up and reduced error, adjustments, and revisions by 50 percent (Boeing). The pilots that fly these and other large airplanes train on sophisticated, computer flight simulators, which enable the pilots to practice dealing with dangerous situations, such as engine fire and wind shear.

3. From the 1960s, numerical **weather prediction** has revolutionized forecasting. "Since then, forecasting has improved side-by-side with the evolution of computing technology, and advances in computing continue to drive better forecasting as weather researchers develop improved numerical models" (Pittsburgh Supercomputing Center 2001).

4. Researchers at the University of Washington's School of Fisheries are employing mathematical modeling to examine the **impact on fish survival of the removal of four dams** on the lower Snake River. Another team at the University of Tennessee's Institute for Environmental Modeling is using computational ecology to study complex options for **ecological management of the Everglades**. Louis Gross, Director of the Institute, says that "computational technology, coupled with mathematics and ecology, will play an ever-increasing role in generating vital information society needs to make tough decisions about its surroundings" (Helly et al.).

5. A group of engineers and computer scientists at Carnegie Mellon University and seismologists from the University of Southern California and the National University of Mexico is building three-dimensional computer simulations of **ground motion during earthquakes** to predict how areas, such as the Greater Los Angeles Basin, will behave during such a disaster. Using powerful parallel-processing computer systems, one simulation indicated a complex pattern of basin ground motion with some sites experiencing nine times greater motion than others. With such information, scientists can predict the damage in an area (Pittsburgh Supercomputing Center 1997). Seattle, Washington is another area prone to earthquakes. The National Tsunami Hazard Mitigation Program has an extensive simulation modeling effort to assess the hazards of **tsunami threats** after earthquakes in the Puget Sound region so that officials can plan and mitigate their dangers (Koshimura and Mofjeld 2001). With computational models, others have studied the **economic impact** of disruption to the water supply caused by an earthquake in the Portland, Oregon region and appropriate responses to minimize the consequences (Rose and Liao).

Such collaboration among scientists, mathematicians, engineers, and computer scientists is indicative of much computational science research and practice. The fruits of these researchers' models and simulations are a deeper understanding of complex systems, a better foundation for important decisions, and a revolution in scientific advances that are helping people all over the world.

Projects

1. Investigate three applications of computational science involving different scientific areas and write at least a paragraph on each. List references.
2. Investigate an application of computational science and write a three-page, typed, double-spaced paper on the topic. List references.

References

Boeing. "Boeing 777 Program Information." http://www.boeing.com/commercial/777family/index.html

Collins, Francis S., et al. 1998. "New Goals for the U.S. Human Genome Project: 1998–2003." *Science*, 282 (October 23): 682–689.

Helly, John, et al., eds. "The State of Computational Ecology." San Diego Supercomputer Center. http://www.sdsc.edu/compeco_workshop/report/helly_publication.html

Koshimura, Shunichi, and Harold O. Mofjeld. 2001. "Inundation Modeling of Local Tsunamis in Puget Sound, Washington, Due to Potential Earthquakes." *ITS 2001 Proceedings*, Session 7, Number 7–18.

Masood, Ehsan. 1998. "UK Seeks Physicists for Environmental Research." *Nature*, 393 (June 4): 400.

Perelson, Alan S., and Patrick W. Nelson. 1999. "Mathematical Analysis of HIV-1 Dynamics in Vivo." *SIAM Review*, 41(1): 3–44.

Pittsburgh Supercomputing Center. 1997. "Getting Ready for the Big One." http://www.psc.edu/science/Bielak97/bielak97.html

———. 2001. "Pittsburgh System Marks a Watershed in Weather Prediction." News Release December 4, 2001. http://www.psc.edu/publicinfo/news/2001/weather-12-04-01.html

Rose, Adam, and Shu-Yi Liao. "Modeling Regional Economic Resiliency to Earthquakes: A Computable General Equilibrium Analysis of Lifeline Disruptions." Public Works Research Institute. http://www.pwri.go.jp/eng/ujnr/joint/34/paper/24rose.pdf

MODULE 1.2

The Modeling Process

Introduction

The process of making and testing hypotheses about models and then revising designs or theories has its foundation in the experimental sciences. Similarly, computational scientists use **modeling** to analyze complex, real-world problems in order to predict what might happen with some course of action. For example, Dr. Jerrold Marsden, a computational physicist at CalTech, models space mission trajectory design (Marsden). Dr. Julianne Collins, a genetic epidemiologist (statistical genetics) at the Greenwood Genetics Center, runs genetic analysis programs and analyzes epidemiological studies using the Statistical Analysis Software (SAS) (Greenwood Genetics Center). Some of the projects on which she has worked involve analyzing data from a genome scan of Alzheimer's disease, performing linkage analyses of X-linked mental retardation families, determining the recurrence risk in nonsyndromic mental retardation, analyzing folic acid levels from a nutritional survey of Honduran women, and researching new methods to detect genes or risk factors involved in autism. Scientists in areas such as cognitive psychology and social psychology at the Human-Technology Interaction Center of The University of Oklahoma perform research on the interaction of people with modern technologies (Human-Technology Interaction Center). Some of the studies involve "strategic planning in air traffic control" and "designing interfaces for effective information retrieval from collections of multimedia." Buried land mines are a serious danger in many areas of the world (Weldon et al. 2001). Scientists are using a combination of mathematics, signal processing, and scientific visualization to model, image, and discover land mines. Lourdes Esteva, Cristobal Vargas, and Jorge Velasco-Hernandez have modeled the oscillating patterns of the disease dengue fever, for which an estimated 50 to 100 million cases occur globally each year (Esteva and Vargas 1999).

> **Definition**   **Modeling** is the application of methods to analyze complex, real-world problems in order to make predictions about what might happen with various actions.

Model Classifications

Several classification categories for models exist. A system we are modeling exhibits **probabilistic** or **stochastic behavior** if an element of chance exists. For example, the path of a hurricane is probabilistic. In contrast, a behavior can be **deterministic**, such as the position of a falling object in a vacuum. Similarly, models can be deterministic or probabilistic. A **probabilistic** or **stochastic model** exhibits random effects, while a **deterministic model** does not. The results of a deterministic model depend on the initial conditions; and in the case of computer implementation with particular input, the output is the same for each program execution. As we see in Module 9.2 on "Simulations" and other modules, we can have a probabilistic model for a deterministic situation, such as a model that uses random numbers to estimate the area under a curve.

> **Definitions**   A system exhibits **probabilistic** or **stochastic behavior** if an element of chance exists. Otherwise, it exhibits **deterministic behavior**. A **probabilistic** or **stochastic model** exhibits random effects, while a **deterministic model** does not.

We can also classify models as static or dynamic. In a **static model**, we do not consider time, so that the model is comparable to a snapshot or a map. For example, a model of the weight of a salamander as being proportional to the cube of its length has variables for weight and length, but not for time. By contrast, in a **dynamic model**, time changes, so that such a model is comparable to an animated cartoon or a movie. For example, the number of salamanders in an area undergoing development changes with time; and, hence, a model of such a population is dynamic. Many of the models we consider in this text are dynamic and employ a static component as part of the dynamic model.

> **Definitions**   A **static model** does not consider time, while a **dynamic model** changes with time.

When time changes continuously and smoothly, the model is **continuous**. If time changes in incremental steps, the model is **discrete**. A discrete model is analogous to a movie. A sequence of frames moves so quickly that the viewer perceives motion. However, in a live play, the action is continuous. Just as a discrete sequence of movie frames represents the continuous motion of actors, we often develop discrete computer models of continuous situations (Voinov 2003).

> **Definitions**   In a **continuous model**, time changes continuously, while in a **discrete model** time changes in incremental steps.

Steps of the Modeling Process

The modeling process is cyclic and closely parallels the scientific method and the software life cycle for the development of a major software project. The process is cyclic because at any step we might return to an earlier stage to make revisions and continue the process from that point.

The steps of the modeling process are as follows:

1. **Analyze the problem**

   We must first study the situation sufficiently to identify the problem precisely and understand its fundamental questions clearly. At this stage, we determine the problem's objective and decide on the problem's classification, such as deterministic or stochastic. Only with a clear, precise problem identification can we translate the problem into mathematical symbols and develop and solve the model.

2. **Formulate a model**

   In this stage, we design the model, forming an abstraction of the system we are modeling. Some of the tasks of this step are as follows:

   a. **Gather data**

      We collect relevant data to gain information about the system's behavior.

   b. **Make simplifying assumptions and document them**

      In formulating a model, we should attempt to be as simple as reasonably possible. Thus, frequently we decide to simplify some of the factors and to ignore other factors that do not seem as important. Most problems are entirely too complex to consider every detail, and doing so would only make the model impossible to solve or to run in a reasonable amount of time on a computer. Moreover, factors often exist that do not appreciably affect outcomes. Besides simplifying factors, we may decide to return to Step 1 to restrict further the problem under investigation.

   c. **Determine variables and units**

      We must determine and name the variables. An **independent variable** is the variable on which others depend. In many applications, time is an independent variable. The model will try to explain the **dependent variables**. For example, in simulating the trajectory of a ball, time is an independent variable; and the height and the horizontal distance from the initial position are dependent variables whose values depend on the time. To simplify the model, we may decide to neglect some variables (such as air resistance), treat certain variables as constants, or aggregate several variables into one. While deciding on the variables, we must also establish their units, such as days as the unit for time.

   d. **Establish relationships among variables and submodels**

      If possible, we should draw a diagram of the model, breaking it into submodels and indicating relationships among variables. To simplify the model, we may assume that some of the relationships are simpler than they really are. For example, we might assume that two variables are related in a linear manner instead of in a more complex way.

e. **Determine equations and functions**

While establishing relationships between variables, we determine equations and functions for these variables. For example, we might decide that two variables are proportional to each other, or we might establish that a known scientific formula or equation applies to the model. Many computational science models involve differential equations, or equations involving a derivative, which we introduce in Module 2.3 on "Rate of Change."

3. **Solve the model**

This stage implements the model. It is important not to jump to this step before thoroughly understanding the problem and designing the model. Otherwise, we might waste much time, which can be most frustrating. Some of the techniques and tools that the solution might employ are algebra, calculus, graphs, computer programs, and computer packages. Our solution might produce an exact answer or might simulate the situation. If the model is too complex to solve, we must return to Step 2 to make additional simplifying assumptions or to Step 1 to reformulate the problem.

4. **Verify and interpret the model's solution**

Once we have a solution, we should carefully examine the results to make sure that they make sense (verification) and that the solution solves the original problem (validation) and is usable. The process of **verification** determines if the solution works correctly, while the process of **validation** establishes if the system satisfies the problem's requirements. Thus, verification concerns "solving the problem right," and validation concerns "solving the right problem." Testing the solution to see if predictions agree with real data is important for verification. We must be careful to apply our model only in the appropriate ranges for the independent data. For example, our model might be accurate for time periods of a few days but grossly inaccurate when applied to time periods of several years. We should analyze the model's solution to determine its implications. If the model solution shows weaknesses, we should return to Step 1 or 2 to determine if it is feasible to refine the model. If so, we cycle back through the process. Hence, the cyclic modeling process is a trade-off between **simplification** and **refinement**. For refinement, we may need to extend the scope of the problem in Step 1. In Step 2, while refining, we often need to reconsider our simplifying assumptions, include more variables, assume more complex relationships among the variables and submodels, and use more sophisticated techniques.

5. **Report on the model**

Reporting on a model is important for its utility. Perhaps the scientific report will be written for colleagues at a laboratory or will be presented at a scientific conference. A report contains the following components, which parallel the steps of the modeling process:

a. **Analysis of the problem**

Usually, assuming that the audience is intelligent but not aware of the situation, we need to describe the circumstances in which the problem arises. Then, we must clearly explain the problem and the objectives of the study.

b. Model design

The amount of detail with which we explain the model depends on the situation. In a comprehensive technical report, we can incorporate much more detail than in a conference talk. For example, in the former case, we often include the source code for our programs. In either case, we should state the simplifying assumptions and the rationale for employing them. Usually, we will present some of the data in tables or graphs. Such figures should contain titles, sources, and labels for columns and axes. Clearly labeled diagrams of the relationships among variables and submodels are usually very helpful in understanding the model.

c. Model solution

In this section, we describe the techniques for solving the problem and the solution. We should give as much detail as necessary for the audience to understand the material without becoming mired in technical minutia. For a written report, appendices may contain more detail, such as source code of programs and additional information about the solutions of equations.

d. Results and conclusions

Our report should include results, interpretations, implications, recommendations, and conclusions of the model's solution. We may also include suggestions for future work.

6. Maintain the model

As the model's solution is used, it may be necessary or desirable to make corrections, improvements, or enhancements. In this case, the modeler again cycles through the modeling process to develop a revised solution.

Definitions   The process of **verification** determines if the solution works correctly, while the process of **validation** establishes if the system satisfies the problem's requirements.

Although we described the modeling process as a sequence or series of steps, we may be developing two or more steps simultaneously. For example, it is advisable to be compiling the report from the beginning. Otherwise, we can forget to mention significant points, such as reasons for making certain simplifying assumptions or for needing particular refinements. Moreover, within modeling teams, individuals or groups frequently work on different submodels simultaneously. Having completed a submodule, a team member might be verifying the submodule while others are still working on solving theirs.

The modeling process is a creative, scientific endeavor. As such, a problem we are modeling usually does not have one correct answer. The problems are complex, and many models provide good, although different, solutions. Thus, modeling is a challenging, open-ended, and exciting venture.

Exercises

1. Compare and contrast the modeling process with the scientific method: Make observations; formulate a hypothesis; develop a testing method for the hypothesis; collect data for the test; using the data, test the hypothesis; accept or reject the hypothesis.
2. Compare and contrast the modeling process with the software life cycle: Analysis, design, implementation, testing, documentation, maintenance.

References

Esteva, Lourdes, and Cristobal Vargas. 1999. "A Model for Dengue Disease with Variable Human Population," *J. Math. Biol.*, 38: 220–240. http://147.46.94.112/e_journals/pdf_full/journal_j/j28_380302.pdf

Giordano, Frank R., Maurice D. Weir, and William P. Fox. 2003. *A First Course in Mathematical Modeling*. 3rd ed. Pacific Grove, Calif.: Brooks/Cole-Thompson Learning.

Greenwood Genetics Center. "Welcome to the GGC." http://www.ggc.org/

Human-Technology Interaction Center. "HTIC." The University of Oklahoma. http://www.ou.edu/HTIC/

Marsden, Jerrold. "Control and Dynamical Systems." California Institute of Technology. http://www.cds.caltech.edu/~marsden/

Voinov, Alexey Arkady. "Model Classifications." Course notes for CS 295 Computer Simulation and Modeling, University of Vermont, Gund Institute for Ecological Economics, 2003. http://www.uvm.edu/giee/AV/CS/class2.1.html (accessed March 20, 2005).

Weldon, T. P., Y. A. Gryazin, and M. V. Kibanov. 2001. "Novel Inverse Methods in Land Mine Imaging." *IEEE International Conference on Acoustics Speech and Signal Processing* 2001 (ICASSP 2001). Salt Lake City, UT, May 7–11, 2001. http://wws2.uncc.edu/tpw/papers/icassp01.pdf

2

FUNDAMENTAL CONSIDERATIONS

MODULE 2.1

Computational Toolbox—Tools of the Trade: Tutorial 1

Download

The text considers two major approaches to computational science problems: system dynamics models and cellular automaton simulations. System dynamics models provide global views of major systems that change with time. For example, one such model considers changes over time in the numbers of predators and prey, such as hawks and squirrels. To model such dynamic systems, students using the text can employ any one of several tools, such as *STELLA®*, *Vensim®* Personal Learning Edition (PLE), or *Berkeley Madonna®*. With the tool, the student can create pictorial representations of models, develop relationships, run simulations, and generate graphs of the results.

In contrast to system dynamics, cellular automaton simulations provide local views of individuals affecting individuals. The world under consideration consists of a rectangular grid of cells, and each cell has a state that can change with time according to rules. For example, the state of one cell could represent a squirrel and the state of an adjacent cell could correspond to a hawk. One rule could be that when adjacent, a hawk gets a squirrel with a probability of 25%. Thus, on the average at the next time step, a 25% chance exists that the particular squirrel will be no more. The text employs a generic approach for cellular automaton simulations, scientific visualizations of results, and calculations, so that students can employ any one of a variety of computational tools, such as *Maple®*, *Mathematica®*, *MATLAB®*, or *Excel®*.

Tutorials, package-specific Quick Review Questions and answers, and files to accompany the text material are available from the text's website in various system dynamics tools—such as *STELLA*, *Vensim* Personal Learning Edition (PLE), and *Berkeley Madonna*—and in several computational tools—such as *Maple*, *Mathematica*, *MATLAB*, and *Excel*. Typically, an instructor picks one system dynamics tool and one computational tool for class use during the term.

From the textbook's website, download Tutorial 1 in the format of your computational tool or in pdf format. We recommend that you work through the tutorial and answer all Quick Review Questions using the corresponding software.

Introduction

Various computer software tools are useful for graphing, numeric computation, and symbolic manipulation. This first computational toolbox tutorial is available for download from the textbook's website for several different software systems. Tutorial 1 in your system of choice gives an introduction to that software and prepares you to use the tool to complete various projects in the first few chapters. The tutorial introduces concepts and functions, such as the following:

- Getting started
- Evaluation
- Saving
- File organization, such as cells
- Styles
- Numbers
- Arithmetic operators
- Built-in functions, such as the natural logarithm, sine, and exponential functions

- Variables
- Assignments
- User-defined functions
- Online documentation
- Printing
- Looping
- Plotting
- Derivatives
- Integrals

The module gives computational examples and Quick Review Questions for you to complete and execute in your desired software system.

MODULE 2.2

Errors

Introduction

Errors can occur in the solution of a computational science problem at any stage, from the earlier stages of data collection and simplifying assumptions to the later stages of computer implementation. The modeler must be aware of possible errors to minimize their occurrence and to avoid drawing incorrect conclusions from flawed solutions. In this module, we discuss various concepts surrounding and the sources of errors.

Data Errors

Unfortunately, the sources for errors in the data upon which we base and verify our models can be numerous. For example, a sensor measuring barometric pressure might malfunction, giving incorrect values or values that are valid in one range but not in another. Moreover, the accuracy of the sensor might not be sufficient. In addition to equipment error, someone can fail to calibrate an instrument properly, misread measurements, or record results incorrectly.

Modeling Errors

Humans can also make errors in formulation of a model. Perhaps the modeler makes simplifying assumptions or determines incorrect equations that cause the model's results to deviate drastically from reality. He or she may not even be aware of crucial factors. For example, Lord Kelvin (William Thomson, Baron Kelvin of Largs), the accomplished nineteenth-century scientist who proposed the absolute temperature scale that bears his name, developed in the mid to late part of that century a mathematical model and calculated the age of the earth to be between 20 and 40 million years old. This computed range is drastically different from its real age, which is about 12 *billion* years. Kelvin based his model on the assumption that the earth was cooling from a molten mass with the sun being its only source of energy. However, his assumption is incorrect. Decaying radioactive elements in the earth's crust also generate heat that helps to maintain the temperature of the planet. Kelvin did not and could not consider the effect of radioactivity, which Becquerel did not discover until 1896. An incorrect assumption led the noted scientist to make a calculation that was off by two orders of magnitude (Darden 1998).

Implementation Errors

In a computer program implementing a model, computational scientists can make logical errors that produce disastrous results. For example, in 1999 NASA's Mars Climate Orbiter spacecraft was lost because the builder of the spacecraft, Lockheed Martin Corp., programmed it to use English units, such as pounds and feet, and NASA's Jet Propulsion scientists employed metric units, such as newtons and meters.

Precision

Other errors we consider in this module also involve computer calculations. In this section, we discuss some basic terms involved with such computations; and in the next section, we define two metrics of error. Then we return to a discussion of other errors encountered in modeling and simulation.

Many computer languages allow floating point numbers to be printed in exponential form as a decimal fraction times a power of 10. For instance, output of $9.843600e02$ means $9.843600 \times 10^2 = 984.36 = 0.98436 \times 10^3$. Usually **floating point numbers**, or numbers with a decimal expansion, are stored in the computer in three parts: 0 or 1 representing the sign + or −, respectively; a **significand, fractional part**, or **mantissa**, such as 98436; and an exponent, such as 3. Every day, we use the **decimal**, or **base 10**, number system with digits 0, 1, 2, ... 9. A computer usually employs the **binary**, or **base 2**, number system with only two binary digits, or **bits**, 0 and 1, but the concepts are the same regardless of base.

> **Definitions**   A **floating point number** is expressed with a decimal expansion. **Exponential notation** represents a floating point number as a decimal fraction times a power of 10. With *a* being a decimal fraction and *n* an integer, the exponential notation ***aen*** represents $a \times 10^n$. The integer formed by dropping the decimal point from *a* is the **significand**, **fractional part**, or **mantissa**; and *n* is the **exponent**.

A **normalized** number in exponential notation has the decimal point immediately preceding the first nonzero digit, as in 0.98436×10^3. This notation is similar to **scientific notation**, which places the decimal point immediately after the first nonzero digit, such as 9.8436×10^2. When a number is expressed in normalized exponential notation, as with 0.98436×10^3, all the digits of the significand, such as 98436, are what we call **significant**. For integers written without a decimal, all the digits except leading and trailing zeros are **significant digits**; for other numbers, all digits are significant except leading zeros. For example, there are four significant digits in $003,704,000 = 0.3704 \times 10^7$, 3, 7, 0, and 4. The **most significant digit** is the leftmost one, 3. The most significant digit of $0.09200 = 0.9200 \times 10^{-1}$ is 9 because the leading zero is not significant. All other digits after the decimal point (9, 2, 0, 0), however, are significant in this number.

> **Definition**   A **normalized** number in exponential notation has the decimal point immediately preceding the first nonzero digit. The **significant digits** of a floating point number are all the digits except the leading zeros. The **significant digits** of an integer are all the digits except the leading and trailing zeros.

Precision is the number of significant digits. Thus, 003,704,000 and 0.09200 each have a precision of 4. **Magnitude** is an indication of the relative size of a number and is 10 to the power when the number is expressed in normalized exponential notation. Therefore, 0.3704×10^7 has a magnitude of 10^7. In C and C++, the precision of a floating point number of type *float*, which we call a **single-precision number**, is about 6 or 7 decimal digits, while the magnitude ranges from about 10^{-38} to 10^{38}. Taking up twice as much computer storage space as a *float* variable, a variable of type *double*, which stores a **double-precision number**, has 14 or 15 significant digits and magnitude from about 10^{-308} to 10^{308}.

> **Definitions**   The **precision** of a number is the number of significant digits. **Magnitude** is 10 to the power when the number is expressed in normalized exponential notation.

Quick Review Question 1

Use 0.0004500 for the following problems.

    a. In normalized exponential notation, give the significand.
    b. In normalized exponential notation, give the exponent of 10.
    c. Give the precision.

Absolute and Relative Errors

To understand the size of a problem, it is helpful to have ways of measuring error. The **absolute error** is the absolute value of the difference between the exact answer and the computed answer. The **relative error** is this difference divided by the absolute value of the exact answer, provided the exact answer is not zero. We often express relative error as a percentage. For example, we can write a relative error of 0.03 as 3%.

Definition   If *correct* is the correct answer and *result* is the result obtained, then

absolute error $= |correct - result|$

$$\textbf{relative error} = \frac{(\text{absolute error})}{|correct|} = \frac{|correct - result|}{|correct|}$$

$$= \frac{(\text{absolute error})}{|correct|} \times 100\% = \frac{|correct - result|}{|correct|} \times 100\%,$$

provided *correct* $\neq 0$.

Example 1

Suppose a computer has a precision of 3, allowing only 3 digits in the significand, and **truncates**, or chops off, the significand to 3 digits. No computer has such limited precision. Limiting the precision to 3, however, simplifies our computations and still illustrates the problem. We evaluate the absolute and relative errors in the computation $(0.356 \times 10^8)(0.228 \times 10^{-3})$. The exact answer is as follows:

$$(0.356 \times 10^8)(0.228 \times 10^{-3}) = (0.356)(0.228)(10^8)(10^{-3})$$
$$= 0.081168 \times 10^5$$

Normalizing, we obtain *correct* $= 0.81168 \times 10^4$.

For our computer with a precision of 3, the result of this computation is *result* $= 0.811 \times 10^4$. Thus, an error has been introduced. The absolute error is as follows:

$$|correct - result| = |0.81168 \times 10^4 - 0.811 \times 10^4| = 0.00068 \times 10^4 = 6.8$$

The relative error is the ratio of the absolute error and the positive correct answer, as shown:

$$(0.00068 \times 10^4)/(0.81168 \times 10^4) = 0.0008378 = 0.08378\%$$

The error is about eight-hundredths of a percent of the exact answer.

Definition   To **truncate** a normalized number to k significant digits, eliminate all digits of the significand beyond the kth digit.

Quick Review Question 2

Using the number 6.239, find the following:

      a. The absolute error as it is truncated to two decimal places.

      b. The relative error for Part a. Express your answer as a percentage.

Round-off Error

Instead of truncating a number to fit in storage, a computer might **round**. To round the significand of 0.81168×10^4 to a precision of 3, we consider the fourth significant digit, here 6. If that digit is less than 5, we **round down**; but if the digit is greater than or equal to 5, we **round up**. Thus, 0.81168×10^4 and 0.81158×10^4 round up to 0.812×10^4, while 0.81138×10^4 rounds down to 0.811×10^4.

> **Definition**   To **round** a normalized number to precision k, consider the $(k + 1)$th significant digit, d. If d is less than 5, **round down** the normalized number by truncating the significand to k significant digits. If d is greater than or equal to 5, **round up** the normalized number by truncating the significand to k significant digits and then adding 1 to the kth significant digit of the significand, carrying as necessary to digits on the left.

Quick Review Question 3

Round each of the following so that the significand has a precision of 2:

      a. 0.93742×10^{-5} **b.** 0.93472×10^{-5} **c.** 0.93572×10^{-5}

Example 2

This example illustrates the difficulty of expressing exact decimal floating point numbers in the computer. Suppose we enter into a cell of a spreadsheet a computation for 1/3, such as follows:

```
= 1/3
```

Alternatively, suppose we make the following **assignment statement** in the programming language C, C++, or Java that gives the value of the expression on the right to the variable, x, on the left:

```
x = 1.0/3.0;
```

The computer stores the floating point representation of $1/3 = 0.333\ldots$ in the spreadsheet cell or in the location for x. If our computer can store only three digits of the significand, the machine rounds or truncates the value to 0.333.

Definition   An **assignment statement** causes the computer to store the value of an expression in a memory location associated with a variable. In most programming languages, the assignment statement has a format similar to the following with the expression always appearing on the right and the variable getting the value always being on the left of an **assignment operator**, here an equal sign:

$$variable = expression$$

Round-off error is the problem of not having enough bits to store an entire floating point number. We use the name "round-off error" whether the computer rounds or truncates the number to fit in a location. If the computer uses a greater number of bits to store the number, the round-off error will not be as serious. For example, if we store the significand with seven digits, the value of x will be 0.3333333; and storage for fifteen significant digits will yield the even more accurate 0.333333333333333.

Definition   **Round-off error** is the problem of not having enough bits to store an entire floating point number and approximating the result to the nearest number that can be represented.

Overflow and Underflow

The problems of overflow and underflow can also ensue from finite storage and binary representation of numbers in a computer. Suppose we are working with a very small computer that uses 16 bits to store an integer. If we ask the computer to perform the sum $20480 + 16384$, the result surprisingly will be a negative number, -28672. The problem arises when the leftmost bit, the sign bit, gets a carry from the addition on the right, converting the result to a negative number. There simply are not enough bits to express the answer, so the final answer has the wrong sign. Overflow also occurs when we add two negative numbers and get a positive result.

Definition   **Overflow** is an error condition that occurs when there are not enough bits to express a value in a computer.

An overflow error caused the European Space Agency's Ariane 5 rocket to explode in 1996. Less than 37 seconds into the launch, the guidance system's computer attempted to convert the rocket's sideways velocity from a 64-bit floating point number to a 16-bit integer. However, because the number was too large, overflow resulted; the guidance system attempted a severe correction for a wrong turn that had not occurred; and very quickly the rocket had to self-destruct. The over-flow of a few bits caused the loss of a rocket that took 10 years and $7 billion to de-velop (Gleick).

Problems can also arise when the result of a computation is too small for a computer to represent in a situation called **underflow**. For example, suppose the smallest floating point number a computer can express has magnitude 10^{-39}. If the correct value for an arithmetic expression, such as 10^{-48}, is smaller than the smallest floating-point value a computer can represent, then underflow occurs, and the computer evaluates the expression as zero.

> **Definition** **Underflow** is an error condition that occurs when the result of a computation is too small for a computer to represent.

Arithmetic Errors

Errors can arise in addition. Consider $(0.684 \times 10^3) + (0.950 \times 10^{-2})$. Unlike in multiplication, the decimal points must be aligned for addition, so we have the following:

$$0.684 \times 10^3 = 684.0000$$
$$0.950 \times 10^{-2} = \underline{+0.0095}$$
$$684.0095$$

If our computer allows for only three significant digits, the normalized result is 0.684×10^3, and the effect of the 0.950×10^{-2} is lost.

Quick Review Question 4

Suppose a particular computer rounds stored floating point numbers to four significant digits. Calculate $(0.1235 \times 10^2) + (0.2499 \times 10^{-1})$. Do not use exponential notation in the answer.

Because of such problems, when adding numbers whose magnitudes are drastically different, we should accumulate smaller numbers first before combining them with larger ones. Thus, the sum of the smaller numbers has a chance of being large enough to make a difference in the final answer.

Similarly, when multiplying and dividing in a term, to avoid loss of precision, we usually should perform all multiplications in the numerator before dividing by the denominator. For example, in our computer that rounds to three significant digits, suppose we are to calculate $(x/y)z$, where $x = 2.41$, $y = 9.75$, and $z = 1.54$. The quotient $x/y = 0.247179$ rounds to 0.247, so that the product is $(x/y)z = (0.247)(1.54) = 0.38038$, or 0.380 after rounding. However, algebraically $(x/y)z = (xz)/y$. Performing multiplication first, we have $xz = 3.7114$, or 3.71 in rounded form. Dividing, we have $(xz)/y = 3.71/9.75 = 0.380513$. The rounded 0.381 is closer to the exact answer of 0.380656. . . .

Of course, whether or not we should perform all numerator multiplications before division depends on the numbers. In the example $(xz)/y$, if the product xz would
cause overflow, we should first divide x by y to obtain a smaller result before multiplying by z.

Quick Review Question 5

Suppose variables r, u, x, y, and z store floating point numbers. Write the following
expression to minimize round-off error upon evaluation, assuming no problems with
overflow or underflow:

$$\left(\frac{3}{z}\right)(r)\left(\frac{x+y}{u}\right)$$

Error Propagation

Looping enables the computer to execute a segment of code several times. Such a
segment that is executed repeatedly is called a **loop**. Performing floating point operations within loops compounds round-off error.

> **Definition**   A **loop** is a segment of code that is executed repeatedly.

Such an accumulation error had disastrous consequences during the Gulf War in
Dharan, Saudi Arabia, when an American Patriot Missile battery failed to intercept a
Scud missile. The Scud hit an American Army barracks, killed 28 soldiers, and injured
more than 100 others. The Patriot's internal computer clock measured time in tenths
of a second, and multiplied the number of ticks by 1/10 to obtain the actual time in
seconds. For example, 15 ticks indicated an elapsed time of (15 ticks) (0.1 sec/tick) =
1.5 sec. The missile's computer used 24 bits to store numbers. However, because 1/10
has an infinite expansion in binary representation (similar to the way 1/3 does in decimal representation), the system could not hold all the number's bits. Each one-tenth
increment produced an error of about 0.000000095 sec. At the time of the disaster, the
Patriot Missile had been operating for 100 hours, causing an error of about (100 hrs)
(60 min/hr)(60 sec/min)(10 ticks/sec)(0.000000095 sec/tick) = 0.34 sec. During that
third of a second, a Scud flew about 1,676 meters, so that the intercepting Patriot Missile missed its target (Arnold 2000).

Example 3

In many simulations of scientific phenomena, such as pollution in a stream, we have
the computer calculate the values of various quantities as time advances in small,
discrete time steps. Suppose time, t, starts at 0.0 seconds and is to end at 10
minutes = 600 seconds. The length of a time step is dt. Perhaps we wish to designate
dt to be 0.1 seconds, so that the number of time steps (*numberOfTimeSteps*) would

be 6000. However, because of conversion to base 2 and finite storage in an exaggeratedly small computer, suppose the actual stored value of dt is $0.09961 = 0.9961 \times 10^{-1}$ seconds. Inside a loop, we compute new values for time and other quantities, such as a simulated amount of mercury in a stream. A **loop variable** or **index**, i, takes on values $1, 2, 3, \ldots, 6000$. Thus, this simulation has the following general algorithm:

> *numberOfTimeSteps* = number of time steps for simulation
> $dt = 0.09961$, the length of a time step in seconds
> $t = 0.0$, the starting time in seconds
> initialize other quantities as necessary
> for i going from 1 through *numberOfTimeSteps* do the following:
>   compute a new value for t
>   compute quantities being simulated

One method of updating time can lead to a serious accumulation of round-off error. Suppose that, each time through the loop, we calculate as follows the new value of time (t on the left-hand side of the assignment statement) as the old value of time (t on the right-hand side of the assignment statement) plus the change in time (dt):

$$t = t + dt$$

Suppose the machine for this example uses the decimal system and rounds to four significant digits. Ignoring problems of having 0.09961 instead of 0.1 for dt and conversion between the decimal and binary number systems, Table 2.2.1 enumerates the absolute and relative errors of t for several iterations of the loop.

As the table illustrates, round-off error increases with the number of loop executions. After the eleventh iteration ($i = 11$), the new value of t should be $(11)(0.09961) = 1.09571$. For a computer with a precision of 4, however, the rounded value is 1.096. The subsequent absolute error is 0.00029, and the relative error is about 0.026%. After iteration $i = 51$, the relative error is about 0.3128% and is more than 30 times the relative error in the loop with $i = 2$.

To avoid the cumulative error of this loop, we should compute t as the index, i, times $dt = 0.09961$, as follows with $*$ indicating multiplication:

$$t = i * dt$$

We still have round-off error, but its effect is minimized because there is no accumulation. For example, in the evaluation with $i = 51$ of $(51)(0.09961) = 5.08011$, the computer stores four significant digits, 5.080. The absolute error is $5.08011 - 5.080 = 0.00011 = 0.011\%$, while the relative error is $0.00011/5.08011 = 0.000022 = 0.0022\%$. In the iteration $i = 51$, the relative error using an accumulated $t = t + dt = t + 0.09961$ is about 140 times greater than the corresponding relative error

Table 2.2.1
Accumulation of Error in Time ($t = t + dt$) for Example 3

Value of i	Correct New Value of t	Accumulated New Value of $t = t + dt$	Absolute Error	Relative Error
1	0.09961 +0.00000 ___ 0.09961	0.09961 +0.00000 ___ 0.09961 is 0.09961 rounded with precision 4	0	0
2	0.09961 +0.09961 ___ 0.19922	0.09961 +0.09961 ___ 0.19922 is 0.1992 rounded with precision 4	0.19922 −0.1992 ___ 0.00002	0.00002/0.19922 ≈ 0.0001 = 0.01%
3	0.09961 +0.19922 ___ 0.29883	0.09961 +0.1992 ___ 0.29881 is 0.2988 rounded with precision 4	0.29883 −0.2988 ___ 0.00003	0.00003/0.29883 ≈ 0.0001 = 0.01%
4	0.09961 +0.29883 ___ 0.39844	0.09961 +0.2988 ___ 0.39841 is 0.3984 rounded with precision 4	0.39844 −0.3984 ___ 0.00004	0.00004/0.39844 ≈ 0.0001 = 0.01%
5	0.09961 +0.39844 ___ 0.49805	0.09961 +0.3984 ___ 0.49801 is 0.4980 rounded with precision 4	0.49805 −0.4980 ___ 0.00005	0.00005/0.49805 ≈ 0.0001 = 0.01%
11	1.09571	1.096	1.096 −1.09571 ___ 0.00029	0.00029/1.09571 ≈ 0.00026 = 0.026%
51	5.08011	5.096	5.096 −5.08011 ___ 0.01589	0.01589/5.08011 ≈ 0.003128 = 0.3128%

using $t = i * dt = (i)(0.09961)$. In some simulations, it is not possible to avoid such repeated additions; but where possible, we should.

> In looping, whenever possible, avoid accumulating floating point values through repeated addition or subtraction.

Quick Review Question 6

Which assignment statement is better (if there is a difference) in a loop with index k whose initial value is 1? Assume that sum is initialized to 0 before the loop.

    A. $sum = sum + 0.00492$
    B. $sum = 0.00492 * k$
    C. It doesn't matter

Violation of Numeric Properties

The section on "Arithmetic Errors" hints at a problem that can lead to errors—numeric properties do not necessarily hold in computer arithmetic. Expressions that are numerically equivalent might not evaluate to equal values on the computer. Mathematically, $(x/y)z = (xz)/y$, but for $x = 2.41$, $y = 9.75$, and $z = 1.54$ on a three-significant-digit machine with rounding, we showed $(x/y)z = 0.380$, while $(xz)/y = 0.381$. Specifically, the following are some of the properties can be violated using computer arithmetic:

Associative Properties: $(a + b) + c = a + (b + c)$ and $(ab)c = a(bc)$

These properties indicate that grouping of numbers in addition or multiplication does not matter. However, a computer can yield different results, such as, if a and b are small numbers and c is very large.

Distributive Property: $a(b + c) = ab + ac$

This property involves the distribution of multiplication over addition. Under certain circumstances, this property may not hold in a computer, such as in the case where a and b are very small in comparison to c.

    In the exercises and projects, we explore these and other properties that can fail using computer arithmetic.

Quick Review Question 7

By evaluating $x(y + z)$ and $x(y) + x(z)$, show that the distributive property does not hold for $x = 2.48$, $y = 9.34$, and $z = 1.55$ on a machine that truncates intermediate and final results to three significant digits.

Comparison of Floating Point Numbers

Conversion back and forth between the decimal system that people usually employ and the binary system of computers can result in the loss of information. For example, with a subscript indicating the base, the decimal number 0.6_{10} is equivalent to the binary number $0.1\overline{1001}_2$, where the line over 1001 indicates that this pattern continues forever. However, a computer cannot store an infinite expansion. If the particular computer truncates to 20 bits to store the significand, the stored value would be 0.10011001100110011001_2, which is equivalent to $0.59999942779541015625_{10}$. Thus, if we request to print this value, the computer might display 0.599999 instead of the expected 0.600000.

This conversion between bases, a finite amount of computer storage, and error propagation are the reasons we should *not* test if floating point numbers are exactly equal in a computer. Thus, in a spreadsheet, we should not test if the floating point contents of cells *B2* and *B3* are equal as follows:

```
= If(B2 = B3, . . .) # Do NOT test floating point
                     # numbers this way
```

Similarly, in a programming language, we should not test if floating point variables x and z are identical, as in the following C/C++/Java statement, which employs two adjacent equal signs, = =, to check for equality:

```
if (x = = z) . . . // Do NOT test floating point numbers
                   // this way
```

Example 4

For ease of computation, let us consider a computer that uses the decimal system but truncates to three digits for the significand. Suppose x has the value 0.536. Now, suppose we multiply x by 7, using $*$ for multiplication, to obtain y and divide this value by 7, assigning the final result to z, as follows:

```
x = 0.536
y = 7 * x
z = y / 7
```

In mathematics x and z are identical; but when we multiply $x = 0.536$ by 7, we obtain $3.752 = 0.3752 \times 10^1$, which truncates to $y = 0.375 \times 10^1$ in a system that has three-digit significands. Division by 7 yields an infinite decimal expansion, $0.535\overline{71428}$, which truncates in three decimal places to 0.535 for z. Thus, $x = 0.536$ and $z = 0.535$ are not identical. We have exaggerated the problem by using only three digits for the significand, but the idea is the same for a larger significand.

To avoid the problem, we should instead test if the difference between two floating point values is "close enough" to zero. For our limited system that uses only three digits for the significand, we might decide that if the difference is within 0.001 of zero, then we will consider the values to be equal. Because the difference could be positive or negative, we take the absolute value of the difference and make the following test:

$$\text{"If } |x - z| < 0.001, \text{ consider } x \text{ and } z \text{ to be the same."} \tag{1}$$

In a more realistic computer system, to determine equality we might test if x and z are within some very small number of each other, say, 0.000001 of each other.

Quick Review Question 8

Write a statement similar to (1) to test if two floating point variables are *not* equal to within 0.000001 of each other.

Truncation Error

Just as a computer cannot store exactly numbers with infinite precision, a computer cannot perform an infinite number of calculations. Let us consider a result of calculus that says e^x has the following infinite series expansion:

$$e^x = 1 + x + \frac{x^2}{1 \cdot 2} + \frac{x^3}{1 \cdot 2 \cdot 3} + \frac{x^4}{1 \cdot 2 \cdot 3 \cdot 4} + \cdots + \frac{x^n}{n!} + \cdots$$

Therefore, $e^1 = e$, which is Euler's number $2.718281828459045\ldots$, is exactly equal to the following infinite series with 1 replacing x above:

$$e = 1 + 1 + \frac{1}{1 \cdot 2} + \frac{1}{1 \cdot 2 \cdot 3} + \frac{1}{1 \cdot 2 \cdot 3 \cdot 4} + \cdots + \frac{1}{n!} + \cdots$$

However, a finite machine is unable to perform such an infinite number of additions. If we wish to use this series to evaluate e, we must truncate the sum. For example, if we perform the additions to $n = 20$ to obtain a partial sum, we have the following close approximation of e:

$$e \approx 1 + 1 + \frac{1}{1 \cdot 2} + \frac{1}{1 \cdot 2 \cdot 3} + \frac{1}{1 \cdot 2 \cdot 3 \cdot 4} + \cdots + \frac{1}{20!} = \frac{6613313319248080001}{2432902008176640000}$$

This finite sum does not include terms from $1/(21!)$ on and results in the following **truncation error**:

$$\frac{1}{21!} + \frac{1}{22!} + \frac{1}{23!} + \cdots \approx 2.05 \times 10^{-20}$$

> **Definition**   A **truncation error** is an error that occurs when a truncated, or finite, sum is used as an approximation for the sum of an infinite series.

Figure 2.1.1 displays graphs of e^x (in color) and the following approximating partial sum functions (in black)

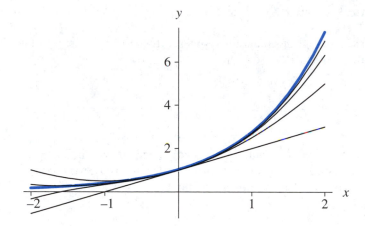

Figure 2.2.1 Graphs of e^x (in color) and four finite series approximations

$$f_n(x) = 1 + x + \frac{x^2}{1 \cdot 2} + \frac{x^3}{1 \cdot 2 \cdot 3} + \frac{x^4}{1 \cdot 2 \cdot 3 \cdot 4} + \cdots + \frac{x^n}{n!} \quad \text{for } n = 1$$

through $n = 4$:

$$f_1(x) = 1 + x$$

$$f_2(x) = 1 + x + \frac{x^2}{1 \cdot 2} = 1 + x + \frac{x^2}{2}$$

$$f_3(x) = 1 + x + \frac{x^2}{1 \cdot 2} + \frac{x^3}{1 \cdot 2 \cdot 3} = 1 + x + \frac{x^2}{2} + \frac{x^3}{6}$$

$$f_4(x) = 1 + x + \frac{x^2}{1 \cdot 2} + \frac{x^3}{1 \cdot 2 \cdot 3} + \frac{x^4}{1 \cdot 2 \cdot 3 \cdot 4} = 1 + x + \frac{x^2}{2} + \frac{x^3}{6} + \frac{x^4}{24}$$

As n becomes larger, the graphs of the partial sum functions $f_n(x)$ approach, or converge to, the graph of $f(x) = e^x$.

Quick Review Question 9

Calculus shows that $\sin(x)$ equals the following infinite series:

$$\sin(x) = x - \frac{x^3}{3!} + \frac{x^5}{5!} - \frac{x^7}{7!} + \cdots$$

a. Obtain a decimal expansion approximation for $\sin(2)$, truncating the infinite series to two terms.
b. Give the error for this approximation.

Exercises

Answers to exercises with numbers in color appear in the appendix on "Answers to Selected Exercises."

Write the numbers of Exercises 1–12 in normalized exponential notation.

1. 63,850 **2.** 29.748 **3.** 0.00032 **4.** 53.7×10^3
5. 0.496 **6.** 0.0000017 **7.** 0.009×10^{-5} **8.** 0.009×10^5
9. −0.82 **10.** −82 **11.** −0.00082 **12.** 4.4

13. Give the magnitude and precision of 0.743621×10^{25}.
14. Give the magnitude and precision of 93.6×10^7.
15. Give the precision and the largest magnitude of numbers in a computer where the significand has 8 digits and the largest power of 10 is 125.

Give the number of significant digits and the most significant digit for the numbers in Exercises 16–21.

16. 63,850 **17.** 29.004 **18.** 0.00074
19. 10^3 **20.** 4×10^{-5} **21.** 0.300500

22. Give the range of the normalized positive numbers where the significand has three digits and the exponent of 10 is from −5 to +5.
23. Give the range of the normalized positive numbers that can be expressed with seven digits in the significand and an exponent of 10 from −78 to +73.

For Exercises 24–26 find the following: (a) the absolute error and (b) the relative error of each number as it is rounded to two decimal places. Then compute (c) the absolute error and (d) the relative error of each number as it is truncated to two decimal places.

24. 6.239 **25.** 6.231 **26.** 1.0/3.0 stored with five significant digits

27. a. Perform the arithmetic $(9.4 \times 10^{-5}) + (3.6 \times 10^4)$, expressing the answer in normalized exponential notation.
 b. Give the final answer if the representation allows only five significant digits, rounded.
 c. Give the absolute error of Part b.
 d. Give the relative error of Part b.
28. Repeat Exercise 27 for $(0.7 \times 10^3) - (0.825 \times 10^2)$. Use three significant digits instead of five significant digits for Parts b–d.
29. Suppose the following sequence is executed: $x = 6.239$; $x = x + x$.
 For a machine that truncates to three significant digits, give the values stored for x after execution of each statement and the relative error for the value of x after the last statement. Compare this error with your answer in Exercise 24.
30. Consider a machine that rounds to four significant digits. Suppose initially $y = 9.649$ and $x = 7.834$. The following assignment statement, which calculates the expression on the right $(y + x)$ and then replaces the value of y on the left, is in a loop that executes four times: $y = y + x$. After each iteration of the loop, give the value stored in y and the absolute and relative errors.

31. Using the computer described in Example 3 and $t = i * dt$, evaluate the computer's value for t, the absolute error, and the relative error for the values of i in Table 2.2.1. Compare your results with those of Table 2.2.1.

32. Mathematics can prove that $1.0 = 0.9\overline{9} = 0.99\overline{9}$. Suppose this value is assigned to x and to y as a series of 9's truncated to four significant digits.

   a. If $x = x + y$ is executed four times in a loop, each time replacing the old value of x with the result of the sum on the right, give the values of x and the absolute and relative errors for the original assignment and after each iteration of the loop.

   b. By observing the results of Part a, give the value of x and the absolute and relative errors after the tenth iteration of the loop.

33. Refer to Example 4 for the proper testing of equality of floating point numbers.

   a. Write an *if* statement that puts 1 in the current cell of a spreadsheet if the floating point values in cells *B2* and *B3* are equal and otherwise puts 0 in the current cell.

   b. In another computational tool, write an *if* statement that displays 1 if the floating point values in x and z are equal and otherwise displays 0.

34. a. Calculus shows that $\cos(x)$ equals the following infinite series:

$$\cos(x) = 1 - \frac{x^2}{2!} + \frac{x^4}{4!} - \frac{x^6}{6!} + \cdots$$

   Obtain a decimal expansion approximation for $\cos(2)$, truncating the infinite series to three terms.

   b. Find finite series approximations for $\cos(2)$ until successive approximations are the same for the first four significant digits and give that approximation.

35. a. With a computational tool, evaluate the following:

$$10000000000000000. + 1. - 10000000000000000.$$

   that is, $1.0 \times 10^{16} + 1. - 1.0 \times 10^{16}$.

   b. What should the result be?

   c. Explain what happened.

36. For a machine that rounds to three significant digits, give an example where a floating point number, x, does not have a multiplicative inverse, y, where $xy = 1$.

37. For a machine that rounds to three significant digits, give an example where the associative property for addition does not hold.

Projects

1. Using a computational tool, evaluate each of the following expressions for $t = 355/113$, $r = 101/113$, and $s = 52/113$: $t - s - r - r - r$, $t - r - r - r - s$, $t - r - r - s - r$, $t - r - 2r - s$, $t - r - s - r - r$, $t - 2r - r - s$, $t - 3r - s$,

$t - r - s - 2r$. Using mathematics, what are the values of the expressions? What numeric property or properties are violated? (Panoff 2004)

2. Using a computational tool, evaluate e^x for a given value of x using the series expansion in the section "Truncation Error" and enough terms so that successive approximations differ by no more than one-hundred-thousandth. Show all steps of the developing expansion on separate rows. Also, evaluate e^x using the built-in exponential function.

3. Do Project 2 for $\sin(x)$ using the series expansion in Quick Review Question 9.

4. Do Project 2 for a function and its Taylor Series expansion. Do not use e^x or $\sin(x)$. Refer to a calculus text for such an expansion.

5. Using a computational tool and the infinite series expansions of $\sin(x)$ (see Quick Review Question 9), define five partial sum approximation functions. Graph $\sin(x)$ and the approximating functions. Evaluate $\sin(x)$ and the approximating functions at several values of x, and compute the absolute and relative errors of the approximations.

6. Do Project 5 for a function and its Taylor Series expansion. Do not use e^x or $\sin(x)$. Refer to a calculus text for such an expansion.

7. This project requires the use of a computer programming language. A chemical is added a drop at a time to a container in which a reaction is occurring. Each drop is measured as precisely 0.xxxx ml. The total amount of the chemical must be computed after each drop is added. Write a program to perform the calculation in two ways: by incrementing the previous total by 0.xxxx (*accumulated*) and by multiplying 0.xxxx (*multiplied*) by the number of drops so far. At the beginning of the program, ask the user for the number of iterations and the reporting interval. Print the iteration number and both ongoing totals at the requested intervals. Use the last four digits of your phone number as the nonzero digits in the measurement of the drop of the chemical. Replace zeros with different odd digits. For example, with a phone number of 555-9389, use 0.9389; and with 555-8090 possibly use 8193. Sample output for 0.xxxx = 0.9389 is as follows:

```
Give the number of iterations: 1000
Give the reporting interval: 200

Iteration 200
Accumulated = 187.781
 Multiplied = 187.78

Iteration 400
Accumulated = 375.561
 Multiplied = 375.56

Iteration 600
Accumulated = 563.342
 Multiplied = 563.34

Iteration 800
Accumulated = 751.123
 Multiplied = 751.12

Iteration 1000
Accumulated = 938.904
 Multiplied = 938.9
```

Print the output and report for each of the following:

a. Run the program with a reporting interval of 1 to determine the first iteration in which there is a difference between the two computations. What are the absolute and relative errors for *accumulated* at that point? Is there an error for *multiplied*?

b. Run the program with a reporting interval of 1,000,000. What are the absolute and relative errors for both totals?

c. Run the program with a reporting interval of 999,999. What are the absolute and relative errors for both totals? Note: To compute the correct answer, take the correct answer for the millionth iteration and subtract 0.xxxx.

d. Run the program with a large enough number of iterations and a large reporting interval so that eventually the value of *accumulated* does not change from one report to the next. What explanation do you have for this phenomenon?

Answers to Quick Review Questions

1. **a.** 4500
   b. −3
   c. 4 because the four digits of 4500 are significant
2. **a.** $6.239 - 6.23 = 0.009$
   b. $(6.239 - 6.23)/6.239 = 0.144\%$
3. **a.** 0.94×10^{-5}
   b. 0.93×10^{-5}
   c. 0.94×10^{-5}
4. 12.37 because the sum is $12.35 + 0.02499 = 12.37499$, rounded to four significant digits.
5. $\dfrac{3r(x + y)}{zu}$ or $3r(x + y)/(zu)$. Note: In the second form, parentheses around the denominator product are essential because of priority of operations. Without the parentheses around zu, z would be divided into $3r(x + y)$ and incorrectly the result would be multiplied by u, which is equivalent to $\left(\dfrac{3r(x+y)}{z}\right) u = \dfrac{3r(x+y)u}{z}$.
6. **B.** $sum = 0.00492 * k$
7. $x(y + z) = 2.48(9.34 + 1.55) = 2.48(10.89)$. However, 10.89 truncates to 10.8. Thus, $2.48(10.8) = 26.784$, which truncates to 26.7.
   $x(y) + x(z) = 2.48(9.34) + 2.48(1.55) = 23.1632 + 3.844$, which after truncation is $23.1 + 3.84 = 26.94$, or 26.9.
   $x(y + z) = 26.7 \neq x(y) + x(z) = 26.9$
8. "If $|x - z| \geq 0.000001$, consider x and z not to be the same."
9. **a.** $2 - 2^3/(3!) = 2 - 8/6 = 2/3$
   b. $\dfrac{2^5}{5!} - \dfrac{2^7}{7!} + \cdots = \sin(2) - 2/3$

References

Arnold, Douglas N. 2000. "The Patriot Missile Failure." University of Minnesota, August 23. http://www.ima.umn.edu/~arnold/disasters/patriot.html

Darden, Lindley. 1998. "The Nature of Scientific Inquiry." University of Maryland, College Park. http://www.philosophy.umd.edu/Faculty/LDarden/sciinq/

Gleick, James. "Sometimes a Bug Is More Than a Nuisance." http://www.around.com/ariane.html

Panoff, Robert. 2004. National Computational Science Institute Workshop on Computational Science at Wofford College, Spartanburg, South Carolina.

MODULE 2.3

Rate of Change

Introduction

This and the next module are for the student who has not had Calculus I or who wants a review of the material to prepare for the remainder of the course. The concept of rate of change from the current module is particularly important. The material in the next module is important but optional to understanding the remainder of the text. The required calculus background for the text is minimal, and students do not need to know formulas to understand the material or develop the models.

Calculus is the mathematics of change. The concept of **instantaneous rate of change**, or **derivative**, from a first course in calculus is used throughout the text. For example, we consider the rates of change of populations of predators and prey to study the dynamics of their populations. The speed with which a radioactive substance decays certainly impacts the amount present at a particular time. In physics or as we drive a car, the rate of change of position with respect to time is the velocity.

Differential calculus, which deals with problems involving the derivative, is one of the two major parts of calculus. The other is **integral calculus**, which we consider in the next module. These two modules consider the concepts of differential and integral calculus that we need to solve modeling problems.

Velocity

We begin the discussion of instantaneous rate of change by considering the average rate of change. Suppose on a windless day someone standing on a bridge holds a ball over the side and tosses the ball straight up into the air. After reaching its highest point, the ball falls, eventually landing in the water. The ball's height above the water (y) is a function (s) of time (t), so we write $y = s(t)$. Figure 2.3.1 gives a plot of a ball's height in meters versus time in seconds, while Table 2.3.1 lists some of the function's values. The graph is not a plot of the ball's trajectory, which is straight up

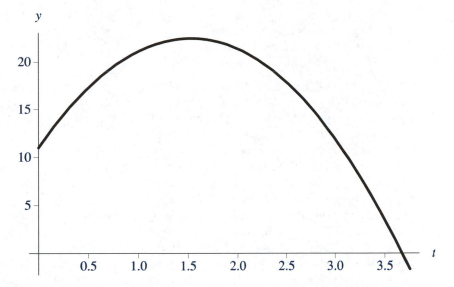

Figure 2.3.1 Height (y) in meters versus time (t) in seconds of a ball thrown straight up from a bridge

Table 2.3.1
Table of Times and Heights for a Ball Thrown Straight Up from a Bridge

Time (t) in Seconds	Height (y) in Meters
0.00	11.0000
0.25	14.4438
0.50	17.2750
0.75	19.4938
1.00	21.1000
1.25	22.0938
1.50	22.4750
1.75	22.2438
2.00	21.4000
2.25	19.9437
2.50	17.8750
2.75	15.1937
3.00	11.9000
3.25	7.9938
3.50	3.4750
3.75	−1.6563

and then straight down again. The height of the ball at water level is $y = 0$ m, and a negative value of y indicates that the ball is under water.

Quick Review Question 1

Consider the ball of Figure 2.3.1 and Table 2.3.1 in approximating the following, giving values and units:

a. The height of the bridge
b. The maximum height of the ball
c. When the ball reaches its maximum height
d. When the ball hits the water

As Figure 2.3.1 and Table 2.3.1 indicate, the ball changes position as time passes. The ball starts at one speed, moving up, and then slows down to a stop because of the effect of gravity. Subsequently, the ball starts to fall, picking up speed as it does, until finally it splashes into the water.

We can approximate the velocity at any particular time t if we know the heights at times shortly before and after t and compute the average velocity over that time period. Thus, an understanding of average velocity is essential to that of instantaneous velocity, or instantaneous rate of change of position with respect to time. The **average velocity** is the ratio of the change in height, or position, to the change in time. For example, if at noon we are 60 km from home and at 2:00 p.m. 260 km from home, then over that time period, we drove at an average speed of $(260 - 60)/2 = 100$ km/hr. For the position function $s(t)$, the average velocity is the average rate of change of s with respect to t. From Table 2.3.1, we see that the average velocity in the first second, which is from time $a = 0$ sec, to time $b = 1$ sec, is as follows:

$$\text{average velocity from 0 to 1 seconds} = \frac{s(1) - s(0)}{1 - 0} = \frac{21.1 - 11.0}{1} = 10.1 \, \text{m/sec}$$

The units for average velocity, meters/second, are the units for the numerator, which are meters, over the units for the denominator, which are seconds.

> **Definition**   Suppose $s(t)$ is the position of an object at time t, where $a \leq t \leq b$. The **average velocity**, or the **average rate of change of s with respect to** t, of the object from time a to time b is
>
> $$\text{average velocity} = \frac{\text{change in position}}{\text{change in time}} = \frac{s(b) - s(a)}{b - a}$$

Quick Review Question 2

Determine the average velocity of the ball from Table 2.3.1

a. From $t = 1$ sec to $t = 2$ sec
b. From $t = 1$ sec to $t = 3$ sec

In Quick Review Question 2, you should have determined the average velocity from 1 to 2 seconds to be 0.3 m/sec. Thus, on the average, the ball is moving faster the second before than the second after $t = 1$ sec. We can approximate the velocity of the ball at the instant $t = 1$ sec by finding the mean of the average velocities during the first second (10.1 m/sec) and the next second (0.3 m/sec), as follows:

$$\text{approximation of velocity at } t = 1 \text{ sec} = \frac{10.1 + 0.3}{2} = 5.2 \text{ m/sec}$$

Equivalently, we can evaluate the average velocity between times on either side of $t = 1$ sec. However, it is best to use known heights for times as close to $t = 1$ sec as possible, in this case for $t = 0.75$ sec and $t = 1.25$ sec.

Quick Review Question 3

Approximate the velocity of the ball at $t = 1$ sec by finding the average velocity from $a = 0.75$ sec to $b = 1.25$ sec.

A slight variation in the notation for determining the average velocity from a to b is advantageous for calculus. Instead of using b, we consider $b = a + \Delta t$, the initial time (a) plus a **change in time**, $\Delta t = b - a$, with Δt pronounced delta-t (Δ is the fourth letter of the Greek alphabet). For example, if $a = 0.75$ and $b = 1.25$, then $\Delta t = 1.25 - 0.75 = 0.50$ sec, which is the change in time; and $b = a + \Delta t = 0.75 + 0.50 = 1.25$ sec. The following definition employs this notation.

Definition   Suppose $s(t)$ is the position of an object at time t, where $a \leq t \leq b$. Then the **change in time**, Δt, is $\Delta t = b - a$; and the **change in position**, Δs, is $\Delta s = s(b) - s(a)$. Moreover, the **average velocity**, or the **average rate of change of s with respect to t**, of the object from time a to time $b = a + \Delta t$ is

$$\text{average velocity} = \frac{\text{change in position}}{\text{change in time}}$$

$$= \frac{\Delta s}{\Delta t}$$

$$= \frac{s(b) - s(a)}{b - a}$$

$$= \frac{s(a + \Delta t) - s(a)}{\Delta t}$$

Quick Review Question 4

Suppose we wish to determine the average velocity of the ball in Table 2.3.1 from time 2.25 sec to time 3.0 sec. Using the notation of the definition of average velocity involving Δt, determine the following, including units:

    a. a
    b. $s(a)$
    c. Δt

d. $a + \Delta t$

e. $s(a + \Delta t)$

f. Δs

g. The average velocity

To obtain the instantaneous velocity at $t = 1$ sec—that is, the *exact* velocity of the ball *precisely* one second after it starts to move, we determine the average velocity with changes in time, Δt, closer and closer to 0. Our answers approach a particular number, the instantaneous velocity at $t = 1$ sec. We say that we are taking the **limit of the average velocities as Δt approaches 0**, and we write the following:

$$\text{instantaneous velocity at 1 sec} = \lim_{\Delta t \to 0} \frac{s(1 + \Delta t) - s(1)}{\Delta t}$$

In the quotient, Δt can be positive or negative, but not zero.

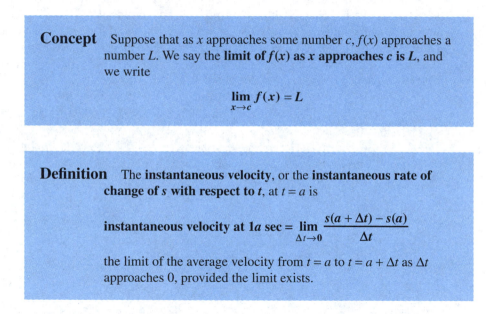

> **Concept**   Suppose that as x approaches some number c, $f(x)$ approaches a number L. We say the **limit of $f(x)$ as x approaches c is L**, and we write
>
> $$\lim_{x \to c} f(x) = L$$

> **Definition**   The **instantaneous velocity**, or the **instantaneous rate of change of s with respect to t**, at $t = a$ is
>
> $$\text{instantaneous velocity at } 1a \text{ sec} = \lim_{\Delta t \to 0} \frac{s(a + \Delta t) - s(a)}{\Delta t}$$
>
> the limit of the average velocity from $t = a$ to $t = a + \Delta t$ as Δt approaches 0, provided the limit exists.

Table 2.3.2 gives additional values of $s(t)$, possibly obtained experimentally, for t close to 1 along with the average velocities between each such time and $t = 1$. The table is in two parts. The left side has values of Δt starting at 0.10 and decreasing to 0.01 along with columns for the corresponding $s(1 + \Delta t)$ and the average velocity $\frac{s(1 + \Delta t) - s(1)}{\Delta t}$. The right side of the table has negative values of Δt from -0.10 to a value closer to 0, namely -0.01, along with the same second and third columns as on the left. Observing the third columns for both sides, the average velocities appear to be converging to 5.20. In fact,

$$\lim_{\Delta t \to 0} \frac{s(1 + \Delta t) - s(1)}{\Delta t} = 5.20$$

so that the instantaneous velocity of this ball at $t = 1$ sec is 5.20 m/sec.

Table 2.3.2
Average Velocities between $(1, s(1)) = (1, 21.1)$ and $(1 + \Delta t, s(1 + \Delta t))$

Δt	$s(1 + \Delta t)$	$\dfrac{s(1 + \Delta t) - s(1)}{\Delta t}$	Δt	$s(1 + \Delta t)$	$\dfrac{s(1 + \Delta t) - s(1)}{\Delta t}$
0.10	21.571	**4.710**	−0.10	20.531	**5.690**
0.09	21.528	**4.759**	−0.09	20.592	**5.641**
0.08	21.485	**4.808**	−0.08	20.653	**5.592**
0.07	21.440	**4.857**	−0.07	20.712	**5.543**
0.06	21.394	**4.906**	−0.06	20.770	**5.494**
0.05	21.348	**4.955**	−0.05	20.828	**5.445**
0.04	21.300	**5.004**	−0.04	20.884	**5.396**
0.03	21.252	**5.053**	−0.03	20.940	**5.347**
0.02	21.202	**5.102**	−0.02	20.994	**5.298**
0.01	21.152	**5.151**	−0.01	21.048	**5.249**

Table 2.3.3
Table for Quick Review Question 5

Δt	$\dfrac{s(2 + \Delta t) - s(2)}{\Delta t}$	Δt	$\dfrac{s(2 + \Delta t) - s(2)}{\Delta t}$
0.0005	27.8072	−0.0005	27.7928
0.0004	27.8058	−0.0004	27.7942
0.0003	27.8043	−0.0003	27.7957
0.0002	27.8029	−0.0002	27.7971
0.0001	27.8014	−0.0001	27.7986

Quick Review Question 5

Using Table 2.3.3, estimate $\displaystyle\lim_{\Delta t \to 0} \frac{s(2 + \Delta t) - s(2)}{\Delta t}$ to one decimal place.

Derivative

The limit $\displaystyle\lim_{\Delta t \to 0} \frac{s(a + \Delta t) - s(a)}{\Delta t}$ above has far more applications than instantaneous velocity at $t = a$. Because of its vast importance, the formula has a special name. In general, when the limit exists, $\displaystyle\lim_{\Delta t \to 0} \frac{s(a + \Delta t) - s(a)}{\Delta t}$ is the **derivative** of s with respect to t at a. We use two notations for the derivative of $y = s(t)$ with respect to t—the first is **dy/dt**, and the second is **$s'(t)$**. For the derivative of s at $t = 1$, which in this case is 5.20, we write $\left.\dfrac{dy}{dt}\right|_{t=1}$ or $s'(1) = 5.20$ m/sec. Notice that the units for the instantaneous and average velocities are the same, m/sec.

> **Definition**   The **derivative of** $y = s(t)$ **with respect to** t at $t = a$ is the instantaneous rate of change of s with respect to t at a:
>
> $$s'(a) = \left.\frac{dy}{dt}\right|_{t=a} = \lim_{\Delta t \to 0} \frac{s(a + \Delta t) - s(a)}{\Delta t}$$
>
> provided the limit exists. If the derivative of s exists at a, we say the function is **differentiable** at a.

Quick Review Question 6

Suppose the population P of a colony of bacteria in millions is a function of time t in hours. Give the units of dP/dt.

Although we will not do so here, we could show for the ball example that the derivative of s at $t = 0$ sec is $s'(0) = 15$ m/sec. Thus, initially, when the ball is at height $s(0) = 11.0$ m (see Table 2.3.1), the ball is increasing its height at a rate of 15 m/sec. With this information, we can estimate the height of the ball one second later at $t = 1$ sec as $11 + 15 = 26$ m. Because of the pull of gravity, the ball does not get quite that high; but the derivative can help us make estimates for the future as well as understand the current situation.

Quick Review Question 7

We know from Table 2.3.1 that $s(2.5) = 17.875$ m. Suppose $s'(2.5) = -9.5$.

    a. Give the units of 2.5.
    b. Give the units of −9.5.
    c. Using this information, estimate $s(3.5)$. Include units.
    d. Interpret this information.

Slope of Tangent Line

In this section, we consider graphically the instantaneous velocity, or the derivative $s'(t)$. Suppose we again wish to examine the velocity of the ball at time $t = 1$ sec. If we keep zooming in on the graph of the height of the ball $y = s(t)$ at time $t = 1$ sec, we observe an interesting phenomenon—the appearance of these graphs as we zoom in is increasingly linear. Figures 2.3.2 through 2.3.4 show plots of y versus t as we zoom in, first from $t = 0.75$ to 1.25 sec, then from $t = 0.9$ to 1.1 sec, and finally from $t = 0.99$ to 1.01 sec. At close range in Figure 2.3.4, the tangent line to $y = s(t)$ at $t = 1$ approximates the graph of $y = s(t)$.

In general, the slope of the tangent line to a curve at a point is the derivative of the function at that point. Figures 2.3.5 through 2.3.7 illustrate that the secant

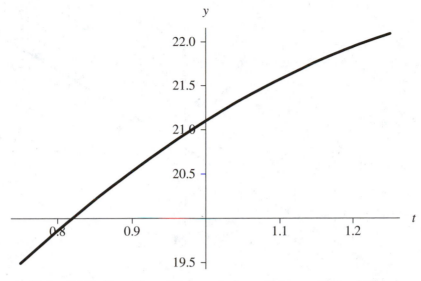

Figure 2.3.2   From Figure 2.3.1, graph of $y = s(t)$ from $t = 0.75$ to 1.25 sec

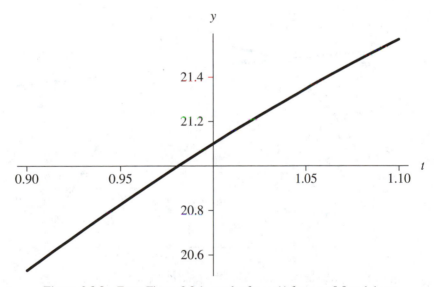

Figure 2.3.3   From Figure 2.3.1, graph of $y = s(t)$ from $t = 0.9$ to 1.1 sec

line through $t = 1$ and $t = 1 + \Delta t$ approaches the tangent line as Δt gets smaller. In Figure 2.3.5, the change in time is $\Delta t = 1.25$, and the slope of the secant line is as follows:

$$\frac{19.9437 - 21.1}{1.25} = -0.925$$

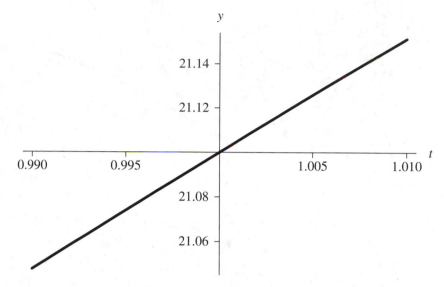

Figure 2.3.4   From Figure 2.3.1, graph of $y = s(t)$ from $t = 0.99$ to 1.11 sec

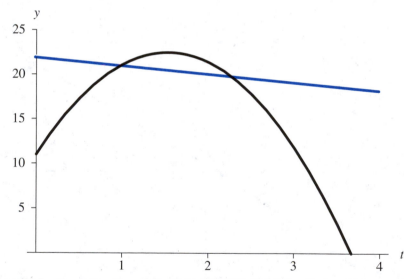

Figure 2.3.5   Secant line through $(1, 21.1)$ and $(2.25, 19.9437)$ with $\Delta t = 1.25$ sec and slope of -0.925 m/sec

As the next Quick Review Question shows, the slopes for $\Delta t = 0.75$ and 0.25 are 1.525 and 3.975, respectively (see Figures 2.3.6 and 2.3.7).

Definition   The **slope** of a nonvertical line through two distinct points (x_1, y_1) and (x_2, y_2) is $(y_2 - y_1)/(x_2 - x_1)$.

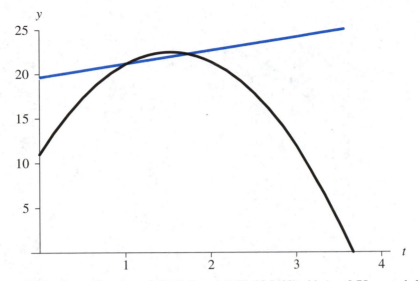

Figure 2.3.6 Secant line through (1, 21.1) and (1,75, 22.2438) with $\Delta t = 0.75$ sec and slope of 1.525 m/sec

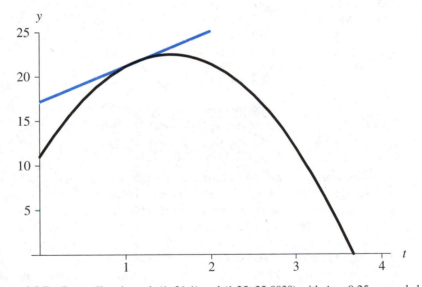

Figure 2.3.7 Secant line through (1, 21.1) and (1.25, 22.0938) with $\Delta t = 0.25$ sec and slope of 3.975 m/sec

Quick Review Question 8

    a. Show the calculation of the slope of the secant line through (1, 21.1) and (1.75, 22.2428) for Figure 2.3.6.

    b. Show the calculation of the slope of the secant line through (1, 21.1) and (1.25, 22.0938) for Figure 2.3.7.

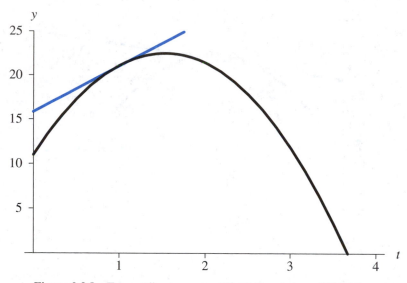

Figure 2.3.8   Tangent line to curve at (1, 21.1) and slope of 5.2 m/sec

As Figures 2.3.5 through 2.3.7 along with the graph of the tangent line in Figure 2.3.8 illustrate, the secant lines approach the tangent line as Δt goes to 0. Thus, the slopes of these secant lines approach the slope of the tangent line at $t = 1$. The slopes of the secant lines through $(1, 21.1)$ and $(1 + \Delta t, s(1 + \Delta t))$ are average velocities, and the limit as Δt approaches 0 is the instantaneous velocity, or the derivative, of the function s at 1. Thus, we can interpret the derivative at a point as the slope of the tangent line to the curve at that point. We also call this slope the **slope of the curve** at that point.

> **Concept**   Geometrically, the **derivative** at a point is the slope of the tangent line to the curve at that point.

The derivative of s at 1 is a number, 5.20, which is the slope of the tangent line to the curve s at $t = 1$ (see Figure 2.3.8). However, the tangent lines, and consequently their slopes, depend on the point on the curve. Table 2.3.4 presents a list of the slopes of some of the tangent lines to the curve s. As expected for this graph, the slopes are positive where the curve is increasing on the left and negative where the curve is decreasing on the right. Because the slope of the tangent line to the curve, and consequently the derivative, depends on t, we can define a **derivative function**, as follows:

$$s'(t) = \frac{dy}{dt} = \lim_{\Delta t \to 0} \frac{s(t + \Delta t) - s(t)}{\Delta t}, \quad \text{provided the limit exists}$$

> **Definition**   The **derivative function of $y = s(t)$ with respect to t** is the instantaneous rate of change of s, provided the limit exists:
>
> $$\frac{dy}{dt} = s'(t) = \lim_{\Delta t \to 0} \frac{s(t + \Delta t) - s(t)}{\Delta t}$$

Table 2.3.4
List of Some Values of t and the Slopes of the Tangent Lines
to s of Figure 2.3.8 at t

t	Slope of Tangent Line at t
0.0	15.0
0.5	10.1
1.0	5.2
1.5	0.3
2.0	−4.6
2.5	−9.5
3.0	−14.4
3.5	−19.3

Quick Review Question 9

Use Table 2.3.4 to evaluate the derivative function at the requested values.

       a. $s'(0.5)$ **b.** $s'(3.0)$

Although we will not verify the result, it can be shown that the derivative function for $y = s(t)$ is $s'(t) = -9.8t + 15$. As we have justified, $s'(1) = 5.20$, which is $-9.8(1) + 15$.

Quick Review Question 10

Using the fact that $s'(t) = -9.8t + 15$, determine the following along with their units:

    a. $s'(1.3)$
    b. The slope of the tangent line to s at $t = 2.9$ sec
    c. The instantaneous rate of change of s at $t = 0.4$ sec

Differential Equations

A **differential equation** is an equation that contains a derivative. For example, if y is a position of a ball above water at time t, then the rate of change, or derivative, of y with respect to t is the velocity. Suppose the velocity function is $v(t) = dy/dt = s'(t) = -9.8t + 15$, and the initial position, or **initial condition**, is $y_0 = s(0) = 11$. Thus, we have the following differential equation with initial condition:

$$dy/dt = -9.8t + 15 \quad \text{and} \quad y_0 = 11$$

or

$$s'(t) = -9.8t + 15 \quad \text{and} \quad s(0) = 11$$

To solve the differential equation means to find a function $y = s(t)$ that satisfies the differential equation and initial condition(s). As we discuss in the next module,

$y = s(t) = -4.9t^2 + 15t + 11$ is the solution to the above differential equation. Verifying the solution, we take the derivative of $y = s(t)$ to obtain $dy/dt = s'(t) = -9.8t + 15$. Moreover, substituting 0 for t in $s(t)$, we find that $y_0 = s(0) = 11$ also holds. The function $y = s(t) = -4.9t^2 + 15\,t + 11$ gives the height above the water as a function of time for the ball example of this module. We can obtain the height values in Tables 2.3.1 and 2.3.2 by substituting appropriate values of t into the function.

Definitions   A **differential equation** is an equation that contains one or more derivatives. An **initial condition** is the value of the dependent variable when the independent variable is zero. A **solution** to a differential equation is a function that satisfies the equation and initial condition(s).

Quick Review Question 11

It can be shown that the derivative of $y = 3t^6 + 7$ is $18t^5$. Why is $y = 3t^6 + 7$ not a solution to the differential equation $dy/dt = 18t^5$ with initial condition $y_0 = 14$?

Second Derivative

Acceleration is the rate of change of velocity with respect to time, and an instantaneous rate of change is a derivative. Thus, the derivative of a velocity function, $v(t)$, is an acceleration function, $a(t) = v'(t)$. However, a velocity function itself is a derivative, the derivative of a position function with respect to time; for $y = s(t)$, $v(t) = s'(t) = dy/dt$. Consequently, acceleration is the derivative of the derivative of position. If we take the derivative of a position function and then the derivative of the result, we obtain the corresponding acceleration function. We say that we have taken the **second derivative** of the position function and write $a(t) = s''(t) = d^2y/dt^2$. Notice the placements of the 2's in the latter stacked notation. This notation elicits the units for the second derivative. If velocity is in m/sec, then the units for acceleration are (m/sec)/sec or, inverting and multiplying, m/sec^2.

Definition   **Acceleration** is the rate of change of velocity with respect to time.

Definition   The **second derivative** of a function $y = s(t)$ is the derivative of the derivative of y with respect to the independent variable t. The notation for this second derivative is $s''(t)$ or d^2y/dt^2.

Quick Review Question 12

a. Suppose $z = f(x) = x^3$; $h(x) = f'(x) = 3x^2$; and $g(x) = h'(x) = 6x$. Evaluate $f''(x)$.
b. Give another notation for $f''(x)$.
c. If velocity is in ft/sec, give the units for acceleration.

Exercises

1. Use the following table of positions (s) of a car at various times (t).

t (hr)	4.0	4.5	5.0	5.5	6.0	6.5	7.0	7.5	8.0	8.5	9.0
s (km)	43.2	31.7	22.3	16.5	15.1	18.5	26.1	36.6	48.5	59.8	68.8

   a. Give the average velocity with units of the car between $t = 5.0$ hr and 9.0 hr.
   b. Estimate the velocity with units of the car at $t = 6.5$ hr.
   c. Estimate the rate of change with units of the car at $t = 4.5$ hr.
2. For the graph in Figure 2.3.9, estimate the following:
   a. The average rate of change of the function from $x = 0.5$ to $x = 1.5$
   b. The average rate of change of the function from $x = 1.5$ to $x = 2$
   c. The slope of the tangent line to the function at $x = 1$
   d. The instantaneous rate of change of the function at $x = 1$
   e. The derivative of the function at $x = 1$
   f. The slope of the tangent line to the function at $x = 0.5$
   g. The instantaneous rate of change of the function at $x = 0.5$
   h. The derivative of the function at $x = 0.5$

3. Table 2.3.5 shows values for $\frac{f(2+\Delta x) - f(2)}{\Delta x}$ as Δx approaches 0 through positive values on the first part of the table and through negative values

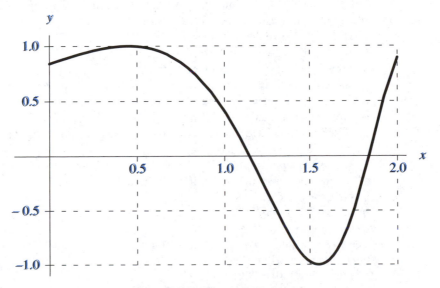

Figure 2.3.9   Graph for Exercise 2

Table 2.3.5
Table for Exercise 3

Δx	0.0051	0.0041	0.0031	0.0021	0.0011	0.0001	−0.0051	−0.0041	−0.0031	−0.0021	−0.0011	−0.0001
$\dfrac{f(2 + \Delta x) - f(2)}{\Delta x}$	3.1955	3.2187	3.2419	3.2649	3.2878	3.3107	3.4275	3.4053	3.3829	3.3605	3.3379	3.3152

on the second part of the table. Estimate the following to two decimal places:

a. $\lim\limits_{\Delta x \to 0} \dfrac{f(2 + \Delta x) - f(2)}{\Delta x}$

b. $f'(2)$

c. $\dfrac{dy}{dx}\Big|_{x=2}$, where $y = f(x)$

d. The slope of the tangent line to the graph of f at 2

e. The instantaneous rate of change of f at 2

4. Suppose $N = f(t)$ is the number of atoms of radium-226 at time t, which is in days. Because radium-226 is radioactive, the substance is decaying.
   a. Give the units for $f'(t)$.
   b. Give the sign for $f'(t)$.
   c. Give the units for $f''(t)$.

5. Suppose the number of tuna $y = T(t)$ in the Mediterranean Sea is a function of time t in years since 1914.
   a. Give the units of the rate of change of tuna numbers with respect to time.
   b. Give two notations for the rate of change of tuna numbers in 1918.
   c. Is it desirable for this rate to be positive or negative?
   d. Give the units for the second derivative of y.
   e. Give two notations for the second derivative of y in 1918.

6. Suppose the time of a chemical reaction, T (in minutes), to oxidize an alcohol is a function of the amount of a catalyst (alcohol dehydrogenase), a (in microliters), that is present. Thus, $T = f(a)$.
   a. If $f(4) = 13$, give the units of 4 and 13.
   b. If $f'(4) = -2$, give the units of 4 and −2.
   c. Interpret these statements taken together.

7. Suppose that on Day 4 of an epidemic in a school the rate at which students are developing influenza is 25 students/day.
   a. Give an interpretation of this rate as the derivative of a function.
   b. If 263 students have influenza on Day 4, estimate the number of students who will have influenza on Day 5.

8. $T = f(t)$ is the temperature in degrees Celsius of a beaker at time t (in hours) after someone places the beaker in a refrigerator.
   a. Give the units for $f'(t)$.
   b. Give the sign for $f'(t)$.
   c. Give the units for $f''(t)$.

9. The size of a drug's dose, S (in milligrams), depends on the weight of the patient, w (in pounds), so that $S = f(w)$.
   a. Interpret $f(150) = 200$.
   b. Interpret $f'(150) = 6$.

c. Using the information from Parts a and b, estimate $f(151)$.

d. Using the information from Parts a and b, estimate $f(155)$.

10. Use a computational tool or calculus to solve the following differential equation:

$$dP/dt = 0.3P - 20, \quad P_0 = 35$$

Project

1. Using a computational tool, such as *Maple*, *Mathematica*, or *MATLAB*, develop a file to explain and illustrate the material of this module. Use different functions than appear in the module for your examples. Employ looping and printing to generate sequences of values as in Tables 2.3.1 and 2.3.2.

Answers to Quick Review Questions

1. a. About 11 m

   b. About 22.5 m

   c. About 1.5 sec

   d. About 3.7 sec

2. a. Average velocity from 1 to 2 seconds $\dfrac{s(2)-s(1)}{2-1} = \dfrac{21.4-21.1}{1} = 0.3\,\text{m/sec}$

   b. Average velocity from 1 to 3 seconds $\dfrac{s(3)-s(1)}{3-1} = \dfrac{11.9-21.1}{2} = -4.6\,\text{m/sec}$

3. Approximation of velocity at $t = 1\,\text{sec} = \dfrac{s(1.25)-s(0.75)}{1.25-0.75} = \dfrac{22.0938-19.4938}{0.5} =$ $5.2\,m/sec$, which in this problem is the same as the mean of the average velocities during the first and second seconds. These values do not necessarily always agree.

4. a. 2.25 sec

   b. 19.9437 m

   c. 0.75 sec

   d. 3.0 sec

   e. 11.9 m

   f. $11.9000 - 19.9437 = -8.0437$ m

   g. $-8.0437/0.75 = -10.725$ m/sec. With up being positive, the average velocity is negative because the ball is falling during this time period.

5. 27.8

6. millions of bacteria/hour

7. a. sec

   b. m/sec

   c. $17.875 + -9.5 = 8.375$ m

   d. At time 2.5 sec, after one additional second (at time 3.5 sec), we estimate that the height of the ball will be 8.375 m.

8. a. $\dfrac{21.1-22.2438}{-0.75} = 1.525$

    b. $\dfrac{21.1 - 22.0938}{-0.25} = 3.975$

9. a. 10.1
    b. -14.4
10. a. $s'(1.3) = -9.8(1.3) + 15 = 2.26$ m/sec
    b. $s'(2.9) = -9.8(2.9) + 15 = -13.42$ m/sec
    c. $s'(0.4) = -9.8(0.4) + 15 = 11.08$ m/sec
11. Substituting 0 for t, $y = 7$, not 14.
12. a. $6x$

    b. $\dfrac{d^2 z}{dx^2}$
    c. ft/sec^2

Reference

Hughes-Hallet, Deborah, Andrew M. Gleason, William G. McCallum, et al. 2004. *Single Variable Calculus*. 3rd ed. New York: John Wiley & Sons.

MODULE 2.4

Fundamental Concepts of Integral Calculus

Introduction

In the last module, we examined some of the fundamental concepts of differential calculus, while in this module we consider several of the high points of the other major part of calculus, **integral calculus**. In a sense, **integration**, or determining the integral of a function, is the reverse of differentiation, or finding the derivative. For example, we can integrate the rate of change function for the amount of a radioactive substance to determine the total amount of decay over a period of time. As we discuss in this module, we can integrate the velocity function for a ball tossed in the air from a bridge—that is, we can integrate the rate of change function for distance of a ball above the water as in Module 2.3 on "Rate of Change"—to obtain the total distance covered from one time to another.

The material in this module is needed for some elective sections in the remainder of the text. While the concept of rate of change from the previous module is essential to understanding the remainder of the text, the concepts of the current module, though extremely important for higher-level computational science, are optional for this introduction to the subject.

Total Distance Traveled and Area

Suppose a car travels along a straight expressway for 2 hours at a constant velocity of 65 km/hr. How far does the car go in that time? Clearly, the answer is (2 hr)(65 km/hr) = 130 km. Figure 2.4.1 gives the graph of this velocity function $v = f(t) = 65$. Notice that the total distance traveled is the area under this curve from $t = 0$ hr to $t = 2$ hr.

Quick Review Question 1

Suppose someone sets the cruise control to drive at 70 km/hr for half an hour. Then, the speed limit changes, and the person resets the cruise control for 60 km/hr for the next hour and a half. Approximate the total distance traveled.

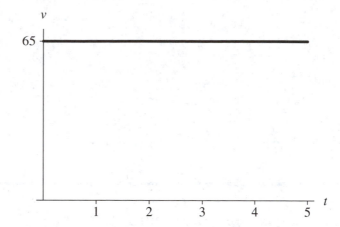

Figure 2.4.1   Graph of velocity function $v(t) = 65$ km/hr with t in hr

As another example, suppose a racecar moves with increasing velocity. The velocities in m/sec are recorded at various times, measured in seconds, in Table 2.4.1 and are displayed in Figure 2.4.2. We can estimate the total distance traveled over the 5 second period in several ways, including calculating under- and overestimates.

One underestimate of the total distance traveled takes the lowest velocity in each 1 second interval. For example, during the first second, from $t = 0$ sec to $t = 1$ sec, the lowest velocity (24 m/sec) occurs initially. If we travel at a constant velocity of 24 m/sec for 1 sec, then during that second we cover a total distance of (24 m/sec) (1 sec) = 24 m. As Figure 2.4.3 illustrates, this result is also the area of the rectangle in the first interval from the t-axis to the leftmost point (0, 24). This rectangle has height 24 and width 1. Because the velocity is increasing, during the next and each subsequent 1 second interval, the minimum velocity also occurs on the left. Summing the underestimates of the distances traveled for the five intervals, we obtain an underestimate of the total distance traveled for the 5 seconds, as follows:

$$\text{underestimate} = (24)(1) + (33)(1) + (40)(1) + (45)(1) + (48)(1) = 190 \text{ m}$$

Although six points appear in Table 2.4.1 and Figure 2.4.2, for the computation, we use only five points, one for each interval. The estimate, 190 m, for the total distance traveled is also the area of the shaded rectangles in Figure 2.4.3.

We can obtain an overestimate of the total distance traveled by using the largest velocity in each of the five intervals. For this increasing function, these velocities occur on the right of each interval. Consequently, we compute the sum of the areas of the rectangles that touch the intervals' rightmost points (see Figure 2.4.4) to obtain an overestimate of the total distance traveled:

$$\text{overestimate} = (33)(1) + (40)(1) + (45)(1) + (48)(1) + (49)(1) = 215 \text{ m}$$

Table 2.4.1
Values for the velocity of a car at certain times

t (sec)	0	1	2	3	4	5
v (m/sec)	24	33	40	45	48	49

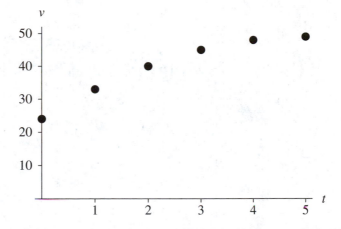

Figure 2.4.2   Plot of velocities from Table 2.4.1 with *t* in sec and *v* in m/sec

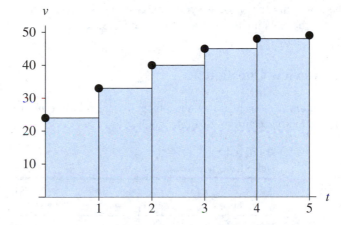

Figure 2.4.3   Underestimate of total distance traveled in m using intervals of 1 sec

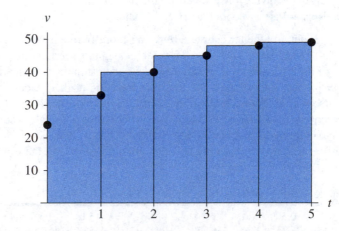

Figure 2.4.4   Overestimate of total distance traveled in m using intervals of 1 sec

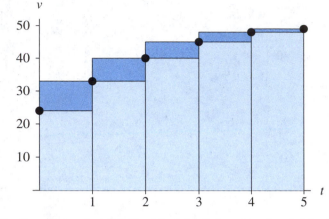

Figure 2.4.5   Over- and underestimates of the total distance in m traveled in m from Figures 2.4.3 and 2.4.4

The actual distance is between the two estimates, 190 m and 215 m. Figure 2.4.5 illustrates these estimates and shows the difference (25 m) between the two with darker shading.

Quick Review Question 2

Suppose after an hour of traveling, a bicyclist starts riding up a mountain with decreasing velocity, according to Table 2.4.2.

Table 2.4.2
Table for Quick Review Question 2

t (hr)	1.0	1.5	2.0	2.5	3.0
v (km/hr)	80.5	68.1	44.9	30.1	25.1

   a. Using intervals of half an hour, determine the best underestimate of the total distance the bicyclist travels up the mountain.
   b. Repeat Part a to obtain an overestimate.

We can obtain a better estimate by using more frequent velocity measurements. Table 2.4.3 and Figure 2.4.6 give the velocities in half-second intervals.

As we did for intervals of width 1 for the underestimate, we employ the minimum velocity in each interval as if the car were traveling at that velocity throughout that small time period. For example, in the first interval, a car traveling at 24.00 m/sec for 0.5 sec covers (24.00 m/sec)(0.5 sec) = 12.00 m during that half second. We must be careful to multiply by the change in time, 0.5 sec; the car does not move 24.00 m during the half-second interval but only half that amount. As above, the estimate of the dis-

Table 2.4.3
Additional Values for the Velocity of a Car at Certain Times

t (sec)	0.0	0.5	1.0	1.5	2.0	2.5	3.0	3.5	4.0	4.5	5.0
v (m/sec)	24.00	28.75	33.00	36.75	40.00	42.75	45.00	46.75	48.00	48.75	49.00

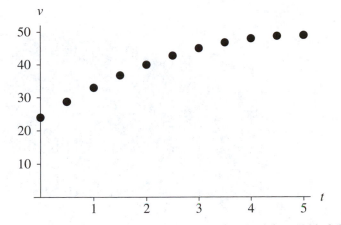

Figure 2.4.6   Plot of velocities (m/sec) versus time (sec) from Table 2.4.3

tance is the area of the rectangle for the width of the interval to the point. Figure 2.4.7 shades the 10 rectangles of width 0.5 whose total area is, as follows:

$$
\begin{aligned}
\text{underestimate} = \ & 24.00(0.5) + 28.75(0.5) + 33.00(0.5) + 36.75(0.5) \\
& + 40.00(0.5) + 42.75(0.5) + 45.00(0.5) + 46.75(0.5) \\
& + 48.00(0.5) + 48.75(0.5) \ \text{m} \\
= \ & 196.875 \ \text{m}
\end{aligned}
$$

This area is an underestimate of the total distance traveled, 196.875 m. To minimize the number of multiplications, we can factor out 0.5, adding the velocities and then multiplying by the length of the interval, as follows:

$$
\begin{aligned}
\text{underestimate} = \ & (24.00 + 28.75 + 33.00 + 36.75 + 40.00 + 42.75 \\
& + 45.00 + 46.75 + 48.00 + 48.75)(0.5) \ \text{m} \\
= \ & 196.875 \ \text{m}
\end{aligned}
$$

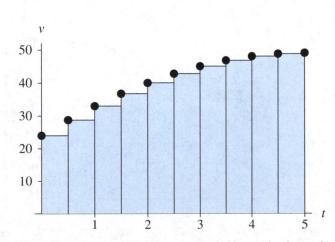

Figure 2.4.7   Underestimate of total distance traveled in m using intervals of 0.5 sec

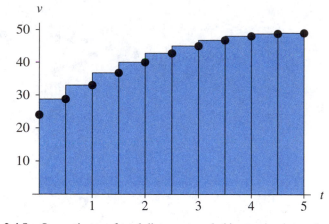

Figure 2.4.8   Overestimate of total distance traveled in m using intervals of 0.5 sec

Figure 2.4.8 shades the corresponding overestimate with the following computation:

$$
\begin{aligned}
\text{overestimate} &= 28.75(0.5) + 33.00(0.5) + 36.75(0.5) + 40.00(0.5) + 42.75(0.5) \\
&\quad + 45.00(0.5) + 46.75(0.5) + 48.00(0.5) + 48.75(0.5) + 49.00(0.5) \\
&= (28.75 + 33.00 + 36.75 + 40.00 + 42.75 + 45.00 + 46.75 \\
&\quad + 48.00 + 48.75 + 49.00)(0.5) \\
&= 209.375 \text{ m}
\end{aligned}
$$

Figure 2.4.9 indicates the difference between these two estimates, 209.375 m − 196.875 m = 12.5 m, which is one-half the difference for intervals of length 1. The estimates are converging as the width of an interval goes to zero and, simultaneously, the number of intervals goes to infinity.

Table 2.4.4 gives the values of Table 2.4.3 along with notations. The times for the velocities are $t_0 = 0.0$, $t_1 = 0.5$, $t_2 = 1.0, \ldots, t_{10} = 5.0$, and the corresponding

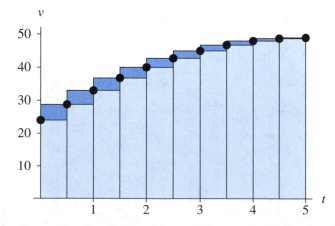

Figure 2.4.9   Over- and underestimates of the total distance traveled in m from Figures 2.4.7 and 2.4.8

Table 2.4.4
Table 2.4.3 with Notation

	t_0	t_1	t_2	t_3	t_4	t_5	t_6	t_7	t_8	t_9	t_{10}
t (sec)	0.0	0.5	1.0	1.5	2.0	2.5	3.0	3.5	4.0	4.5	5.0
	$f(t_0)$	$f(t_1)$	$f(t_2)$	$f(t_3)$	$f(t_4)$	$f(t_5)$	$f(t_6)$	$f(t_7)$	$f(t_8)$	$f(t_9)$	$f(t_{10})$
v (m/sec)	24.00	28.75	33.00	36.75	40.00	42.75	45.00	46.75	48.00	48.75	49.00

velocities are $f(t_0) = f(0.0) = 24.00$, $f(t_1) = f(0.5) = 28.75$, $f(t_2) = f(1.0) = 33.00, \ldots$, $f(t_{10}) = f(5.0) = 49.00$. The total segment goes from $a = 0.0$ to $b = 5.0$, and we have $n = 10$ intervals. We write the width of an interval, or the **change in t**, as $\Delta t = \frac{b-a}{n} = \frac{5-0}{10} = 0.5$ sec. For this example, the underestimate is a **left-hand sum**, where we use the velocity value on the left of each interval, as follows:

$$\text{left-hand sum} = 24.00(0.5) + 28.75(0.5) + 33.00(0.5) + \cdots + 48.75(0.5)$$
$$= f(t_0)\Delta t \quad + f(t_1)\Delta t \quad + f(t_2)\Delta t \quad + \cdots + f(t_9)\Delta t$$

Similarly, the overestimate is a **right-hand sum**, where we use the velocity value on the right of each interval, as follows:

$$\text{right-hand sum} = 28.75(0.5) + 33.00(0.5) + \cdots + 48.75(0.5) + 49.00(0.5)$$
$$= f(t_1)\Delta t \quad + f(t_2)\Delta t \quad + \cdots + f(t_9)\Delta t \quad + f(t_{10})\Delta t$$

Perhaps we know the function f from a model or by estimation from the data. For this particular function, if the number of intervals approaches infinity with Δt going to 0, the left- and right-hand sums approach $203\frac{1}{3}$. Thus, the total distance traveled from $t = 0$ to $t = 5$ sec, is $203\frac{1}{3}$, and as the figures indicate, $203\frac{1}{3}$ is also the area under the velocity curve in Figure 2.4.10.

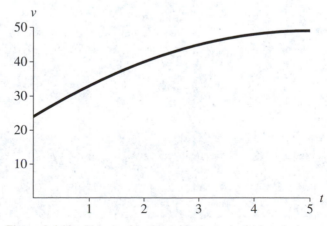

Figure 2.4.10 Velocity function with t in sec and v in m/sec

Quick Review Question 3

Using Quick Review Question 2, give values for the following:

    a. a
    b. b
    c. n
    d. Δt
    e. Times $t_0, t_1, \ldots, t_n$
    f. Velocities $f(t_0), f(t_1), \ldots, f(t_n)$, where $v = f(t)$ is the velocity function

Definite Integral

The limit of such a sum has so many applications other than computation of the total distance from velocity and the area under the curve that it has a special name and notation, the **definite integral**, as follows:

$$\int_a^b f(t)\,dt = \lim_{n \to \infty} (\text{left-hand sum}) = \lim_{n \to \infty} (f(t_0)\Delta t + f(t_1)\Delta t + \cdots + f(t_{n-1})\Delta t)$$

and

$$\int_a^b f(t)\,dt = \lim_{n \to \infty} (\text{right-hand sum}) = \lim_{n \to \infty} (f(t_1)\Delta t + f(t_2)\Delta t + \cdots + f(t_n)\Delta t),$$

where the width of an interval $\Delta t = (b - a)/n$ gets smaller as n gets larger.

> **Definitions**   If f is continuous (unbroken) for $a \le t \le b$, then the **left-hand sum** is
>
> $$\text{left-hand sum} = f(t_0)\Delta t + f(t_1)\Delta t + \cdots + f(t_{n-1})\Delta t$$
>
> and the **right-hand sum** is
>
> $$\text{right-hand sum} = f(t_1)\Delta t + f(t_2)\Delta t + \cdots + f(t_n)\Delta t$$
>
> where $\Delta t = (b - a)/n$.

> **Definitions**   The **definite integral** of f from a to b is
>
> $$\int_a^b f(t)\,dt = \lim_{n \to \infty} (\text{left-hand sum}) = \lim_{n \to \infty} (f(t_0)\Delta t + f(t_1)\Delta t + \cdots + f(t_{n-1})\Delta t$$

and

$$\int_a^b f(t)\,dt = \lim_{n \to \infty} (\text{right-hand sum}) = \lim_{n \to \infty} (f(t_1)\Delta t + f(t_2)\Delta t + \cdots + f(t_{n-1})\Delta t)$$

The function f is the **integrand**, and a and b are the **upper** and **lower limits of integration**, respectively.

Quick Review Question 4

Suppose $v = f(t)$ is the continuous (unbroken) velocity function for the cyclist in Quick Review Question 2. Give the following:

    a. The definite integral for the total distance the cyclist travels during the indicated time up the mountain

    b. The definite integral for the area under the curve during that period

    c. The limits of integration

Total Change

Above, we saw that the definite integral of a velocity function from time $t = 0$ to time $t = 5$ sec gives the total change in distance during that period. Velocity is the rate of change of position with respect to time, so the definite integral of the rate of change of position yields the total change in position. In general, the definite integral of a rate of change of a function is the total change in that function. Because a rate of change is a derivative, the following expresses this fact symbolically:

$$\int_a^b F'(t)\,dt = \left(\begin{array}{c} total\ change\ in\ F(t) \\ from\ t = a\ to\ t = b \end{array} \right) = F(b) - F(a)$$

For example, suppose the instantaneous rate of change of the number of **disintegrations per minute (dpm)** per gram (A) of radioactive carbon-14 in a gram of a dead tree is $dA/dt = -15.3e^{-0.000121t}$ dpm/g/year from the time t the tree dies (Higham; Mahaffy). An estimate of the total change in the number of particles of carbon-14 between years 10 and 20 is as follows:

$$\int_{10}^{20} A'(t)\,dt = \int_{10}^{20} (-15.3e^{-0.000121t})\,dt = A(20) - A(10)$$

Using a computational tool that integrates or knowledge of integration, we can calculate the answer as approximately -152.723 dpm/g. From year 10 to year 20, a gram of carbon from the dead tree loses about 153 dpm per gram of carbon-14.

Quick Review Question 5

Suppose at time $t = 5$ hr the rate of change of a population of bacteria is $417(2^t)$.

a. Give the appropriate notation to estimate the increase in the number of bacteria from $t = 5$ to $t = 9$ hr.

b. Using the answer from Part a in a computational tool that integrates, we can calculate this increase in the population as about 288,770 bacteria. If the number of bacteria at time $t = 5$ hr is about 18,650, estimate the number of bacteria at time $t = 9$ hr.

Fundamental Theorem of Calculus

We have observed that the definite integral of a rate of change, or derivative, of a function gives the total change in that function. This result is the essential connection between differential and integral calculus, and the name of the theorem that explicitly states the relationship, **The Fundamental Theorem of Calculus**, indicates its significance.

> **The Fundamental Theorem of Calculus**   If f is continuous (unbroken) on the interval from a to b and $f(t) = F'(t)$ is the derivative, or rate of change, of F with respect to t, then
>
> $$\int_a^b f(t)dt = F(b) - F(a)$$
>
> or
>
> $$\int_a^b F'(t)dt = F(b) - F(a)$$
>
> That is, the definite integral of a derivative, or a rate of change, of a function is the total change in the function from a to b.

If the derivative of F is f, or $F'(t) = f(t)$, we call the function F an **antiderivative** of f. For example, we can show that one antiderivative of the velocity function $f(t) = -t^2 + 10t + 24$, whose graph is in Figure 2.4.10, is $F(t) = -t^3/3 + 5t^2 + 24t$. An infinite number of antiderivatives exist for $f(t) = -t^2 + 10t + 24$, because the derivative of $-t^3/3 + 5t^2 + 24t + C$, where C is any constant, is also $-t^2 + 10t + 24$. Thus, $-t^3/3 + 5t^2 + 24t + 1$, $-t^3/3 + 5t^2 + 24t + 37.8$, $-t^3/3 + 5t^2 + 24t - 3$, etc., are all antiderivatives of $-t^2 + 10t + 24$. We call the most general antiderivative of $f(t) = -t^2 + 10t + 24$, namely, $-t^3/3 + 5t^2 + 24t + C$ for arbitrary constant C, the **indefinite integral** of f and employ a similar notation to that of the definite integral, as follows:

$$\int (-t^2 + 10t + 24)dt = -\frac{t^3}{3} + 5t^2 + 24t + C$$

> **Definition** F is an **antiderivative** of f if $F'(t) = f(t)$, or the derivative of
> F is f.

> **Definition**   The **indefinite integral** of $f(t)$ is $F(t) + C$, where $F'(t) = f(t)$
> and C is an arbitrary constant. The notation for the indefinite inte-
> gral is as follows:
>
> $$\int f(t)dt = F(t) + C$$

Quick Review Question 6

The derivative of $3x^6$ is $18x^5$. Using these functions, complete the following statements:

    a. _____ is an antiderivative of _____.

    b. $\int$ _____ = _____.

The velocity function $f(t) = -t^2 + 10t + 24$ is a rate of change, or derivative, of position with respect to time. The definite integral of f from $t = 0$ to $t = 5$ hr is the total
change in the position. $F(t) = -t^3/3 + 5t^2 + 24t$ is an antiderivative of f, or $F'(t) = f(t)$.
Thus,

$$\int_0^5 (-t^2 + 10t + 24)dt = F(5) - F(0)$$

We substitute 5 for t in $F(t)$ and subtract the substitution of 0 for t in $F(t)$, as
follows:

$$F(5) - F(0) = (-(5^3)/3 + 5(5^2) + 24(5)) - (-(0^3)/3 + 5(0^2) + 24(0)) = 203\tfrac{1}{3}$$

This value is the total change in position indicated above.

To recap, if a function f has an antiderivative F, then to calculate the definite integral of f from a to b, we compute $F(b) - F(a)$. However, not all functions have antiderivatives. In such cases, we estimate the definite integral using a technique of
numeric integration. We consider several such methods in the text. Many computational tools employ numeric integration techniques in their computations of definite integrals.

Quick Review Question 7

Using the fact that $3x^6$ is an antiderivative of $18x^5$, compute

$$\int_1^2 18x^5 dx$$

Differential Equations Revisited

Module 2.3 on "Rate of Change" defines a differential equation as an equation that contains a derivative. For example, suppose y is the position of a bicyclist at time t, and the following differential equation for the rate of change of y with respect to t gives the bicyclist's velocity function:

$$dy/dt = -t^2 + 10t + 24$$

We take the indefinite integral to find a general position function, as follows:

$$\int (-t^2 + 10t + 24)dt = -\frac{t^3}{3} + 5t^2 + 24t + C$$

Thus, the general position function is $y = -t^3/3 + 5t^2 + 24t + C$. We must have additional information to determine C. Frequently, we know the initial value of the function. For example, suppose initially, at time $t = 0$, the bicyclist is at position $y_0 = 30$ km from a starting location. Substituting 0 for t and 30 for y, we obtain a specific solution, namely, $y = -t^3/3 + 5t^2 + 24t + 30$, to the differential equation.

Quick Review Question 8

Using the fact that $\int 18x^5\, dx = 3x^6 + C$ from Quick Review Question 7, solve the differential equation $dy/dx = 18x^5$ with initial condition $y_0 = 14$.

Exercises

1. Suppose someone standing on a bridge throws a ball straight up over the water. With up being positive, suppose the velocity function for the ball is $v(t) = 15 - 9.8t$ in m/sec.
   a. When is the velocity zero? At this instant, the ball is at its highest point.
   b. Over what time period is the ball going up?
   c. Generate a table of values from $t = 0$ to $t = 4$ similar to Table 2.4.3 with $\Delta t = 0.5$.
   d. Using these values, under- and overestimate the total change in position (height) from $t = 0$ to $t = 1.5$ sec.
   e. Using these values, under- and overestimate the total change in position from $t = 2$ to $t = 4$ sec. Why is your result negative?
   f. Use a computational tool that integrates or an integration formula from calculus to solve the differential equation $dy/dt = 15 - 9.8t$ with initial condition $y_0 = 11$ m for the position (height) function $y(t)$. The solution is the height function whose graph is in Figure 2.3.4 of the module "Fundamental Concepts of Differential Calculus."
   g. Using the position function in Part f, determine when the ball hits the water. At this instant, the position of the ball is at 0 m.

h. Graph the velocity function from time $t = 0$ to $t = 4$ sec.

i. Using the formula for the area of a triangle and your answer from Part a, determine the area under the velocity curve from $t = 0$ sec to the time at which the velocity is zero. The area of a triangle is $0.5bh$, where b is the base and h is the height.

j. Using your answer from Part i, determine the total change in position of the ball from the time it is thrown until the time it reaches its highest point.

k. Using the formula for the area of a triangle (see Part i) and your answers from Parts a and g, determine the area between the velocity curve and the t-axis from the time at which the ball is at its highest point until it hits the water.

l. Using your answer from Part k, determine the total change in position of the ball from the time at which the ball is at its highest point until it hits the water. Why should your answer be negative?

m. Using your answers from Parts j and l, determine the ball's total change in position from $t = 0$ until it hits the water. How is your answer related to the initial condition $y_0 = 11$ m?

n. Using your answer for the position function from Part f, determine the ball's total change in position from $t = 0$ until it hits the water. Do your answers from this and Part m agree?

2. Use the facts that 1 meter = 3.281 feet and 1 mile = 5280 feet, to compute the following:

   a. The velocity 65 km/hr from Figure 2.4.1 in miles per hour (mph)

   b. The velocities 70 km/hr and 60 km/hr from Quick Review Question 1 in mph

   c. The velocities 24 m/sec and 49 m/sec from Table 2.4.1 in ft/sec

   d. The total distance traveled ($203\frac{1}{3}$ m) for that example in feet

   e. The velocities 80.5 km/hr and 25.1 km/hr from Quick Review Question 2 in mph

   f. Your answer for an underestimate total distance traveled from Quick Review Question 2a in miles

3. For Quick Review Question 5, use a computational tool that integrates or an integration formula from calculus to obtain the increase in the number of bacteria from $t = 5$ to $t = 9$ hr.

4. Suppose that Q is the total quantity of salt in pounds in a reservoir. During a certain period of time, the amount of salt is increasing due to runoff from rains at the rate $dQ/dt = 10e^{-0.01t}$ pounds/day.

   a. Generate a table of values from $t = 100$ to $t = 250$ days similar to Table 2.4.3 with $\Delta t = 50$ days.

   b. Using these values, under- and overestimate the total change in salt from $t = 100$ to $t = 250$ days.

   c. Repeat Part a using $\Delta t = 25$ days.

   d. Repeat Part b using $\Delta t = 25$ days.

   e. Use integration with an appropriate computational tool or an integration formula from calculus to determine the total change in salt from $t = 100$ to $t = 250$ days.

f. Repeat Part e for $t = 0$ to $t = 250$ days.

g. Use a computational tool that integrates or knowledge of calculus to solve the differential equation with initial condition $Q_0 = 0$ pounds.

Project

1. Using a computational tool that integrates, develop a document to explain and illustrate the material of this module. Use different functions than appear in the module for your examples.

Answers to Quick Review Questions

1. $70(0.5) + 60(1.5) = 125$ km. The velocity is multiplied by the length of time for each segment.

2. a. $(68.1)(0.5) + (44.9)(0.5) + (30.1)(0.5) + (25.1)(0.5) = 84.1$ km. Each interval lasts 0.5 hours. There are four half-hour periods, so the sum consists of four terms. Because the function is decreasing, we use the velocities at the end (on the right) of each interval for the underestimate.

   b. $(80.5)(0.5) + (68.1)(0.5) + (44.9)(0.5) + (30.1)(0.5) = 111.8$ km. Because the function is decreasing, we use the velocities on the left of each of the four intervals for the overestimate.

3. a. $a = 1$

   b. $b = 3$

   c. $n = 4$

   d. $\Delta t = 0.5$

   e. $t_0 = 1.0, t_1 = 1.5, t_2 = 2.0, t_3 = 2.5, t_4 = 3.0$

   f. $f(t_0) = 80.5, f(t_1) = 68.1, f(t_2) = 44.9, f(t_3) = 30.1, f(t_4) = 25.1$

4. a. $\int_1^3 f(t)dt$

   b. $\int_1^3 f(t)dt$

   c. 1 and 3

5. a. $\int_5^9 417(2^t)dt$

   b. 307,420 bacteria

6. a. $3x^6$ is an antiderivative of $18x^5$

   b. $\int 18x^5 dx = 3x^6 + C$

7. $189 = 3(2)^6 - 3(1)^6$

8. $y = 3x^6 + 14$

References

Higham, Thomas. "The ^{14}C Method." http://www.c14dating.com/int.html

Hughes-Hallet, Deborah, Andrew M. Gleason, William G. McCallum, et al. 2004. *Single Variable Calculus*. 3rd ed. New York: John Wiley & Sons: 623.

Mahaffy, Joseph M. "Math 122—Calculus for Biology II." San Diego State University. http://www-rohan.sdsu.edu/~jmahaffy/courses/f00/math122/labs/labe/q3v1.htm

3

SYSTEM DYNAMICS PROBLEMS WITH RATE PROPORTIONAL TO AMOUNT

MODULE 3.1

System Dynamics Tool: Tutorial 1

Download

From the textbook's website, download Tutorial 1 in pdf format for your system dynamics tool. We recommend that you work through the tutorial and answer all Quick Review Questions using the corresponding software.

Introduction

Dynamic systems, which change with time, are usually very complex, having many components with involved relationships. Two examples are systems involving competition among different species for limited resources and the kinetics of enzymatic reactions.

With a system dynamics tool, we can model complex systems using diagrams and equations. Thus, such a tool helps us perform Step 2 of the modeling process—formulate a model—by helping us document our simplifying assumptions, variables, and units; establish relationships among variables and submodels; and record equations and functions. Then, a system dynamics tool can help us solve the model—Step 3 of the modeling process—by performing simulations using the model and generating tables and graphs of the results. We use this output to perform Step 4 of the modeling process—verify and interpret the model's solution. Often such examination leads us to change a model. With its graphical view and built-in functions, a system dynamics tool facilitates cycling back to an earlier step of the modeling process to simplify or refine a model. Once we have verified and validated a model, the tool's diagrams and equations from the design and the results from the simulation should be part of our report, which we do in Step 5 of the modeling process. The tool can even help us as we maintain the model (Step 6) by making corrections, improvements, or enhancements.

This first tutorial is available for download from the textbook's website for several different system dynamics tools. Tutorial 1 in your system of choice prepares you to perform basic modeling with such a tool, including the following:

- Diagramming a model
- Entering equations and values
- Running a simulation
- Constructing graphs
- Producing tables

The module gives examples and Quick Review Questions for you to complete and execute with your desired tool.

MODULE 3.2

Unconstrained Growth and Decay

Introduction

Many situations exist where the rate at which an amount is changing is proportional to the amount present. Such might be the case for a population, say of people, deer, or bacteria. When money is compounded continuously, the rate of change of the amount is also proportional to the amount present. For a radioactive element, the amount of radioactivity decays at a rate proportional to the amount present.

Differential Equation

For example, suppose we have a population in which no individuals arrive or depart; the only change in the population comes from births and deaths. No constraints, such as competition for food or a predator, exist on growth of the population. When no limiting factor exists, we have the **Malthusian model** for unconstrained population growth, where the rate of change of the population is **directly proportional** ($\propto$) to the number of individuals in the population. If P represents the population and t represents time, then we have the following proportion:

$$\frac{dP}{dt} \propto P$$

For a positive growth rate, the larger the population, the greater the change in the population. With the same positive growth rate in two cities, say New York City and Spartanburg, S.C., the population of the larger New York City increases more in magnitude in a year than that of Spartanburg. In a later section in this module, "Unconstrained Decay," we consider a situation in which the rate is negative. We write the above proportion in equation form as follows:

$$\frac{dP}{dt} = rP$$

The constant r is the **growth rate** or **instantaneous growth rate** or **continuous growth rate**, while dP/dt is the **rate of change of the population**.

In "System Dynamics Tool: Tutorial 1" (Module 3.1), we started with a bacterial population of size 100, an instantaneous growth rate of $10\% = 0.10$, and time measured in hours. Thus, we had

$$\frac{dP}{dt} = 0.10P$$

with $P_0 = 100$. The equation $\frac{dP}{dt} = 0.10P$ with the **initial condition** $P_0 = 100$ is a **differential equation** because it contains a derivative. We begin by reconsidering this example from Tutorial 1 for reinforcement and a more in-depth examination of the concepts.

Difference Equation

Each diagram in Figure 3.2.1, developed with a choice of modeling tools and with the generic format employed in text, depicts the situation with stock (box variable) *population* indicating P, converter (variable) *growth_rate* representing r, and flow *growth* meaning dP/dt. Both population and growth rate are necessary to determine the growth. The growth is the additional number of organisms that joins the population.

For a simulation with a system dynamics tool or a program we write, we consider time advancing in small, incremental steps. For time t and length of a time step, Δt, the **previous time** is $t - \Delta t$. Thus, if t is 7.75 sec and Δt is 0.25 sec, the previous time is 7.50 sec. A system dynamics tool might call the change in time dt, DT, or something else instead of Δt. As some tools do to avoid confusion, we replace each blank in a diagram component name with an underscore when using the name in equations and discussions. For example, we employ "growth rate" in the diagrams of Figure 3.2.1 and the corresponding "growth_rate" in the following discussion. Regardless of the notation, with initial *population* $= 100$, *growth_rate* $= 0.1$, and *growth* $=$ *growth_rate* $*$ *population*, as in Figure 3.2.1, a system dynamics tool generates an equation similar to the following where *population*(t) is the population at time t and *population*($t - \Delta t$) is the population at time $t - \Delta t$:

$$population(t) = population(t - \Delta t) + (growth) * \Delta t$$

This equation, called a **finite difference equation**, indicates that the population at one time step is the population at the previous time step plus the change in population over that time interval:

$$(\text{new population}) = (\text{old population}) + (\text{change in population})$$

or

$$population(t) = population(t - \Delta t) + \Delta population$$

where $\Delta population$ is a notation for the **change in population**. We approximate the change in the population over one time step, $\Delta population$ or (*growth*) $* \Delta t$, as the

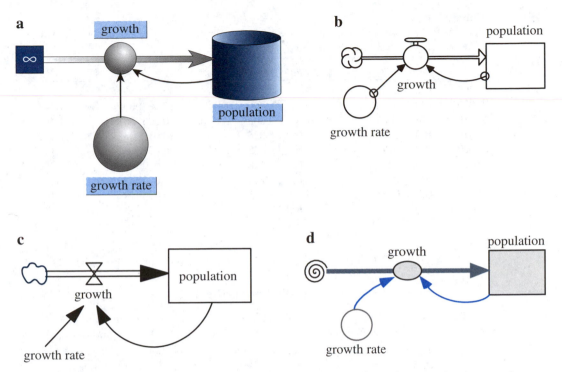

Figure 3.2.1   Diagrams of population models where growth rate is proportional to population: (a) Berkeley Madonna (b) *STELLA* (c) *Vensim PLE* (d) Text's format

finite difference of the populations at one time step and at the previous time step, *population*(t) − *population*($t − \Delta t$). Thus, solving for *growth*, we have an approximation of the derivative *dP/dt* as follows:

$$growth = \frac{\Delta population}{\Delta t} = \frac{population(t) - population(t - \Delta t)}{\Delta t}$$

Computer programs and system dynamics tools employ such finite difference equations to solve differential equations.

> **Definition**   A **finite difference equation** is of the following form:
>
> (new value) = (old value) + (change in value)
>
> Such an equation is a discrete approximation to a differential equation.

Quick Review Question 1

Consider the differential equation $dQ/dt = -0.0004Q$ with $Q_0 = 200$.

a. Using delta notation, give a finite difference equation corresponding to the differential equation.

Table 3.2.1

Table of Estimated Populations, Where the Initial Population is 100, the Continuous Growth Rate is 10% per Hour, and the Time Step is 0.005 hr

t	$population(t)$	$=$	$population(t - \Delta t)$	$+$	$(growth)$	$*$	Δt
0.000	100.000000						
0.005	100.050000	=	100.000000	+	10.000000	*	0.005
0.010	100.100025	=	100.050000	+	10.005000	*	0.005
0.015	100.150075	=	100.100025	+	10.010003	*	0.005
0.020	100.200150	=	100.150075	+	10.015008	*	0.005
0.025	100.250250	=	100.200150	+	10.020015	*	0.005
0.030	100.300375	=	100.250250	+	10.025025	*	0.005
0.035	100.350525	=	100.300375	+	10.030038	*	0.005
0.040	100.400701	=	100.350525	+	10.035053	*	0.005

b. At time $t = 9.0$ sec, give the time at the previous time step, where $\Delta t = 0.5$ sec.

c. If $Q(t - \Delta t) = 199.32$ and $Q(t) = 199.28$, give ΔQ.

The *growth* is the *growth_rate* (*r* above) times the current *population* (*P* above). For example, we can show that the population at time $t = 0.025$ hr is approximately *population*(0.025) = 100.250250 bacteria, so that *growth* is about *growth_rate* * *population*(0.025) = 0.1 * 100.250250 = 10.025025 bacteria per hour at that instant. For $\Delta t = 0.005$ hr, the change in the population of bacteria to the next time step, 0.025 + 0.005 = 0.030 hr, is approximately *growth* * Δt = 10.025025 * 0.005 = 0.050125 bacteria.* We calculate the population at time 0.030 hr as follows:

$$population(0.030) = population(0.025) + (growth \text{ at time } 0.025 \text{ hr}) * \Delta t$$
$$= 100.250250 + 10.025025 * 0.005$$
$$= 100.250250 + 0.050125$$
$$= 100.300375$$

Thus, we compute the value at the line $t = 0.030$ hr of Table 3.2.1 using the previous line.

Quick Review Question 2

Evaluate to six decimal places *population*(0.045), the population at the next time interval after the end of Table 3.2.1.

Because of the compounding, the number of bacteria at $t = 1$ hr is slightly more than 10% of 100, namely, 110.51. Table 3.2.2 lists the growth and the population on the hour for 20 hr, and Figure 3.2.2 graphs the population versus time. The model states and the table and figure illustrate that as the population increases, the growth does, too.

---

*Computations in this module use Euler's Method for estimating values of a function. In Chapter 5, we examine this and two other techniques for numeric integration.

Table 3.2.2
Table of Estimated Growths and Populations, Reported on the Hour,
Where the Initial Population is 100, the Growth Rate is 10%, and the
Time Step is 0.005 hr

t (hr)	growth	population
0.000	10.00	100.00
1.000	11.05	110.51
2.000	12.21	122.13
3.000	13.50	134.98
4.000	14.92	149.17
5.000	16.49	164.85
6.000	18.22	182.18
7.000	20.13	201.34
8.000	22.25	222.51
9.000	24.59	245.90
10.000	27.18	271.76
11.000	30.03	300.33
12.000	33.19	331.91
13.000	36.68	366.81
14.000	40.54	405.38
15.000	44.80	448.00
16.000	49.51	495.11
17.000	54.72	547.16
18.000	60.47	604.69
19.000	66.83	668.27
20.000		738.54

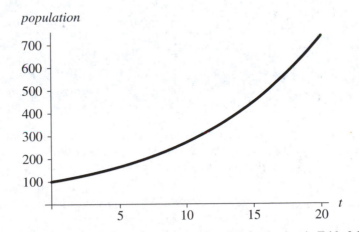

Figure 3.2.2   Graph of population versus time (hr) for the data in Table 3.2.2

The model gives an estimate of the population at various times. If the model is
analytically correct, a simulation estimates the values for *growth* and *population*.
Until round-off error causes the step size to be zero, it is usually the case that the
smaller the step size, the more accurate will be the results. (In Exercise 9, we explore a situation where the smaller step size does not produce better results.)

Simulation Program

In developing a simulation program, we use statements similar to the above finite difference equations. We initialize constants, such as *growthRate*, *population*, Δt, and the length of time the simulation is to run (*simulationLength*). The calculation for the total number of iterations (*numIterations*) of the loop is *simulationLength*/Δt. For example, if the simulation length is 10 hr and Δt is 0.25 hr, then the number of loop iterations is *numIterations* = 10/0.25 = 40. We have a loop index (*i*) go from 1 through *numIterations*. Inside the loop, we calculate time *t* as the product of *i* and Δt. For example, if Δt is 0.25 hr, during the first iteration, the index *i* becomes 1 and the time is 1 $*$ Δt = 0.25 hr. On loop iteration *i* = 8, the time gets the value 8 $*$ Δt = 8 $*$ 0.25 h = 4.00 hr.

Algorithm 1 contains the design for generating and displaying the time, growth, and population at each time step. In the algorithm, a **left-facing arrow** ($\leftarrow$) indicates assignment of the value of the expression on the right to the variable on the left. For example, *numIterations* $\leftarrow$ *simulationLength*/Δt represents an assignment statement in which *numIterations* gets the value of *simulationLength*/Δt.

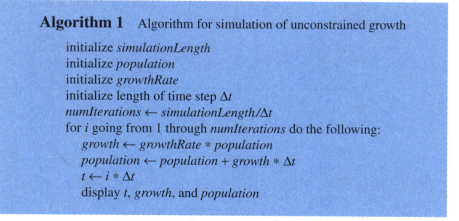

Algorithm 1   Algorithm for simulation of unconstrained growth

initialize *simulationLength*
initialize *population*
initialize *growthRate*
initialize length of time step Δt
numIterations $\leftarrow$ *simulationLength*/Δt
for *i* going from 1 through *numIterations* do the following:
   growth $\leftarrow$ *growthRate* $*$ *population*
   population $\leftarrow$ *population* + *growth* $*$ Δt
   t $\leftarrow$ *i* $*$ Δt
   display *t*, *growth*, and *population*

If we do not need to display *growth* (derivative) at each step and the length of a step (Δt) is constant throughout the simulation, we can calculate the constant growth rate per step (*growthRatePerStep*) before the loop, as follows:

$$growthRatePerStep \leftarrow growthRate * \Delta t$$

Within the loop, we do not compute *growth* but estimate *population* as follows:

$$population \leftarrow population + growthRatePerStep * population$$

Thus, within the loop, we have two assignments instead of three and two multiplication instead of three, saving time in a lengthy simulation. The revised algorithm appears as Algorithm 2.

Algorithm 2 Alternative algorithm to Algorithm 1 for simulation of unconstrained growth that does not display *growth*.

> initialize *simulationLength*
> initialize *population*
> initialize *growthRate*
> initialize Δt
> **growthRatePerStep ← growthRate * Δt**
> *numIterations ← simulationLength / Δt*
> for *i* going from 1 through *numIterations* do the following:
> > **population ← population + growthRatePerStep * population**
> > $t ← i * \Delta t$
> > display *t* and *population*

Analytical Solution Introduction

We can solve analytically the above model for unconstrained growth, which is the differential equation $dP/dt = 0.10P$ with initial condition $P_0 = 100$, as follows:

$$P = 100 \, e^{0.10t}$$

The next three sections develop the analytical solution. The first section starts the explanation using indefinite integrals, while the second section begins the discussion without using integrals. Thus, you may select the section that matches your calculus background. The third section completes the development of the analytical solution for both tracks.

When it is possible to solve a problem analytically, we should usually do so. We have employed simulation of unconstrained growth with a system dynamic tool as an introduction to fundamental concepts and as a building block to more complex problems for which no analytical solutions exist.

Analytical Solution: Explanation with Indefinite Integrals

We can solve the differential equation $dP/dt = 0.10P$ using a technique called **separation of variables**. First, we move all terms involving P to one side of the equation and all those involving t to the other. Leaving 0.10 on the right, we have the following:

$$\frac{1}{P} dP = 0.10 dt$$

Then, we integrate both sides of the equation, as follows:

$$\int \frac{1}{P} dP = \int 0.10 dt$$

$$\ln |P| = 0.10t + C \quad \text{for an arbitrary constant } C$$

We solve for $|P|$ by taking the exponential function of both sides and using the fact that the exponential and natural logarithmic functions are inverses of each other.

$$e^{\ln|P|} = e^{0.10t + C}$$

$$|P| = e^{0.10t} e^C = Ae^{0.10t}$$

where $A = e^C$. Solving for P, we have

$$P = (\pm A) e^{0.10t}$$

or

$$P = k e^{0.10t}$$

where $k = (\pm A)$ is a constant.

Analytical Solution: Explanation without Indefinite Integrals

We can solve the differential equation $dP/dt = 0.10P$ for P analytically by finding a function whose derivative is 0.10 times the function itself. The only functions that are their own derivative are exponential functions of the following form:

$$f(t) = ke^t, \quad \text{where } k \text{ is a constant}$$

For example, the derivative of $5e^t$ is $5e^t$. To obtain a factor of 0.10 through use of the chain rule, we have the general solution

$$P = ke^{0.10t}$$

For example, if $P = 5e^{0.10t}$, we have

$$\frac{dP}{dt} = \frac{d(5e^{0.10t})}{dt} = 5\frac{d(e^{0.10t})}{dt} = 5(0.10e^{0.10t}) = 0.10(5e^{0.10t}) = 0.10P$$

Completion of Analytical Solution

Using the initial condition that $P_0 = 100$, we can determine a particular value of k and, thus, a particular solution of the form $P = ke^{0.10t}$. Substituting 0 for t and 100 for P, we have the following:

$$100 = ke^{0.10(0)} = ke^0 = k(1) = k$$

The constant is the initial population. For this example,

$$P = 100e^{0.10t}$$

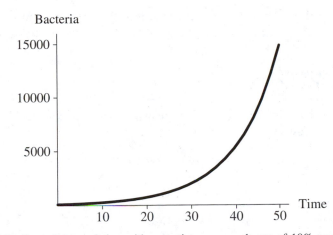

Figure 3.2.3 Bacterial population with a continuous growth rate of 10% per hour and an initial population of 100 bacteria

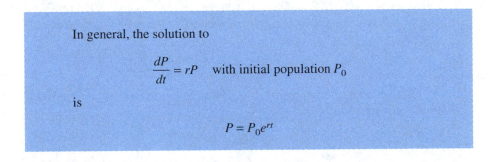

In general, the solution to

$$\frac{dP}{dt} = rP \quad \text{with initial population } P_0$$

is

$$P = P_0 e^{rt}$$

Figure 3.2.3 displays the dramatic increase in the bacterial population as time advances.

Quick Review Question 3

Give the solution of the differential equation

$$\frac{dP}{dt} = 0.03P, \quad \text{where } P_0 = 57$$

The simulated values for the bacterial population are slightly less than those the model $P = 100e^{0.10t}$ determines. For example, after 20 hours, a simulation may display in two decimal places a population of 738.54. However, $100e^{0.10(20)}$, expressed to two decimal places, is 738.91. The simulation compounds the population every step, and, in this case, the step size is $\Delta t = 0.005$ hr. The analytic model compounds the population continuously; that is, the analytic solution is the limit of simulated values as the step size goes to zero and the number of steps goes to infinity.

Both the analytic model and simulation produce valid estimates of the population of bacteria. After 20 hours, the number of bacteria will be an integer, not a decimal number, such as 738.54 or 738.91. Moreover, the population probably does not grow in an ideal fashion with a 10% per hour growth rate at every instant. Both the ana-lytic model and the simulation produce estimates of the population at various times.

Further Refinement

We can refine the model further by having separate parameters for birth rate and death rate instead of the combined growth rate. Thus,

$$growth\_rate = birth\_rate - death\_rate$$

Unconstrained Decay

The rate of change of the mass of a radioactive substance is proportional to the mass of the substance, and the constant of proportionality is negative. Thus, the mass decays with time. For example, the constant of proportionality for radioactive carbon-14 is ap-proximately -0.000120968. The continuous decay rate is about 0.0120968% per year, and the differential equation is as follows, where Q is the quantity (mass) of carbon-14:

$$\frac{dQ}{dt} = -0.000120968Q$$

As indicated in the "Completion of Analytical Solution" section, the analytical solu-tion to this equation is

$$Q = Q_0 \, e^{-0.000120968t}$$

After 10,000 years, only about 29.8% of the original quantity of carbon-14 remains, as the following shows:

$$Q = Q_0 \, e^{-0.000120968(10,000)} = 0.298293Q_0$$

Figure 3.2.4 displays the decay of carbon-14 with time.

Carbon dating uses the amount of carbon-14 in an object to estimate the age of an object. All living organisms accumulate small quantities of carbon-14, but accu-mulation stops when the organism dies. For example, we can compare the propor-tion of carbon-14 in living bone to that in the bone of a mummy and estimate the age of the mummy using the model.

Example 1

Suppose the proportion of carbon-14 in a mummy is only about 20% of that in a liv-ing human. To estimate the age of the mummy, we use the above model with the in-formation that $Q = 0.20Q_0$. Substituting into the analytical model, we have

$$0.20Q_0 = Q_0 \, e^{-0.000120968t}$$

After canceling Q_0, we solve for t by taking the natural logarithm of both sides of the equation. Because the natural logarithm and the exponential function are

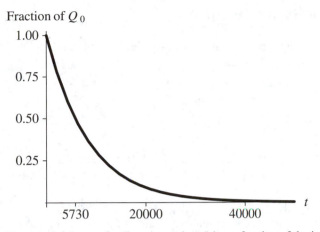

Figure 3.2.4   Exponential decay of radioactive carbon-14 as a fraction of the initial quantity Q_0, with time in years

inverses of each other, we have the following:

$$\ln(0.20) = \ln(e^{-0.000120968t}) = -0.000120968t$$
$$t = \ln(0.20)/(-0.000120968) \approx 13{,}305 \text{ yr}$$

We often express the rate of decay in terms of the half-life of the radioactive substance. The **half-life** is the period of time that it takes for the substance to decay to half of its original amount. Figure 3.2.5 illustrates that the half-life of radioactive carbon-14 is about 5730 years. We can determine this value analytically as we did in Example 1 using 50% instead of 20%; $Q = 0.50Q_0$.

Definition   The **half-life** is the period of time that it takes for a radioactive substance to decay to half of its original amount.

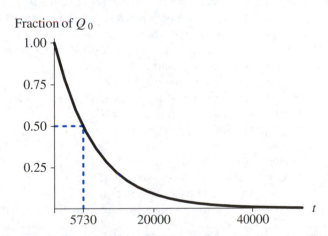

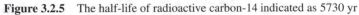

Figure 3.2.5   The half-life of radioactive carbon-14 indicated as 5730 yr

Quick Review Question 4

Radium-226 has a continuous decay rate of about 0.0427869% per year. Determine its half-life in whole years.

Exercises

1. **a.** For an initial population of 100 bacteria and a continuous growth rate of 10% per hour, determine the number of bacteria at the end of one week.
   b. How long will it take the population to double?
2. **a.** Suppose the initial population of a certain animal is 15,000, and its continuous growth rate is 2% per year. Determine the population at the end of 20 years.
   b. Suppose we are performing a simulation of the population using a step size of 0.083 yr. Determine the population at the end of the first three time steps.
3. Adjust the model in Figure 3.2.1 to accommodate birth rate and death rate instead of just growth rate.
4. **a.** **Newton's Law of Heating and Cooling** states that the rate of change of the temperature (T) with respect to time (t) of an object is proportional to the difference between the temperatures of the object and of its surroundings. Suppose the temperature of the surroundings is 25°C. Write the differential equation that models Newton's Law.
   b. Solve this equation for T as a function of time t.
   c. Suppose cold water at 6°C is placed in a room that has temperature 25°C. After one hour, the temperature of the water is 20°C. Determine all constants in the equation for T.
   d. What is the temperature of the water after 15 minutes?
   e. How long will it take for the water to warm to room temperature?
5. **a.** Suppose someone, whose temperature is originally 37°C, is murdered in a room that has constant temperature 25°C. The temperature is measured as 28°C when the body is found and at 27°C one hour later. How long ago was the murder committed from discovery of the body? See Exercise 4 for Newton's Law of Heating and Cooling.
   b. Suppose we are performing a simulation using a step size of 0.004 hr. Using the decay rate from Part a, determine the temperature at the end of the first three time steps after discovery of the body.
6. **a.** What proportion of the original quantity of carbon-14 is left after 30,000 years?
   b. If 60% is left, how old is the item?
7. **a.** The half-life of radioactive strontium-90 is 29 years. Give the model for the quantity present as a function of time.
   b. What proportion of strontium-90 is present after 10 years?
   c. After 50 years?
   d. How long will it take for the quantity to be 15% of the original amount?

8. Suppose an investment has approximately a continuous growth rate of 9.3%. Calculate analytically the value of an initial investment of $500 after

    a. 10 yr   **b.** 20 yr   **c.** 30 yr   **d.** 40 yr

    d. How long will it take for the value to double?
    e. Quadruple?

9. Suppose the amount of deposited ash, A, in millimeters (mm) is a function of time t in days. Suppose the model states that the rate of change of ash with respect to time is 4 mm/day and the initial quantity is 3 mm.
    a. Using a step size of 0.5 days, estimate the amount of ash when $t = 1$ day.
    b. Repeat Part a using a step size of 0.25 days.
    c. Does the smaller step size change the result?
    d. Solve the model for A.
    e. What kind of function do you obtain?

Projects

For additional projects, see Module 7.1 on "Radioactive Chains—Never the Same Again," Module 7.2 on "Turnover and Turmoil—Blood Cell Populations," Module 7.3 on "Deep Trouble—Ideal Gas Laws and Scuba Diving," Module 7.4 on "What Goes Around Comes Around—The Carbon Cycle," and Module 7.9 on "Mercury Pollution—Getting on Our Nerves."

For all model development, use an appropriate system dynamics tool.

1. Develop a model for Newton's Law of Heating and Cooling (see Exercise 4). Using this model, answer the questions of Exercises 4 and 5.
2. In 1854, Dr. John Snow, the Father of Epidemiology, identified a particular London water pump as the point source of the Broad Street Cholera Epidemic, which spread in a radial fashion from the pump. Model such a spread of disease assuming that the rate of change of the number of cases of cholera is proportional to the square root of the number of cases.
3. Develop a model for Exercise 8. Calculate the absolute and relative errors of the simulated values in comparison to the analytical values.
4. A young professional would like to save enough money to pay cash for a new car. Develop a model to determine when such a purchase will be possible. Take into account the following issues: The price of a new car is rising due to inflation. The buyer plans to trade in a car, which is depreciating. This person already has some savings and plans to make regular monthly payments. Thus, use a Δt value of 1 month. Assume appropriate rates and values.

Develop a computational tool file for each of Projects 5–11.

5. Algorithm 1 to perform the simulation discussed in the "Simulation Program" section
6. Algorithm 2 adjusted so that the user enters the report interval, which should be an integer multiple of Δt
7. Algorithm 1 to perform a simulation for Exercise 2

8. Algorithm 1 to perform a simulation for Exercise 3
9. Model for Newton's Law of Heating and Cooling (see Exercise 4). Using this model, answer the questions of Exercises 4 and 5.
10. Project 2
11. Project 4

Develop a spreadsheet for each of Projects 12–15.

12. Exercise 2
13. Exercise 4
14. Exercise 5
15. Exercise 8. Calculate the absolute and relative errors of the simulated values in comparison to the analytical values.

Answers to Quick Review Questions

1. a. $Q(t) = Q(t - \Delta t) + \Delta Q$, where $\Delta Q = -0.0004Q(t - \Delta t) \cdot \Delta t$ and $Q(0) = 200$
  b. $t - \Delta t = 9.0 - 0.5 = 8.5$ sec
  c. $\Delta Q = 199.28 - 199.32 = -0.04$
2. 100.450901

$$growth = 100.400701 * 0.10 = 10.040070$$

Thus, $population(0.045) = 100.400701 + 10.040070 * 0.005 = 100.450901$

3. $P = 57e^{0.03t}$
4. 1620.
  Reasoning:

  $Q = Q_0\, e^{-0.000427869t}$
  For $Q = 0.50Q_0$, $0.50Q_0 = Q_0\, e^{-0.000427869t}$ or $0.50 = e^{-0.000427869t}$
  $\ln(0.50) = -0.000427869t$
  $t = \ln(0.50)/(-0.000427869) = 1620$

Reference

Zill, Dennis G. 2001. *A First Course in Differential Equations with Modeling Applications*. 7th ed. Pacific Grove, Calif: Brooks/Cole Publishing Co.

MODULE 3.3

Constrained Growth

Introduction

If you introduce an animal to a new environment, it will often reproduce at a very high rate. That is what happened when the Australian Bureau of Sugar Experimental Stations introduced about 3000 marine toads (*Bufo marinus*) to sugar cane planta-tions in northern Queensland, Australia in 1935. The toads were supposed to feed on two species of beetle grubs that fed on roots of sugar cane, killing it. Unfortunately, the population of toads exploded, and they extended their range well beyond the cane fields. They had no natural predators and plenty of food. Female toads can lay up to 35,000 eggs twice a year, and adults can reach sexual maturity in less than two years. Forty years after their introduction, they numbered almost 600,000. They are considered quite a nuisance, poisoning and feeding on various native animals. Other parts of Australia and New Zealand maintain constant surveillance to prevent further spread of this unwelcome creature.

Because births exceed the numbers maturing and reproducing, all populations, the-oretically, have the potential for exponential growth. Endemic populations increase rapidly at first, but they eventually encounter resistance from the environment—competitors, predators, limited resources, disease. Thus, the environment tends to limit the growth of populations, so that they usually increase only to a certain level and then do not increase or decrease drastically unless a change in the environment occurs. This maximum population size that the environment can support indefinitely is termed the **carrying capacity**. Many introduced species that become pests, such as the marine toad in Australia, have a very high carrying capacity because there are few competitors, no predators, vast resources, and little disease.

Carrying Capacity

In Module 3.2 on "Unconstrained Growth and Decay," we considered a population growing without constraints, such as competition for limited resources. For such a population P with instantaneous growth rate r, the rate of change of the population has the following differential equation model:

$$\frac{dP}{dt} = rP$$

With initial population P_0, we saw that the analytical solution is $P = P_0 e^{rt}$. In that module, we also developed the following finite difference equation for the change in P from one time to the next, which we used in simulations:

$$\Delta P = P(t) - P(t - \Delta t)$$
$$= (r\, P(t - \Delta t))\Delta t$$

The simulation and analytical solution graphs in Figures 3.2.2 and 3.2.3, respectively, of Module 3.2 display the exponential growth of unconstrained growth.

After developing such a model in Step 2 of the modeling process and solving the model (Step 3) as above, we should verify that the solution (Step 4) agrees with real data. However, as examples in the introduction to the current module illustrate, no confined population can grow without bound. Competition for food, shelter, and other resources eventually limits the possible growth. For example, suppose a deer refuge can support at most 1000 deer. We say that the carrying capacity (M) for the deer in the refuge is 1000.

> **Definition**   The **carrying capacity** for an organism in an area is the maximum number of organisms that the area can support.

Quick Review Question 1

Cycling back to Step 2 of the modeling process, this question begins refinement of the population model to accommodate descriptions of population growth from the "Introduction" of this module.

   a. Determine any additional variable and its units.
   b. Consider the relationship between the number of individuals (P) and carrying capacity (M) as time (t) increases. List all the statements below that apply to the situation where the population is much smaller than the carrying capacity.
      A. P appears to grow almost proportionally to t.
      B. P appears to grow almost without bound.
      C. P appears to grow faster and faster.
      D. P appears to grow more and more slowly.
      E. P appears to decline faster and faster.
      F. P appears to decline more and more slowly.
      G. P appears to grow almost linearly with slope M.
      H. P is appears to be approaching M asymptotically.
      I. P appears to grow exponentially.
      J. dP/dt appears to be almost proportional to P.

K. dP/dt appears to be almost zero.
L. The birth rate is about the same as the death rate.
M. The birth rate is much greater than the death rate.
N. The birth rate is much less than the death rate.
 c. List all the choices from Part b that apply to the situation where the population is close to but less than the carrying capacity.
 d. List all the choices from Part b that apply to the situation where the population is close to but greater than the carrying capacity.

Revised Model

In the revised model, for an initial population much lower than the carrying capacity, we want the population to increase in approximately the same exponential fashion as in the earlier unconstrained model. However, as the population size gets closer and closer to the carrying capacity, we need to dampen the growth more and more. Near the carrying capacity, the number of deaths should be almost equal to the number of births so that the population remains roughly constant. To accomplish this dampening of growth, we could compute the number of deaths as a changing fraction of the number of births, which we model as rP. When the population is very small, we want the fraction to be almost zero, indicating that few individuals are dying. When the population is close to the carrying capacity, the fraction should be almost $1 = 100\%$. For populations larger than the carrying capacity, the fraction should be even larger so that the population decreases in size through deaths. Such a fraction is P/M. For example, if the population P is 10 and the carrying capacity M is 1000, then $P/M = 10/1000 = 0.01 = 1\%$. For a population $P = 995$ close to the carrying capacity, $P/M = 995/1000 = 0.995 = 99.5\%$; and for the excessive $P = 1400$, $P/M = 1400/1000 = 1.400 = 140\%$.

Thus, we can model the instantaneous rate of change of the number of deaths (D) as the fraction P/M times the instantaneous rate of change of the number of births (rP), as the following differential equation indicates:

$$\frac{dD}{dt} = \left(r\frac{P}{M} \right)P$$

The differential equation for the instantaneous rate of change of the population subtracts this value from the instantaneous rate of change of the number of births, as follows:

$$\frac{dP}{dt} = \underbrace{(rP)}_{\text{births}} - \underbrace{\left(r\frac{P}{M} \right)P}_{\text{deaths}}$$

or

$$\frac{dP}{dt} = r\left(1 - \frac{P}{M} \right)P \tag{1}$$

For the discrete simulation, where $P(t-1)$ is the population estimate at time $t-1$, the number of deaths from time $t-1$ to time t is

$$\Delta D = \left(r\frac{P(t-1)}{M} \right)P(t-1) \quad \text{for } \Delta t = 1$$

We approximate the number of deaths from time $(t-\Delta t)$ to time t by multiplying the corresponding value by Δt, as follows:

$$\Delta D = \left(r\frac{P(t-\Delta t)}{M} \right)P(t-\Delta t)\Delta t$$

where $P(t-\Delta t)$ is the population estimate at $(t-\Delta t)$. Thus, the change in population from time $(t-\Delta t)$ to time t is the difference of the number of births and the number of deaths over that period:

$$\Delta P = \text{births} - \text{deaths}$$

$$\Delta P = \underbrace{(rP(t-\Delta t))\Delta t}_{\text{births}} - \underbrace{\left(r\frac{P(t-\Delta t)}{M} \right)P(t-\Delta t)\Delta t}_{\text{deaths}}$$

$$= (r\Delta t)\left(1 - \frac{P(t-\Delta t)}{M} \right)P(t-\Delta t)$$

or

$$\Delta P = k\left(1 - \frac{P(t-\Delta t)}{M} \right)P(t-\Delta t), \quad \text{where } k = r\Delta t \qquad (2)$$

Differential equation (1) and difference equation (2) are called **logistic equations**. Figure 3.3.1 displays the S-shaped curve characteristic of a logistic equation, where the initial population is less than the carrying capacity of 1000. Figure 3.3.2 shows how the population decreases to the carrying capacity when the initial population is 1500. Thus, the model appears to match observations from the "Introduction" qualitatively. To verify a particular model, we should estimate parameters, such as birth rate, and compare the results of the model to real data.

Quick Review Question 2

a. Complete the difference equation to model constrained growth of a population P with respect to time t over a time step of 0.1 units, given that the population at time $t-\Delta t$ is $p \le 1000$, the carrying capacity is 1000, the instantaneous rate of change of births is 105%, and the initial population is 20.

$$\Delta P = \underline{\quad}(\underline{\quad}\ \underline{\quad}\ \underline{\quad})(p)(0.1)$$

b. What is the maximum population?

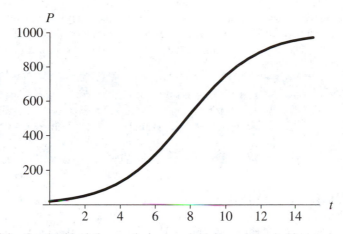

Figure 3.3.1 Graph of logistic equation, where initial population is 20, carrying capacity is 1000, and instantaneous rate of change of births is 50%, with time (t) in years

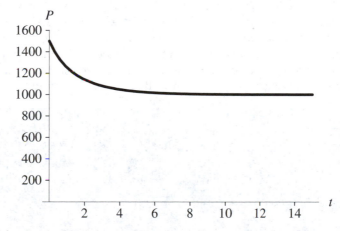

Figure 3.3.2 Graph of logistic equation, where initial population is 1500, carrying capacity is 1000, and instantaneous rate of change of births is 50%, with time (t) in years

c. Suppose the population at time $t = 5$ yr is 600 individuals. What is the population, rounded to the nearest integer, at time 5.1 years?

Equilibrium and Stability

The logistic equation with carrying capacity $M = 1000$ has an interesting property. If the initial population is less than 1000, as in Figure 3.3.1, the population increases to a limit of 1000. If the initial population is greater than 1000, as in Figure 3.3.2, the population decreases to the limit of 1000. Moreover, if the initial population is 1000, we see from Equation (1) that $P/M = 1000/1000 = 1$ and $dP/dt = r(1 - 1)P = 0$. In discrete terms, $\Delta P = 0$. A population starting at the carrying capacity remains there. We say that $M = 1000$ is an **equilibrium** size for the population because the population remains steady at that value or $P(t) = P(t - \Delta t) = 1000$ for all $t > 0$.

> **Definition**   An **equilibrium solution** for a differential equation is a solution where the derivative is always zero. An **equilibrium solution** for a difference equation is a solution where the change is always zero.

Quick Review Question 3

Give another equilibrium size for the logistic differential equation (1) or logistic difference equation (2).

Even if an initial positive population does not equal the carrying capacity $M = 1000$, eventually, the population size tends to that value. We say that the solution $P = 1000$ to the logistic equation (1) or (2) is **stable**. By contrast, for a positive carrying capacity, the solution $P = 0$ is **unstable**. If the initial population is close to but not equal to zero, the population does not tend to that solution over time. For the logistic equation, any displacement of the initial population from the carrying capacity exhibits the limiting behavior of Figures 3.3.1 or 3.3.2. In general, we say that a solution is stable if for a small displacement from the solution, P tends to the solution.

> **Definition**   Suppose that q is an equilibrium solution for a differential equation dP/dt or a difference equation ΔP. The solution q is **stable** if there is an interval (a, b) containing q, such that if the initial population $P(0)$ is in that interval then
>
> 1. $P(t)$ is finite for all $t > 0$
> 2. $\lim_{t \to \infty} P(t) = q$
>
> The solution q is **unstable** if no such interval exists.

Exercises

1. Using calculus or a computational tool, solve the following:
   a. The differential equation (1),

$$\frac{dp}{dt} = r\left(1 - \frac{P}{M}\right)P$$

   where the carrying capacity M is 1000, $P_0 = 20$, and the instantaneous rate of change of the number of births r is 50%.
   b. The differential equation (1) in general.

2. Consider $dy/dt = \cos(t)$.
  a. Give all the equilibrium solutions.
  b. Give a function $y(t)$ that is a solution.
  c. Give the most general function y that is a solution.

3. It has been reported that a mallard must eat 3.2 ounces of rice each day to remain healthy. On the average, an acre of rice in a certain area yields 110 bushels per year; and a bushel of rice weighs 45 pounds. Assuming that in the area 100 acres of rice are available for mallard consumption and mallards eat only rice, determine the carrying capacity for mallards in the area (Reinecke).

4. The **Gompertz differential equation**, which follows, is one of the best models for predicting the growth of cancer tumors:

$$\frac{dN}{dt} = kN \ln\!\left(\frac{M}{N}\right), \quad N(0) = N_0$$

where N is the number of cancer cells and k and M are constants.
  a. As N approaches M, what does dN/dt approach?
  b. Make the substitution $u = \ln(M/N)$ in the Gompertz equation to eliminate N and convert the equation to be in terms of u.
  c. Solve the transformed differential equation for u.
  d. Using the relationship between u and N from Part b, convert your answer from Part c to be in terms of N. The result is the solution to the Gompertz differential equation.
  e. Verify that $N(t) = Me^{\ln(N_0/M)e^{-kt}}$ is the solution to the Gompertz differential equation.
  f. Using the solution in Part e, what does N approach as t goes to infinity?

5. a. Graph $y = e^{-t}$.
Match each of the following scenarios to a differential equation that might model it.

A. $dP/dt = 0.05P$ **B.** $dP/dt = 0.05P + e^{-t}$
C. $dP/dt = 0.05(1 - e^{-t})P$ **D.** $dP/dt = 0.05P - 0.0003P^2 - 400$
E. $dP/dt = 0.05e^{-t}P$ **F.** $dP/dt = 0.05P - 0.0003P^2$

  b. At first, a bacterial colony appears to grow without bound; but because of limited medium, the population eventually approaches a limit.
  c. Because of degradation of its medium, the growth of a bacterial colony becomes dampened.
  d. A bacterial colony has unlimited medium and grows without bound.
  e. Because of adjustment to its new medium, a bacterial colony grows slowly at first before appearing to grow without bound.
  f. Each day, a scientist removes a constant amount from the colony.

Projects

For additional projects, see Module 7.4 on "What Goes Around Comes Around— The Carbon Cycle" and Module 7.5 on "A Heated Debate—Global Warming."

For all model development, use an appropriate system dynamics tool.

1. Develop a model for constrained growth.
2. Develop a simulation for constrained growth in a programming language.
3. Develop a model for the mallard population in Exercise 3. Have a converter or variable for the number of acres of rice available for mallard consumption, and from this value, have the model compute the carrying capacity. Report on the effect of decreasing the number of acres of rice available (Reinecke).
4. In some situations, the carrying capacity itself is dynamic. For example, the performance of airplanes had one carrying capacity with piston engines and a higher limit with the advent of jet engines. Many think that human population growth over a limited period of time follows such a pattern as technological changes enable more people to live on the available resources. In such cases, we might be able to model the carrying capacity itself as a logistic. Suppose M_1 is the first carrying capacity and M_2 is the second with $M_1 < M_2$. The differential equation for the carrying capacity $M(t)$ as a function of time t would be as follows:

$$\frac{dM(t)}{dt} = a(M(t) - M_1)\left(1 - \frac{M(t) - M_1}{M_2}\right) \quad \text{for some constant } a > 0$$

By using $M(t)$, we have a logistic for the carrying capacity as well as a logistic for the population. Figure 3.3.3 displays $P(t)$ in black and $M(t)$ in color with the first carrying capacity $M_1 = 20$, the second $M_2 = 70$, and an inflection point for M at $t = 450$.

Develop a model for the scenario below. First, generate an appropriate logistic carrying capacity $M(t)$. Then, use this dynamic carrying capacity to limit the population.

In a population study of England from 1541 to 1975, starting with a population of about 1 million, early islanders appear to have a carrying capacity of around 5 million people. However, beginning about 1800 with the advent

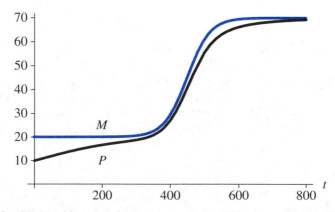

Figure 3.3.3 Graphs of functions for carrying capacity, $M(t)$, and population, $P(t)$, with time (t) in years.

of the Industrial Revolution, the carrying capacity appears to have increased to about 50 million people. The change in the concavity from concave up to concave down for this new logistic appears to occur in about 1850 (Meyer and Ausubel 1999).

5. Refer to Project 4 for a description of a logistic carrying-capacity function. Using that information, develop a model for the Japanese population from the year 1100 to 2000. With an initial population of 5 million, the island population was mainly a feudal society that leveled off to about 35 million. The industrial revolution came to Japan in the latter part of the nineteenth century, and the population rose rapidly over a 77-year period with the inflection point occurring about 1908 (Meyer and Ausubel 1999).

6. Develop a model for the number of trout in a lake. The lake is initially stocked with 400 trout. They increase at a rate of 15%, and the lake has a carrying capacity of 5000 trout. However, vacationers catch trout at a rate of 8%.

7. It has been estimated that for the Antarctic fin whale, $r = 0.08$, $M = 400,000$, and $P_0 = 70,000$ in 1976. Model this population. Then, revise the model to consider harvesting the whales as a percentage of rM. Give various values for this percentage that lead to extinction and other values that lead to increases in the population. Estimate the **maximum sustainable yield**, or the percentage of rM that gives a constant population in the long term (Zill 2001).

8. Army ants on a 17 km^2 island forage at a rate of 1500 m^2/day, clearing the area almost completely of other insects. Once the ants have departed, it takes about 150 days for the number of other insects to recover in the area. Assume an initial number of 1 million army ants and a growth rate of 3.6%, where the unit of time is a week. Model the population.

Answers to Quick Review Questions

1. **a.** Carrying capacity, say M, in units of the population, such as deer or bacteria.

   b. B. P appears to grow almost without bound.
   C. P appears to grow faster and faster.
   I. P appears to grow exponentially.
   J. dP/dt appears to be almost proportional to P.
   M. The birth rate is much greater than the death rate.

   c. D. P appears to grow more and more slowly.
   H. P is appears to be approaching M asymptotically.
   K. dP/dt appears to be almost zero.
   L. The birth rate is about the same as the death rate.

   d. F. P appears to decline more and more slowly.
   H. P is appears to be approaching M asymptotically.
   K. dP/dt appears to be almost zero.
   L. The birth rate is about the same as the death rate.

2. **a.** $\Delta P = 1.05(1 - p/1000)\,(p)(0.1)$
   b. 1000 individuals

c. 625 individuals because $P + \Delta P = 600 + 1.05(1 - 600/1000)600(0.1) = 625.2$ individuals

3. 0 because $dP/dt = r(1 - P/M)P = r(1 - 0)0 = 0$

References

Meyer, Perrin S., and Jesse H. Ausubel. 1999. "Carrying Capacity: A Model with Logistically Varying Limits." *Technological Forecasting and Social Change*, 61(3): 209–214.

Reinecke, Kenneth J. Personal communication. USGS, Pantuxent Wildlife Research Center.

Zill, Dennis G. 2001. *A First Course in Differential Equations with Modeling Applications*. 7th ed. Pacific Grove, Calif.: Brooks/Cole Publishing Co.: 134–135.

MODULE 3.4

System Dynamics Tool: Tutorial 2

Prerequisite: Module 3.1 on "System Dynamics Tool: Tutorial 1"

Download

From the textbook's website, download Tutorial 2 in pdf format and the *unconstrained* file for your system dynamics tool. We recommend that you work through the tutorial and answer all Quick Review Questions using the corresponding software.

Introduction

This tutorial introduces the following functions and concepts, which subsequent modules employ for model formulation and solution using your system dynamics tool:

- Built-in functions and constants, such as the *if-then-else* construct, absolute value, initial value, exponential function, sine, pulse function, time, time step, and π
- Relational and logical operators
- Comparative graphs
- Graphical input
- Conveyors, an optional topic useful for some of the later projects

MODULE 3.5

Drug Dosage

Downloads

The text's website has available for download in various system dynamics systems *OneCompartAspirin* and *OneCompartDilantin* files, which contain models for Examples 1 and 2.

Introduction

Errors in the dispensing and administration of medications occur frequently. Although most do not result in great harm, some do. For instance, a Colorado pharmacy dispensed five times the prescribed dose of gentamycin to a child, which resulted in bilateral hearing loss ((Fitzgerald and Wilson 1998). In other tragedies, a ten-month-old infant died after receiving a tenfold overdose of the chemotherapy agent Cisplatin (Hospital Pharmacy Reporter 1998), and three nurses have been prosecuted for administering a tenfold (fatal) overdose of penicillin to an infant (Virtual Nurse).

According to the Institute of Medicine, in a study of two well-respected teaching hospitals, two percent of admissions were subjected to avoidable, adverse drug events. They estimated that the costs of these errors would result in almost three million dollars in increased hospital costs for a 700-bed facility. Nationwide, this would translate into billions in increased medical care costs, not to mention the human suffering.

Not only healthcare professionals make mistakes in drug administration. On June 28, 2003, an Oklahoma teenager died from an overdose of Tylenol (acetaminophen). Suffering from a migraine headache, she took 20, 500-mg capsules, two and one-half times the maximum dosage recommended in 24 hours. Apparently, the quantity was enough of the drug to cause liver and kidney failure. Assuming that an over-the-counter analgesic was safe, she apparently did not read the label and made a fatal dosage error (Robert 2004).

There are prescribed dosages for various drugs, but how do we determine what the correct/effective dosage is? There are quite a number of factors that are considered, including drug **absorption**, **distribution**, **metabolism**, and **elimination**. These factors are components of the quantitative science of **pharmacokinetics**.

One-Compartment Model of Single Dose

Metabolism of a drug in the human body is a complex system to represent in a model. Thus, in Step 2 of the modeling process, particularly for our first attempt, we should make simplifying assumptions about the drug and the body. A **one-compartment model** is a simplified representation of how a body processes a drug. In this model, we consider the body to be one homogeneous compartment, where distribution is instantaneous, the **concentration** of the drug in the system (amount of drug/volume of blood) is proportional to the drug dosage, and the rate of elimination is proportional to the amount of drug in the system. The concentration of a drug instead of the absolute quantity is important because a quantity that might be appropriate for a small child might be ineffective for a large adult. A drug has a **minimum effective concentration** (**MEC**), which is the least amount of drug that is helpful, and a **maximum therapeutic concentration** or **minimum toxic concentration** (**MTC**), which is the largest amount that is helpful without having dangerous or intolerable side effects. The **therapeutic range** for a drug consists of concentrations between the MEC and MTC. A drug's **half-life**, or the amount of time for half the drug to be eliminated from the system, is useful for modeling as well as patient treatment. Often concentrations and half-life are expressed in relationship to the drug in the plasma or blood serum. The total amount of blood in an adult's body is approximately 5 liters; while the amount of **plasma**, or fluid that contains the blood cells, is about 3 liters. Blood **serum** is the clear fluid that separates from blood when it clots, and an adult human has about 3 liters of blood serum.

Example 1

We begin by modeling the concentration in the body of aspirin (acetylsalicylic acid). For adults and children over the age of 12, the dosage for a headache is one or two 325 mg tablets every four hours as necessary up to 12 tablets/day. Analgesic effectiveness occurs at plasma levels of about 150 to 300 μg/ml, while toxicity may occur at plasma concentrations of 350 μg/ml. The plasma half-life of a dose from 300 to 650 mg is 3.1 to 3.2 hr, with a larger dose having a longer half-life.

For simplicity, we assume a one-compartment model with the aspirin immediately available in the plasma. A stock (box variable), *aspirin_in_plasma*, represents the mass of aspirin in the compartment, which is the person's system, and has an initial value of the mass of two aspirin, (2)(325 mg)(1000 μg/mg), where 1 milligram (mg) is equivalent to 1000 micrograms (μg).

The flow from *aspirin_in_plasma* (*elimination*) is proportional to the amount present in the system, *aspirin_in_plasma*. Thus, the rate of change of the drug leaving the system is proportional to the quantity of drug in the system (*aspirin_in_plasma*, or Q in the following equation):

$$dQ/dt = -KQ$$

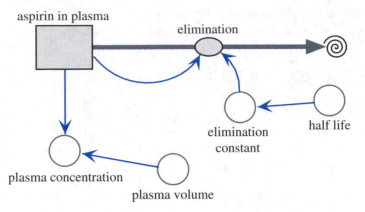

Figure 3.5.1   One-compartment model of aspirin

As Module 3.2 on "Unconstrained Growth" shows, the solution to this differential equation is as follows:

$$Q = Q_0 e^{-Kt}$$

Using this solution, as Exercise 1 shows, the constant of proportionality K above and *elimination_constant* in the system dynamics software model has the following relationship to the drug's half-life ($t_{1/2}$):

$$K = -\ln(0.5)/t_{1/2}$$

Pharmaceutical sources widely report a drug's half-life.

Quick Review Question 1

Determine the elimination constant with units for aspirin assuming a half-life of 3.2 hr.

To compute aspirin's plasma concentration (*plasma_concentration*) in a converter (variable), we have another converter for the volume of the system (*plasma_volume*) with a value of 3000 ml and appropriate connectors and equation. Figure 3.5.1 contains a one-compartment model for one dose of a drug, where the initial value of *plasma_concentration* is the dosage; and Equation Set 3.5.1 gives the corresponding equations and values explicitly entered for the model of aspirin.

Quick Review Question 2

In terms of the variables in the model of Figure 3.5.1, give the equation for *plasma_concentration*.

Equation Set 3.5.1

Explicitly entered equations and values for one-compartment model of aspirin

half_life = 3.2 hr
plasma_volume = 3000 ml

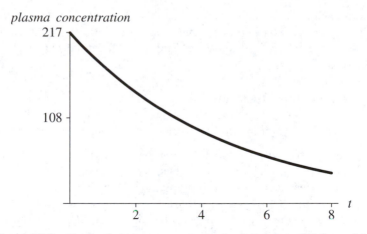

Figure 3.5.2   Graph of *plasma_concentration* (µg/ml) versus *t*(hr) for aspirin

aspirin_in_plasma(0) = 2 * 325 * 1000 µg
elimination_constant = −ln(0.5)/*half_life*
elimination = *elimination_constant* * *aspirin_in_plasma*
plasma_concentration = *aspirin_in_plasma*/*plasma_volume*

Running the simulation for 8 hr and plotting *plasma_concentration*, the resulting graph in Figure 3.5.2 indicates that the concentration of the drug in the plasma is initially approximately 217 µg/ml, which is a safe, therapeutic dose. Subsequently, the concentration decreases exponentially.

One-Compartment Model of Repeated Doses

Example 2

In this example, we model the concentration in the body of the drug Dilantin, a treatment for epilepsy that the patient takes on a regular basis. Adult dosage is often one 100-mg tablet three times daily. The effective serum blood level is 10 to 20 µg/ml, which may take seven to ten days to achieve. Although individual variations occur, serious side effects can appear at a serum level of 20 mg/ml = 20,000 µg/ml. The half-life of Dilantin ranges from 7 to 42 hours but averages 22 hours.

For simplicity, we assume a one-compartment model with instantaneous absorption. A stock (box variable), *drug_in_system*, represents the mass of Dilantin in the compartment, which is the person's blood serum. A flow, *ingested*, into *drug_in_system* is for the drug absorbed into the system. Because of the periodic nature of the dosage, we employ a pulse function with converters/variables for the dose (*dosage*), time of the initial dose (*start*), and time interval between doses (*interval*). Presuming that only a fraction (*absorption_fraction*) actually enters the system, we multiply this constant (say, 0.12 from experimental evidence) and the pulse value together for the equation of *injested*. We can estimate the value of *absorption_fraction* by plotting actual data of drug concentration versus time and employing techniques of curve fitting, which Module 8.3 on "Empirical Models" discusses.

Quick Review Question 3

Give the equation for *injested*.

The flow from *drug_in_system* (*elimination*) is proportional to the amount present in the system, *drug_in_system*. Thus, between doses of a drug, the rate of change of the drug leaving the system is proportional to the quantity of drug in the system. As for aspirin in Example 1, we use a constant of proportionality (*elimination_constant*) of $-\ln(0.5)/t_{1/2}$, where $t_{1/2}$ is Dilantin's half life.

For comparison purposes, we have converters (variables) for *MEC*, *MTC*, and the concentration of the drug in the system (*concentration*). To compute the latter, we have a converter (variable) for the volume of the blood serum (*volume*) with a possible value of 3000 ml and appropriate connectors and equation. Figure 3.5.3 contains a one-compartment model, and Equation Set 3.5.2 the corresponding explicitly entered equations and constants for Dilantin.

Note that, except for name changes, the middle and right side of the diagram agree with those of aspirin in Figure 3.5.1. The inflow for Figure 3.5.3 models the multiple doses of Dilantin in contrast to no inflow for Figure 3.5.1 because of the assumption that exactly one dose of aspirin is immediately available in the plasma.

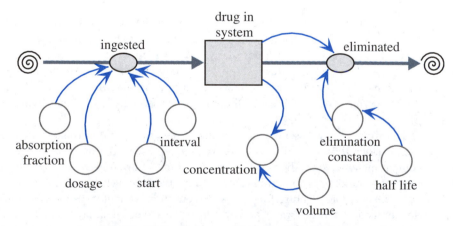

Figure 3.5.3   One-compartment model of Dilantin

Equation Set 3.5.2

Explicitly entered equations and constants for one-compartment model of Dilantin

half_life = 22 hr; *interval* = 8 hr; *MEC* = 10 μg/ml; *MTC* = 20 μg/ml; *start* = 0 hr;
volume = 3000 ml; *dosage* = 100 * 1000 μg; *absorption_fraction* = 0.12
elimination_constant = −ln(0.5)/*half_life*
drug_in_system(0) = 0
entering = *absorption_fraction* * (pulse of amount *dosage* beginning at *start*
    every *interval* hours)
elimination = *elimination_constant* * *drug_in_system*
concentration = *drug_in_system*/*volume*

Running the simulation and plotting the various concentrations that occur over 168 hr (7 days), the resulting Figure 3.5.4 indicates that the concentration of the drug in the system between doses fluctuates. In less than two days, the concentration remains within the therapeutic range; and after about five days, the drug reaches a steady state.

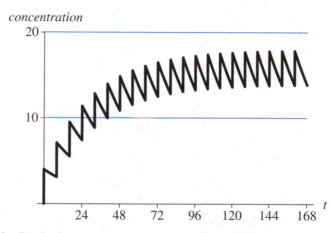

Figure 3.5.4 Graph of concentrations *MEC* = 10 μg/ml, *MTC* = 20 μg/ml, and *concentration* (μg/ml) versus time (hr)

Mathematics of Repeated Doses

Let us show the mathematics of why the drug concentration in Example 2 tends to a fixed value, in this case about 18 μg/ml, immediately after a dose. Suppose that the patient takes a 100-mg tablet every 8 hours. In the model, we assumed an absorption level of 0.12, so that the effective dosage is $Q_0 = (0.12)(100) = 12$ mg. With an elimination rate of −ln(0.5)/22, which is about 0.0315, the amount of drug in the system after 8 hours is $Q = Q_0 e^{-0.0315(8)} \approx (12)(0.7772) = 9.3264$ mg $= 9326.4$ μg. Thus, at the end of 8 hours, about 77.72% of the drug remains in the system. The analytical value (9326.4 μg) for the mass of drug in the system is close to the simulated value

(9327.91 μg) of *drug_in_system* at time 8.00 hr (using a time step of 0.01 hr and Runge-Kutta 4 numeric integration, which Module 5.4 discusses).

Suppose Q_n is the quantity (in mg) in the system immediately after the nth tablet. Thus, assuming 77.72% of the drug remains in the system at the end of an 8-hour interval immediately before a dose, we have the following:

$$Q_1 = 12 \text{ mg}$$

$$Q_2 = \underbrace{(12 \text{ mg})(0.7772)}_{\text{remainder of tablet 1}} + \underbrace{12 \text{ mg}}_{\text{tablet 2}} = 21.3264 \text{ mg}$$

$$Q_3 = \underbrace{Q_2(0.7772)}_{\text{remainder of tablets 1 \& 2}} + \underbrace{12 \text{ mg}}_{\text{tablet 3}}$$

$$= (12(0.7772) + 12)(0.7772) + 12$$

$$= 12(0.7772)^2 + 12(0.7772) + 12 = 28.5749 \text{ mg}$$

$$Q_4 = \underbrace{Q_3(0.7772)}_{\text{remainder of tablets 1–3}} + \underbrace{12 \text{ mg}}_{\text{tablet 4}}$$

$$= (12(0.7772)^2 + 12(0.7772) + 12)(0.7772) + 12$$

$$= 12(0.7772)^3 + 12(0.7772)^2 + 12(0.7772) + 12 = 34.2084 \text{ mg}$$

Continuing in the same pattern, we determine that the general form of the quantity of the drug in the system immediately after the fifth tablet is as follows:

$$Q_5 = 12(0.7772^4) + 12(0.7772^3) + 12(0.7772^2) + 12(0.7772) + 12$$

$$= 12(0.7772^4) + 12(0.7772^3) + 12(0.7772^2) + 12(0.7772^1) + 12(0.7772^0)$$

$$= 12(0.7772^4 + 0.7772^3 + 0.7772^2 + 0.7772^1 + 0.7772^0)$$

Similarly, the quantity of the drug immediately after the nth tablet, Q_n, follows:

$$Q_n = 12(0.7772^{n-1} + \cdots + 0.7772^2 + 0.7772^1 + 0.7772^0)$$

Quick Review Question 4

Suppose a patient takes a 200-mg tablet once a day, and within 24 hours 75% of the drug is eliminated from the body. With Q_n being the quantity of the drug in the body after the nth dose, determine the following:

   a. Q_1
   b. Q_2 expressed as a sum
   c. Q_3 expressed as a sum
   d. Q_4 expressed as a sum
   e. Q_n expressed as a sum

We would like to determine what happens to the quantity of the drug in the system over a long period of time. To do so, we need a formula for the sum $0.7772^{n-1} + \cdots + 0.7772^2 + 0.7772^1 + 0.7772^0$ for positive integer n. This sum is a **finite geometric series**, and its general form is as follows:

$$a^{n-1} + \cdots + a^2 + a^1 + a^0 \quad \text{for } a \neq 1 \quad \text{and positive integer } n$$

As we verify in the next section, this sum is the following ratio:

$$a^{n-1} + \cdots + a^2 + a^1 + a^0 = \frac{(1-a^n)}{(1-a)} \quad \text{for } a \neq 1$$

Thus, for $a = 0.7772$ and $n = 5$, we can compute the value of Q_5:

$$Q_5 = 12(0.7772^4 + 0.7772^3 + 0.7772^2 + 0.7772^1 + 0.7772^0)$$

$$= 12 \cdot \frac{1 - 0.7772^5}{1 - 0.7772} = 38.5868 \text{ mg} = 38,586.8 \ \mu g$$

Within simulation error, this value agrees with *drug_in_system* (38,580.92) after the fifth dose, at time 32.01 hr. In general, the quantity of the drug immediately after the nth tablet, Q_n, is as follows:

$$Q_n = 12(0.7772^{n-1} + \cdots + 0.7772^2 + 0.7772^1 + 0.7772^0)$$

$$= 12 \cdot \frac{1 - (0.7772)^n}{1 - 0.7772}$$

Definition $a^{n-1} + \cdots + a^2 + a^1 + a^0$ for $a \neq 1$ and positive integer n is a **finite geometric series** with **base** a.

Quick Review Question 5

Using the drug of Quick Review Question 4 and the formula for the sum of a finite geometric series, evaluate the following:

a. Q_{10}
b. Q_n

Using the formula for the sum of a finite geometric series, we can compute the quantity of drug after the nth tablet. To determine the long-range affect, we let n go to infinity and see that Q_n approaches 53.8600 mg, as follows:

$$Q_n = 12 \cdot \frac{1 - (0.7772)^n}{1 - 0.7772} \rightarrow 12 \cdot \frac{1 - 0}{1 - 0.7772} \approx 53.8600 \text{ mg}$$

Thus, the serum concentration is about (53.8600 mg)/(3000 ml) = 0.0179533 mg/ml = 17.95 μg/ml, which agrees closely with the peak value of the concentration in Figure 3.5.4.

Quick Review Question 6

Using the drug of Quick Review Questions 4 and 5, determine the quantity of drug after the nth tablet when the patient has been taking the drug for a long time.

Sum of Finite Geometric Series

To derive the formula for the sum of a finite geometric series, we start by considering a particular example, Q_5 above. Let s be equal to the sum of the powers from 0 through 4 of 0.7772, as follows:

$$s = 0.7772^4 + 0.7772^3 + 0.7772^2 + 0.7772^1 + 0.7772^0 \tag{1}$$

Multiplying both sides by 0.7772, we have the following:

$$0.7772s = (0.7772)(0.7772^4 + 0.7772^3 + 0.7772^2 + 0.7772^1 + 0.7772^0)$$

$$0.7772s = 0.7772^5 + 0.7772^4 + 0.7772^3 + 0.7772^2 + 0.7772^1 \tag{2}$$

Subtracting Equation 2 from Equation 1, we subtract off all but two terms on the right:

$$
\begin{array}{l}
s = \qquad\qquad 0.7772^4 + 0.7772^3 + 0.7772^2 + 0.7772^1 + 0.7772^0 \\
-0.7772s = -0.7772^5 - 0.7772^4 - 0.7772^3 - 0.7772^2 - 0.7772^1 \\
\hline
s - 0.7772s = -0.7772^5 \qquad\qquad\qquad\qquad\qquad\qquad\quad + 0.7772^0
\end{array}
$$

With 0.7772^0 being 1, we factor out s on the left as follows:

$$s(1 - 0.7772) = -0.7772^5 + 1$$

or

$$s(1 - 0.7772) = 1 - 0.7772^5$$

Dividing both sides by the factor $(1 - 0.7772)$, we obtain the following formula:

$$s = \frac{1 - 0.7772^5}{1 - 0.7772}$$

By the same reasoning, we have the general formula for the sum of a finite geometric series.

> The formula for the sum of a finite geometric series is as follows:
>
> $$a^{n-1} + \cdots + a^2 + a^1 + a^0 = \frac{(1 - a^n)}{(1 - a)} \quad \text{for } a \neq 1$$

Two-Compartment Model

The one-compartment model is more appropriate for an injection of a drug into the system than for a pill, which takes time to dissolve, be absorbed, and be distributed within the system. In such cases, a **two-compartment model** might yield better results. The first compartment represents the digestive system (stomach and/or intestines), while the second might indicate the blood, plasma, serum, or a particular organ that the drug targets. A flow pumps the drug from one compartment to the other in the model. One option for modeling the rate of change of absorption from the intestines to blood serum has the rate proportional to the amount of drug in the intestines. Probably a more accurate representation has the rate of change of absorption from the intestines to blood serum be proportional to the volume of the intestines and to the difference of the drug concentrations in the intestines and serum.

Although the one- or two-compartment model is appropriate for most situations, a drug dosage problem could benefit from more compartments in a **multicompartment model**. Various projects employ more than one compartment.

Quick Review Question 7

This question applies to the rate of change of absorption of a drug from the intestines to blood serum in a two-compartment model. Suppose k is a constant of proportionality; i and b are the masses of the drug in the intestines and blood serum, respectively; vi and vb are the volumes of the intestines and blood serum, respectively; ci and cb are the drug concentrations in the stomach and blood serum, respectively; and time t is in hours.

   a. Give the differential equation for this rate if the rate of absorption is proportional to the mass of drug in the intestines.
   b In this case, give the units of k.
   c. Give the differential equation for this rate if the rate of absorption is proportional to the volume of the intestines and to the difference of the drug concentrations in the intestines and blood serum.
   d. In this case, give the units of k.

Exercises

   1. Assuming that the rate of change of a quantity of a drug (Q) is $Q = Q_0 e^{Kt}$, show that $K = -\ln(0.5)/t_{1/2}$, where $t_{1/2}$ is the drug's half-life.
   2. a. In Figure 3.5.4, what are the units for MEC and MTC?
   b. What are the units for *dosage*?
   c. With a dosage of Dilantin being 100 mg, why is the value of *dosage* 100 * 1000?
   3. Prove the general formula for the sum of a finite geometric series.
   4. a. In Example 2, describe the effect a longer half-life has on *elimination_constant*.
   b. Evaluate *elimination_constant* for $t_{1/2} = 7$ hr.
   c. Evaluate *elimination_constant* for $t_{1/2} = 22$ hr.
   d. Evaluate *elimination_constant* for $t_{1/2} = 42$ hr.
   5. a. Suppose a patient taking Dilantin decides for convenience to take 300 mg once a day instead of 100 mg every 8 hr. Adjusting the model in *OneCompartDilantin*, determine the results of such a decision. Is the decision advisable?

    b. Mathematically, determine the long-term value of Q_n, the quantity of Dilantin in the system immediately after the nth dose, assuming absorption of only $(0.09)(300 \text{ mg})$.

6. a. Determine mathematically the quantity of Dilantin in the system immediately before the fifth dose. Use the same assumptions as in the sections on "Mathematics of Repeated Doses."

    b. Determine mathematically the long-term value of the quantity of Dilantin in the system immediately before the nth dose.

    c. Compare your answers to the values in *OneCompartDilantin*.

7. How should Example 1 be adjusted to incorporate the weight of a male patient? Assume that 1 kg of body liquid has a volume of 1 liter. Assume that about 8% of a human's weight is blood.

Projects

For additional projects, see Module 7.6 on "Cardiovascular System—A Pressure-Filled Model."

    For all model development, use an appropriate system dynamics tool.

1. Develop a two-compartment model for one dose of aspirin (see Example 1).

2. Develop a two-compartment model for aspirin (see Example 1), where someone with a headache takes three aspirin tablets and two hours later takes two more aspirin tablets. Consider using a step function.

3. In attempt to raise the concentration of a drug in the system to the minimum effective concentration quickly, sometimes doctors give a patient a **loading dose**, which is an initial dosage that is much higher than the maintenance dosage. A loading dose for Dilantin is three doses—400 mg, 300 mg, and 300 mg two hours apart. Twenty-four hours after the loading dose, normal dosage of 100 mg every eight hours begins. Develop a model for this dosage regime.

4. Develop a two-compartment model for Dilantin (see Example 2), where the rate of change of absorption from the stomach to the blood serum is proportional to the amount of drug in the stomach.

5. Develop a two-compartment model for Dilantin (see Example 2), where the rate of change of absorption from the stomach to the blood serum is proportional to the volume of the stomach and to the difference of the drug concentrations in the stomach and serum. Assume the volume of the stomach is 500 ml.

6. Develop a two-compartment model for a pediatric dosage of Dilantin. The initial dose is 5 mg/kg/day in two or three equally divided doses. The maintenance dosage is usually 4 to 8 mg/kg/day.

7. Develop a model for vancomycin HCl, which is a treatment for serious infections by susceptible strains of methicillin-resistant staphylococci in penicillin-allergic patients. The drug is administered by IV infusion. The intravenous dose is usually 2 g divided either as 500 mg every 6 hr or 1 g every 12 hr, and the rate is no more than 10 mg/min or over a period of at least

50 minutes, whichever is longer. When kidney function is normal, multiple intravenous dosing of 1 g results in mean plasma concentrations of about 63 μg/ml immediately after infusion, 23 μg/ml in 2 hr, and 8 μg/ml in 11 hr after infusion. In such patients, the mean elimination half-life from plasma is 4 to 6 hours. The mean plasma clearance is approximately 0.058 l/kg/hr, while the mean renal clearance is about 0.048 l/kg/hr.

8. Repeat Project 7 for patients with renal dysfunction in which the average half-life of elimination is 7.5 days, and the distribution coefficient is from 0.3 to 0.43 l/kg.

9. Develop a model for Vancocin HCl in which the patient initially has normal kidney function (see Exercise 7). However, at the start of the third day, one of the patient's kidneys stops functioning; and the elimination rate becomes half its previous value. Consider using a step function.

10. Do Project 7 for children, where the dosage is 10 mg/kg every 6 hours, and the rate of administration is over a period of at least 60 minutes.

11. Do Project 7 for neonates and young infants. The initial dose is 15 mg/kg. Thereafter, the dosage is 10 mg/kg every 12 hours for neonates in their first week of life and afterward up to age of 1 month, every 8 hours. Administration is over 60 minutes.

12. Model drug dosage of aspirin for arthritis, where the initial dose is 3 g/day in divided doses. The dosage can be increased. Relief usually occurs at plasma levels of 20 to 30 mg per 100 ml. The plasma half-life of aspirin increases with dosage, so that a dose of 1 g has a half-life of about 5 hours and a dose of 2 g has a half-life of about 9 hours.

13. Considering the information about mass in Exercise 7, do any of the previous projects except one involving children or infants, accounting for the mass of a male patient.

14. By consulting a pharmacy reference or website, such as http://www.nlm.nih.gov/medlineplus/druginformation.html, obtain relevant information about some drug. Model the dosage of this drug.

Answers to Quick Review Questions

1. $K = -\ln(0.5)/3.2$ per hr $= 0.22$/hr
2. *plasma_concentration = aspirin_in_plasma/plasma_volume*
3. *absorption_fraction* $\ast$ (pulse of amount *dosage* beginning at *start* every *interval* hours), where the pulse function depends on the particular system dynamics tool
4. a. 200 mg
   b. $(200 \text{ mg})(0.25) + 200 \text{ mg}$
   c. $(200 \text{ mg})(0.25)^2 + (200 \text{ mg})(0.25) + 200 \text{ mg}$
   d. $(200 \text{ mg})(0.25)^3 + (200 \text{ mg})(0.25)^2 + (200 \text{ mg})(0.25) + 200 \text{ mg}$
   e. $(200 \text{ mg})(0.25)^{n-1} + \cdots + (200 \text{ mg})(0.25)^2 + (200 \text{ mg})(0.25) + 200 \text{ mg}$
5. a. $(200 \text{ mg})(1 - (0.25)^{10})/(1 - 0.25) = 266.67 \text{ mg}$
   b. $(200 \text{ mg})(1 - (0.25)^n)/(1 - 0.25) = (200 \text{ mg})(1 - (0.25)^n)/(0.75)$
6. $(200 \text{ mg})(1 - 0)/(0.75) = 266.67 \text{ mg}$

7. a. $db/dt = ki$
   b. 1/hr
   c. $db/dt = k(vi)(ci - cb)$
   d. 1/hr

References

Fisher, Diana M. 2001. *Lessons in High School Mathematics, A Dynamic Approach.* Hanover, N.H.: High Performance Systems.

Fitzgerald, Walter L., Jr., and Dennis B. Wilson. 1998. "Medication errors: Lessons in law." *Drug Topics*, January 19, 1998. http://www.drugtopics.com/be_core/MVC?mag=d&action=viewArticle&y=1998&m=01&d=19&article=d1b084.html&path=/be_core/content/journals/d/data/1998/0119&title=Medication@errors:@Lessons@in@law&template=past_issues_show_article.jsp&navtype=d (accessed June 24, 2003).

Hospital Pharmacy Reporter. February 1998.

Khorasheh, Farhad, Amir Mahbod Ahmadi, and Abbas Gerayeli. 1999. "Application of Direct Search Optimization for Pharmacokinetic Parameter Estimation." *Journal of Pharmacological and Pharmaceutical Science* 2(3): 92–98. http://www.ualberta.ca/~csps/JPPS2(3)/F.Khorasheh/application.htm.

Makoid, Michael C. *Basic Pharmacokinetics*, 1996–2000. http://pharmacy-creighton.edu/pha443/pdf/ (accessed June 25, 2003).

Pharmacy Network Group. "Pharmacy Network Group Home Page." http://www.pharmacynetworkgroup.com/ (accessed June 24, 2003).

Robert, Teri. 2004. "Teenager Dies from Acetaminophen Overdose." *Your Guide to Headaches/Migraine*, About.com. http://headaches.about.com/cs/medicationsusage/a/acet_death.htm. Source for this web article: "Teenager Accidentally Overdoses On Over-The-Counter Analgesic," July 2, 2003. http://TheAssociatedPress.Channel Oklahoma.com

Virtual Nurse Communications. "Virtual Nurse Communications Home Page." http://www.virtualnurse.com (accessed June 24, 2003).

4

FORCE AND MOTION

MODULE 4.1

Modeling Falling and Skydiving

Introduction

What is it like to skydive? Imagine ascending in a small plane to, say, 10,000 feet, when the jumpmaster opens the door. The jumpmaster asks you if you are ready to jump. You head for the door and walk out onto a step under the wing, holding on to a strut. You experience lots of wind and noise. Your heart is pounding wildly. The jumpmaster yells, "Go!" You arch your body and release your grip on the strut. Your adrenalin levels have never been higher as you plunge toward earth at 120 mph. Nevertheless, you are in control. For the next 50 seconds, simple body movements can alter your speed, direction and position. At three thousand feet, the landscape is fast approaching, and you pull your cord. As it deploys, your descent slows, and the mad rush of wind ceases, replaced by the rustling sounds of your canopy. Soon you gently settle to the ground.

The use of parachutes or parachute-like devices to slow the descent of jumpers from positions of considerable height may have begun with the twelfth-century Chinese. However, the first evidence of a parachute in the western world appeared in the late-fifteenth-century drawings of Leonardo da Vinci. His pyramid-shaped design was to be constructed of linen and a wooden frame. There is no record of Leonardo experimenting with his invention, but late last century it was demonstrated successfully.

Not much development of parachutes took place until late in the eighteenth century, when hot-air balloons were being shown across Europe. Andres-Jacques Garnerin, a French balloonist of dubious reputation, was one of the first persons to demonstrate a parachute without a rigid frame. He successfully descended from his balloon (which exploded) at about 3000 feet using a gondola suspended by an umbrella-shaped parachute.

Jumps using parachutes from airplanes began in the early twentieth century but were primarily for rescuing observation balloon pilots. Barnstormers performed parachute-jumping demonstrations at air shows during the time between the world wars. During World War II, both sides exploited the capabilities of parachutes for dispersing men and supplies.

Sport parachuting (skydiving) probably has its roots in the first free-fall conducted in 1914, but the sport really gained popularity only in the 1950s and 1960s (Bates; Cislunar 1997).

In this module, using a system dynamics tool we model the motion of someone skydiving. Such a jump has two phases, a free-fall stage followed by a parachute stage with greater air friction. In preparation for development of this model, we reconsider the main example in Module 2.3 on "Rate of Change" and a follow-up exercise (Exercise 1) in Module 2.4 on "Fundamental Concepts of Integral Calculus" involving the motion of a ball thrown straight up from a bridge. We model this motion, first ignoring air friction and then refining the model to consider this additional force.

Acceleration, Velocity, and Position

The above-mentioned modules on calculus discuss how the instantaneous rate of change, or derivative, of position (s) with respect to time (t) is velocity (v), and the instantaneous rate of change of velocity with respect to time is acceleration (a). In derivative notation, we have the following:

$$v(t) = \frac{ds}{dt}$$

$$a(t) = \frac{dv}{dt}$$

In Example 1, we use these derivatives in modeling the main illustration from the module on "Rate of Change" (Module 2.3).

Quick Review Question 1

This question reflects on Step 2 of the modeling process—formulating a model—for developing a model for a falling object. We simplify this first attempt at a model by ignoring friction. After completing this question and before continuing in the text, we suggest that you develop a model for a falling object.

    a. Determine four variables for the model and their units in the metric system.
    b. Give a differential equation relating time (t), position (s), and velocity (v).
    c. Give a differential equation relating time (t), velocity (v), and acceleration (a).
    d. Ignoring friction, give any of the following that is constant in a fall: time, distance, velocity, acceleration. In a model diagram, we will store such a value in a converter/variable.

e. In a model diagram, list the components that will be in stocks (box variables): *t*, *s*, *v*, *a*, *ds/dt*, *dv/dt*.

f. In a model diagram, give the value(s) that will flow into the position stock (box variable) for change in position: *t*, *s*, *v*, *a*, *ds/dt*, *dv/dt*.

g. In a model diagram, give the value(s) that will flow into the velocity stock (box variable) for change in velocity: *t*, *s*, *v*, *a*, *ds/dt*, *dv/dt*.

Example 1

To model with a system dynamics tool the motion of a ball that someone throws straight up from a bridge, we have stocks for the quantities that accumulate, the height (*position*) and velocity (*velocity*) of the ball. During the simulation, we can observe their changing values in a graph and table. A flow representing the change goes into velocity (*change_in_velocity*). Change in velocity is acceleration, and in this case, the acceleration is due to gravity. Therefore, a converter/variable (*acceleration_due_to_gravity*) contains the constant for **acceleration due to gravity**, which with up being the positive direction is approximately **−9.81 m/sec²**. The converter connects to *change_in_velocity*, which has this constant as its equation. Also, the flow for the change in height (*change_in_position*) is identical to the current velocity, *velocity*. Thus, we have a connector from *velocity* to *change_in_position*, and define the value of this flow to be *velocity*. Because velocity can be positive, zero, or negative, we specify that the flow can go into or out of *position*. For flexibility in models that we derive from this one, we also make *change_in_velocity* a biflow. Moreover, we specify that *velocity* and *position* can take on negative as well as positive values. To match the example in the earlier modules, we initialize *velocity* to be 15 m/sec and *position* to be 11 m, which is the height of the bridge. Figure 4.1.1 presents a diagram for a model of motion of the ball with a white arrowhead on each flow, indicating the secondary biflow direction.

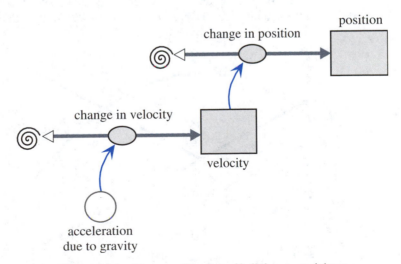

Figure 4.1.1   Diagram of motion of ball thrown straight up

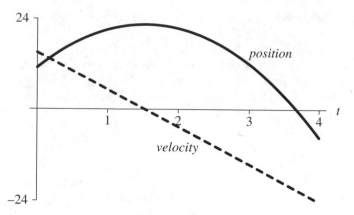

Figure 4.1.2 Graph of velocity (m/sec) and position (m) of ball versus time (sec)

Quick Review Question 2

Give the formula in metric units for each of the following components in Figure 4.1.1:

    a. The converter *acceleration_due_to_gravity*
    b. The flow *change_in_velocity*
    c. The flow *change_in_position*

Output consists of a graph and a table of velocity and height versus time. With $\Delta t = 0.25$ sec and the Runge-Kutta 4 integration technique, which Module 5.4 discusses, we obtain a table of values that matches Table 2.3.1 in Module 2.3 on "Rate of Change." The graph of velocity in Figure 4.1.1 agrees with a similar graph (Figure 2.3.2) in that module. As Figure 4.1.2 shows, the graph of velocity versus time is the line $v(t) = 15 - 9.8t$.

For some of the models, it is more convenient to consider speed than velocity. The **speed** gives the magnitude of the change in position with respect to time, while

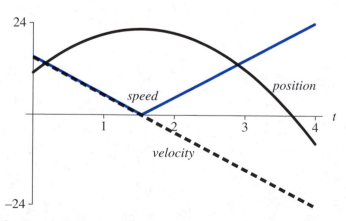

Figure 4.1.3 Graph of velocity (m/sec), position (m), and speed (m/sec) of ball versus time (sec)

the velocity expresses the magnitude with the direction. Thus, speed is the absolute value of the velocity. To incorporate speed, we have a connector/arrow from the *velocity* stock to a new converter/variable, *speed*, which stores the equation for the absolute value of velocity. The graph in Figure 4.1.3 shows speed and velocity decreasing in a linear fashion to 0 m/sec at about time 1.5 sec. Afterward, speed steadily increases.

Physics Background

Before developing additional examples of falling and skydiving, we need to consider some formulas from physics—Newton's Second Law and approximations of friction. Newton's Second Law concerns force applied to a mass imparting acceleration. So that we can refine models to account for air friction, we also consider several approximations of such a force.

Newton's Second Law has far-reaching significance. In this text, we employ the law in modeling situations from the motion of skydivers to the motion of the planets. The law states that a force F acting on a body of mass m gives the body acceleration a. Moreover, as the following models indicate, the acceleration is directly proportional to the force and inversely proportional to the mass:

$$a = F/m$$

or

$$F = ma$$

> **Newton's Second Law**   A force F acting on a body of mass m gives the body acceleration a according to the following formula:
>
> $$F = ma$$

We can apply this formula to obtain the relationship between weight and mass. **Weight** is a force and is not the same as mass. The acceleration involved is acceleration due to gravity, which is about -9.81 m/sec^2 or -32 ft/sec^2 for up being the positive direction. For example, an object that has mass of 20 kilograms (kg) has a weight of -196.2 newtons, as the following shows:

$$\text{weight} = F = (20 \text{ kg})(-9.81 \text{ m/sec}^2) = -196.2 \text{ kg m/sec}^2 = -196.2 \text{ newtons}$$

The metric unit for force is a **newton (N)** or kg m/sec^2.

> **Definition**   A **newton (N)** is a measure of force, and **1 N = 1 kg m/sec^2**.

Quick Review Question 3

Determine the following including units:

   a. The mass of an object that weighs 981 N.
   b. The acceleration that results when a net force of 10 N is applied to an object with mass 5 kg.

Kinetic friction or **drag**, too, is a force. This force between objects is in the opposite direction to a moving object and tends to slow motion. Thus, kinetic friction dampens motion of an object. When an object moves through a fluid, such as air or water, the fluid friction is a function of the object's velocity. For example, the faster we pedal a bicycle, the harder it is for us to do so. As our velocity increases, so does the friction of the air on our bodies.

Several models that estimate friction exist. In Module 8.3 on "Empirical Models," we study how to derive our own model, such as a model for drag, from data. In this module, we consider two models for drag on a body traveling through a fluid.

For a small object traveling slowly, such as a dust particle floating through the air, we usually employ **Stokes' friction**, which states that friction on the particle is approximately proportional to its velocity,

$$F = kv$$

where k in kg/sec is a constant of proportionality for the particular object and fluid, and v in m/sec is the velocity.

For a larger object moving faster through a fluid, we usually employ **Newtonian friction**, which states that the drag is approximately as follows:

$$F = 0.5CDAv^2$$

where C is a dimensionless constant of proportionality (the **coefficient of drag** or **drag coefficient**) related to the shape of the object, D is the density of the fluid, and A is the object's projected area in the direction of movement. For a particular situation, C, D, and A are constants, so that the drag is approximately proportional to the velocity squared. **At 0°C, the density of air at sea level is 1.29 kg/m³.** For shapes that are hydrodynamically good, $C < 1$; for spheres, C is about 1; and for shapes that are hydrodynamically inefficient, $C > 1$. Many objects have a coefficient of drag of about 1. Thus, through air with $C = 1$, Newtonian friction is approximately the following:

$$F = 0.65Av^2$$

The **density of water at 3.98°C**, where the fluid achieves its maximum density, is **1.00000 g/cm³**, yielding a formula with a different coefficient. Table 4.1.1 summarizes the three models for fluid friction considered here.

The drag force is in the opposite direction of motion, and the sign of velocity indicates the direction. On the upward portion of a trajectory, drag and gravity both act downward; while on the downward part, drag is upward, and gravity downward. Thus, for the general formula for Newtonian friction, we take the absolute value of only one of the velocity terms and multiply the entire formula by −1, yielding

Table 4.1.1
Summary of Several Models for Magnitude of Fluid Friction

Name	Formula	Meanings of Symbols	When to Use
Stokes' friction	$F = kv$	k constant v velocity	Very small object moving slowly through fluid
Newtonian friction	$F = 0.5CDAv^2$	C coefficient of drag D density of fluid A object's projected area in direction of movement v velocity	Larger objects moving faster through fluid
Newtonian friction through air	$F = 0.65Av^2$	A object's projected area in direction of movement v velocity	Larger objects with $C = 1$ moving faster through sea-level air

$-0.5CDAv|v|$. If *ABS* is the absolute value function, the translation of this formula into a system dynamics tool is as follows:

```
-0.5 * drag_coefficient * density * projected_area * velocity
    * ABS(velocity)
```

Quick Review Question 4

Calculate the following:

a. The density of 3.98°C water in kg/m^3
b. The magnitude of friction in newtons of a ball falling through 3.98°C water, where the coefficient of drag is 0.9, the cross-sectional area of the ball is 0.03 m^2, and its velocity is −20 m/sec
c. Write the formula for Newtonian friction for a system dynamics tool, where the coefficient of drag is 1 and the air density is 1.29 kg/m^3, namely, $-0.65Av|v|$, A and v are appropriate variables, and *ABS* is the absolute value function.

Quick Review Question 5

This question reflects on refinement of the model of an object falling through sea-level air to account for friction. After completing this question and before continuing in the text, we suggest that you revise the model in Example 1 to account for drag friction for practice in model development.

a. Give the inputs to compute drag friction.
b. Give a formula for air friction in a system dynamics tool's model with v for velocity, A for projected area, and *ABS* for the absolute value function.
c. Give the force(s) acting on the object.
d. Give a formula for an object's weight in a system dynamics tool's model, where g is the acceleration due to gravity and m is the mass of the object.

e. Give a formula for an object's acceleration in a system dynamics tool's model, where F is the total force on the object (weight + air friction) and m is the mass of an object.

Friction During Fall

Example 2

Example 1 to model the motion of a ball thrown straight up does not account for air friction. To do so, we consider two forces on the ball, gravity and drag friction. The force due to gravity is its weight, which by Newton's Second Law is $F = ma$. Thus, adjusting the model diagram in Figure 4.1.1, we include a converter/variable for *weight* with connections from converters/variables for *mass* and *acceleration_due_to_gravity* (see Figure 4.1.4). Newtonian friction for the air friction including direction is $F = -0.65Av|v|$. In the diagram, connectors/arrows go from *velocity* and

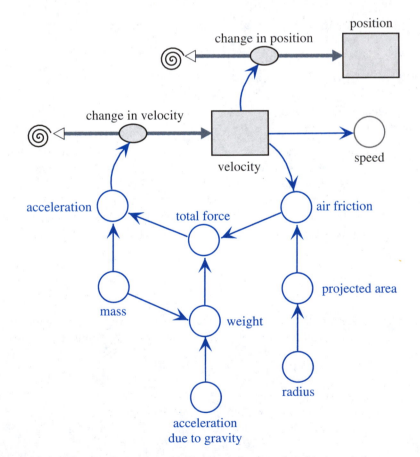

Figure 4.1.4 Diagram for motion of ball under influence of air friction; changes to converters/variables from Figure 4.1.1 in color

from a new *projected_area* converter to a new converter for *air_friction*. The *projected_area* converter/variable stores the cross-sectional, or projected, area of the object in the direction of motion. Assuming spherical objects, another converter/variable stores the *radius*; and the equation in *projected_area* is *pi* ∗ *radius*^2, where *pi* is built in or an approximated constant 3.15169, depending on the system dynamics tool. Both forces, *weight* and *air_friction*, connect to a new converter/variable for *total_force*, which is the sum of the individual forces. Employing Newton's Second Law again with $a = F/m$, *acceleration* is *total_force/mass*. This acceleration provides the change in velocity for the flow into *velocity*.

Figure 4.1.4 contains a **feedback loop**. The initial value of air friction employs the initial velocity, here 0 m/sec; and *air_friction* contributes to the *total_force*, which *acceleration* uses. Acceleration is the *change_in_position*, which contributes to *velocity*. Then, the current value of *velocity* "feeds back" into *air_friction* for a new computation of that force.

To detect the influence of drag, we consider a ball of mass 0.5 kg and radius 0.05 m dropped (initial velocity = 0 m/sec) from a height of 400 m. Equation Set 4.1.1 presents various underlying equations for the model.

Equation Set 4.1.1

Various underlying equations to accompany diagram in Figure 4.1.4

mass = 0.5 kg
acceleration_due_to_gravity = −9.81 m/sec²
radius = 0.05 m
weight = *mass* ∗ *acceleration_due_to_gravity*
projected_area = 3.14159 ∗ *radius*^2
air_friction = −0.65 ∗ *projected_area* ∗ *velocity* ∗ ABS(*velocity*)
total_force = *weight* + *air_friction*
acceleration = *total_force/mass*
change_in_velocity = *acceleration*
change_in_position = *velocity*
speed = ABS(*velocity*)
velocity(0) = 0 m/sec
velocity(t) = *velocity*(t − Δt) + (change_in_velocity) ∗ Δt
position(0) = 400 m
position(t) = *position*(t − Δt) + (change_in_position) ∗ Δt

Running the simulation for 15 seconds, we see in Figure 4.1.5 that the ball reaches a constant, or **terminal speed**, of about 31 m/sec. From about time 6 seconds on, the position graph is almost linear, so that acceleration is approximately 0 m/sec².

Quick Review Question 6

At the terminal velocity, give the relationship between *weight* and *air_friction:* (A) *weight* < *air_friction*; (B) *weight* = *air_friction*; (C) *weight* > *air_friction*

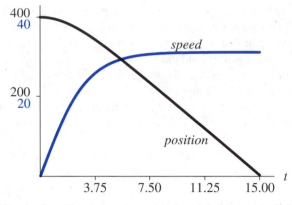

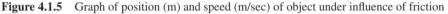

Figure 4.1.5   Graph of position (m) and speed (m/sec) of object under influence of friction

Quick Review Question 7

This question reflects on refinement of the model of Example 2 to incorporate sky-diving. After completing this question and before continuing in the text, we suggest that you revise the model.

a. Give the phases of the fall during the simulation.
b. Give the variable whose value we can use to trigger the change in phase: *acceleration*, *mass*, *position*, *velocity*, *weight*.
c. Give the value(s) that change upon opening of the parachute: *acceleration_due_to_gravity*, *mass*, *projected_area*, *weight*.
d. Describe anticipated changes to the graphs in Figure 4.1.5 after deployment of a parachute.

Modeling a Skydive

Example 3

To model a skydive, we build heavily on Example 2 of a falling object. For sim-plicity, we consider someone jumping out of a stationary helicopter at 2000 m (about 6562 ft), and we ignore changes in air density. Project 5 considers para-chuting out of a moving plane, which imparts a horizontal velocity to the jumper. The model for a skydive out of a helicopter has two phases, one where the person is in a free-fall and the other after the parachute opens when the larger surface area results in more air resistance. For our model, the main difference in these two phases is the projected area in the direction of motion, down. The cross-sectional area of a jumper in the stable arch position with arms arched back and legs bent at the knees is approximately 0.4 m² (about 4.3 ft²). Parachutes vary in their designs, but 28 m² (about 301 ft²) is a reasonable value. We trigger the pull of the ripcord by the height (*position*) above the ground, say, 1000 m (about 3281 ft). Thus, the diagram contains a converter/variable (*position_open*) for this quantity and

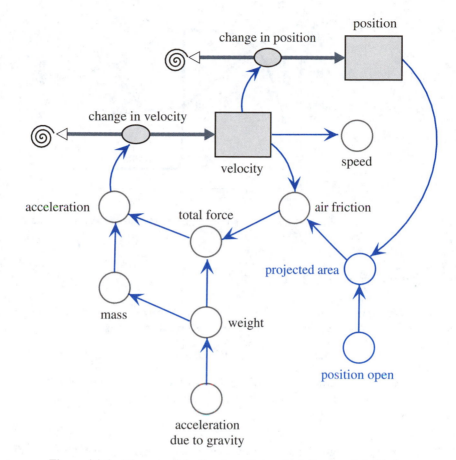

Figure 4.1.6   Diagram of skydiver's motion under influence of air friction

connectors/arrows from *position* to *projected_area* and from *position_open* to *projected_area*. Figure 4.1.6 presents a model diagram for this example with changes in color from Figure 4.1.4 on a ball's fall. Assuming the parachute fully opens instantaneously, the equation in *projected_area* is no longer a constant but employs the following logic:

   if (*position* > *position_open*)
     projected_area ← 0.4
   else
     projected_area ← 28

Figure 4.1.7 shows graphs of the position and speed of a 90-kg (comparable to about 198-lb) skydiver versus time. Until a height of 1000 m, which occurs at about 21.3 seconds into the fall from 2000 m, the skydiver is in a free-fall approaching a terminal speed of about 58.2 m/sec (about 130 mph). At 1000 m, the person pulls the ripcord, and in a very short amount of time, the parachutist's speed slows to a new terminal speed of 6.96 m/sec (about 15.6 mph).

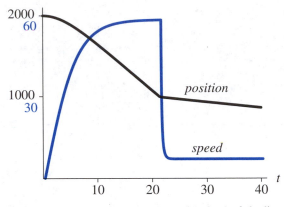

Figure 4.1.7 Position (m) and speed (m/sec) of skydiver

Quick Review Question 3

    a. How does the terminal speed of a skydiver who curls into a ball compare to that of the same skydiver who is in a stable arch position?

        A. Less   **B.** Equal   **C.** Greater

    b. Referring to Figure 4.1.7, approximately how long does it take for the sky-diver in free-fall to be close to terminal speed?

        A. 13 sec   **B.** 21 sec   **C.** 40 sec

    c. Referring to Figure 4.1.7, at approximately what time does the skydiver pull the ripcord?

        A. 13 sec   **B.** 21 sec   **C.** 40 sec

Assessment of the Skydive Model

The shapes of the graphs of position and velocity in Figure 4.1.7 match the opening description of a skydive. However, our model exhibits a terminal speed of about 130 mph (about 58.2 m/sec), while actual, measured speeds of 110–120 mph are common. The drag coefficient of a jumper is probably larger than the assumed $C = 1$ of the model. Also, the example employs the sea-level density of air, while the air density at 10,000 ft (about 3,048 m) is about 73.8% (0.952 kg/m³) of sea-level density. Adjusting the initial position to be 3048 m and using an air density of 0.952 kg/m³ with the Newtonian friction of $F = 0.5CDAv^2$, the model indicates a terminal velocity of 68 m/sec (about 152 mph) for the free-fall for less than 35 sec-onds. However, the air density changes as the skydiver descends. Projects 4 and 7 explore refinements of the model to account for this variation. Project 5 also considers the skydiver jumping from a moving plane as opposed to a stationary helicopter.

Exercises

1. a. Using the equations and values of Example 1, write differential equations with initial conditions for acceleration and velocity.

   b. Using calculus or an appropriate computational tool, solve the differential equations of Part a to obtain velocity and position as functions of time.

2. Adjust *Fall* of Example 1 so that the object falls with an initial velocity of zero and initial position of 400 m. Compare the results with those in *Fall-Friction* of Example 2, which accounted for friction.

3. a. Using the equations and values of Example 2, write a differential equation involving the derivative of velocity for when an object reaches terminal velocity. At terminal velocity, the forces acting on the body are equal.

   b. Solve the equation of Part a using calculus or a computational tool.

4. Give the adjustments to the diagram in Figure 4.1.6 along with equations so that graphs of new converters/variables *adjusted_position* and *adjusted_speed* become horizontal lines at position 0 m after the parachutist lands.

5. Repeat Exercise 3 using Stokes' friction instead of Newtonian friction.

6. Suppose a raindrop evaporates as it falls but maintains its spherical shape. Assume that the rate at which the raindrop evaporates (that is, the rate at which it loses mass) is proportional to its surface area, where the constant of proportionality is -0.01. The density (mass per volume) of water at 3.98°C is 1 g/cm^3. The surface area of a sphere is $4\pi r^2$, and its volume is $4\pi r^3/3$, where r is the radius. Assume no air resistance. (Project 8 models the motion of this raindrop under the influence of air resistance.)

   a. Assume that the initial radius is 0.3 cm. Determine the raindrop's initial mass.

   b. Write a differential equation for the rate of change of mass with respect to time as a function of r.

   c. Write an equation for r as a function of mass.

7. Adjust Example 3 so that the parachute opening depends on time, not height above the ground.

8. Write a system of differential equations to represent Example 3.

9. Using the models in your system dynamics tool's *Fall* and *FallFriction* files (see "Download"), compare position graphs for a dropped object with and without consideration of friction. Also, consider the velocity graphs. Discuss the results.

Projects

For all model development, use an appropriate system dynamics tool.

1. Develop a model to estimate the total change in position of the car with velocities from Table 2.4.3 of Module 2.4, "Fundamental Concepts of Integral Calculus." Employ an input graph instead of an equation to record Table 2.4.3's values for the change in position. Give the absolute and relative errors of your estimate in comparison to the exact value of $203\frac{1}{3}$ m.

Table 4.1.2
Approximate Air Densities at Various Altitudes

Altitude (m)	Density (kg/m³)
0	1.290
610	1.216
1219	1.146
1829	1.078
2438	1.014
3048	0.952
3658	0.894
4267	0.839
4877	0.786

2. Using Stokes' friction, develop a model for the motion of a dust particle floating down from a height of 50 m. Using comparative plots, determine its terminal speeds for various values of Stokes' constant of proportionality.

3. A bathysphere is a pressurized metal vessel in the shape of a sphere that allows people to explore the ocean to much greater depths than are possible by skin diving. A ship lowers and raises the sphere using a steel cable and communicates with its two occupants by telephone. In the 1930s, explorers William Beebe and Otis Barton developed the first bathysphere, which weighed 4500 pounds and had a diameter of 4′9″. In a subsequent one, they descended to about 3000 ft in the ocean. Ignoring currents but not drag, model the sinking motion of a bathysphere. Assume that the boat reels out the steel cable fast enough so as not to affect the bathysphere's motion (Col 2000; Uscher 2000).

4. Table 4.1.2 contains air densities at various altitudes. Using these values on an input graph, refine the model for Example 3 (Aber and Aber 2003).

5. Suppose an airplane is traveling in a straight line horizontally at 130 m/sec at a height of 600 m when a parachutist jumps out of the plane at an angle of 30° with the horizon. Model the motion of the skydive.

6. Model the motion of a meteor falling to the earth. Assume an initial height of 100,000 m, initial velocity of −10,000 m/sec, coefficient of drag of 2, mass of 500 kg, and density of 8000 kg/m³ for iron or 3500 kg/m³ for stone (Schecker 1996). Give graphs for position, velocity, and acceleration versus time. Give comparison graphs for velocity versus height for meteors of various masses. Similarly, give comparison graphs for acceleration versus height. NASA's Glenn Research Center gives the following model for air density using variables D for density (slugs/ft³), P for pressure (lbs/ft²), T for temperature (°F), and h for altitude (ft):

$$D = \frac{P}{1718(T + 459.7)}, \quad \text{where}$$

for $h > 82,345$ ft, $T = -205.05 + 0.00164\,h$ and $P = 51.97 \left(\frac{T + 459.7}{389.98} \right)^{-11.388}$

for $36{,}152 < h < 82{,}345$ ft, $T = -70$ and $P = 473.1e^{(1.73 - 0.000048h)}$; and

for $h < 36{,}152$ ft, $T = 59 - 0.00356h$ and $P = 2116\left(\dfrac{T + 459.7}{518.6}\right)^{5.256}$

Note, if you wish to use metric instead of English units, you can use the following: 1 slug = 14.5939 kg and 1 ft = 0.3048 m (Benson).

7. Using NASA's Glenn Research Center model for air density at heights less than 36,152 ft (see Project 6), refine the model in Example 3.

8. **a.** Model the change in mass of the raindrop that Exercise 5 describes.
   b. Model the motion of this raindrop taking into account air resistance.

9. Develop a model to compare the terminal velocities of objects of different masses, such as a mouse, cat, human, horse, elephant, etc. With the density of living protoplasm being almost constant across a wide variety of species, assume mass is proportional to the cube of a linear dimension, such as length or circumference; but surface area is proportional to the square of a linear dimension. How do the terminal velocities of more massive objects compare to those of less massive objects? Can a cat survive a fall from a tall building (Diamond 1989)?

Answers to Quick Review Questions

1. **a.** time, perhaps in seconds; distance, perhaps in meters; velocity, perhaps in m/sec; acceleration, perhaps in m/sec²

   b. $v(t) = \dfrac{ds}{dt}$

   c. $a(t) = \dfrac{dv}{dt}$

   d. Acceleration, which is acceleration due to gravity, -9.81 m/sec²

   e. s and v

   f. ds/dt

   g. dv/dt or a, which is the constant acceleration due to gravity without friction

2. **a.** $acceleration\_due\_to\_gravity = -9.81$ m/sec²
   b. $change\_in\_velocity = acceleration\_due\_to\_gravity$
   c. $change\_in\_position = velocity$

3. **a.** $m = F/a = 981$ N/(9.81 m/sec²) = 100 kg
   b. $a = F/m = 10$ N/(5 kg) = 2 m/sec²

4. **a.** $\dfrac{1\,\text{g}}{\text{cm}^3} \times \dfrac{1\,\text{kg}}{10^3\,\text{g}} \times \left(\dfrac{10^2\,\text{cm}}{1\,\text{m}}\right)^3 = 10^3\,\dfrac{\text{kg}}{\text{m}^3}$

   b. $F = -0.5CDAv|v| = -0.5(0.9)(10^3)(0.03)(-20)|-20| = 5400$ N

   c. `-0.65 * A * v * ABS(v)`

5. **a.** velocity and projected area
   b. $-0.65 * A * v * ABS(v)$
   c. weight and air friction

     d. $m * g$

     e. F/m

  6. (B) *weight = air_friction*

  7. a. Before and after opening of the parachute

     b. *position*

     c. *projected_area*

     d. The position curve should continue to decrease but not as steeply. The speed curve should suddenly drop and then level off to a new terminal velocity.

  8. a. C. Greater because *projected_area* is less, causing *air_friction* to be less, making the absolute values of *total_force*, *acceleration*, *change_in_velocity*, *velocity*, and *speed* more.

     b. A. 13 sec

     c. B. 21 sec

References

Aber, James S., and Susan W. Aber. 2003. "High-altitude Kite Aerial Photography," *Kite Aerial Photography*, April. http://www.geospectra.net/kite/weather/h_altit.htm

Bates, Jim. "The World of Parachutes, Parachuting, and Parachutists." Aero.com. http://www.aero.com/publications/parachutes/9511/pc1195.htm

Benson, Tom. "Earth Atmosphere Model." NASA Glenn Learning Technologies Project, Glenn Research Center. http://www.grc.nasa.gov/WWW/K-12/

Cislunar Aerospace, Inc. 1997. "Vehicles—Parachutes—Advanced." The National Business Aviation Association. http://wings.avkids.com/Book/Vehicles/advanced/parachutes-01.html

Col, Jeananda. 2000. "Undersea-Related Inventors and Inventions." Enchanted Learning.com. http://www.enchantedlearning.com/inventors/undersea.shtml

Diamond, J. 1989. "How Cats Survive Falls from New York Skyscrapers." *Natural History* (August): 20–26.

Glenn Elert, ed. 2002. "Speed of a Skydiver (Terminal Velocity)." *The Physics Factbook*. http://hypertextbook.com/facts/JianHuang.shtml

Schecker, H.P. 1996. "Modeling Physics, System Dynamics in Physics Education." *The Creative Learning Exchange* (newsletter), 5(2): 1–8.

Uscher, Jennifer. 2000. "Deep-Sea Machines." NOVA Online, October. http://www.pbs.org/wgbh/nova/abyss/frontier/deepsea.html

Weisstein, Eric. 2003. "Eric Weisstein's World of Physics." Wolfram Research. http://scienceworld.wolfram.com/physics/

Zill, Dennis G. 2001. *A First Course in Differential Equations with Modeling Applications*. 7th ed. Pacific Grove, Calif.: Brooks/Cole Publishing Co.

MODULE 4.2

Modeling Bungee Jumping

Downloads

The text's website has available for download with various system dynamics tools the following files containing the models in this module: *VerticalSpring* and *Bungee*.

Introduction

On April Fool's Day in 1979, four members of Oxford University's Dangerous Sports Club, dressed in tails and top hats, climbed out onto the Clifton Suspension Bridge in Bristol, U.K. Each attached one end of a nylon-braided, rubber shock cord to themselves and the other to the bridge. Then, they jumped off toward the 250-foot Avon Gorge. Voila! The sport of bungee jumping had begun in the western world.

What in the world possessed these men to do such a thing? The story goes that they watched a film on "land divers" from Pentecost Island in the South Pacific and became inspired to try diving themselves.

What are "land divers"? These divers are the male inhabitants of Pentecost who dive from platforms at various heights along a wooden tower. For these dives, lianas (vines) attached to the tower are tied to their ankles. Divers may be as young as seven years of age. Naturally, the lianas have to be selected very carefully. They must be just the right length and elasticity for the height of the platform and the weight of the diver. Consideration must be given to the length of the platform (which collapses and absorbs some shock), the slope of the land, and the swaying of the tower. A perfect dive will have the hair of the diver just brushing the ground. A miscalculation might be fatal. Land diving is part of ceremonies that ensure the yam harvest and fertility. Now, extreme-sports enthusiasts come from all over the world to experience "land diving."

How did this practice get started? Why would men choose to jump from platforms with vines tied to their ankles? The annual land dives are based on local lore about a young girl betrothed to a much older man. The frightened young girl, attempting to escape her new husband, climbed high into a banyan tree. The angry husband pursued her up the tree. As he ascended, she tied vines to her ankles and jumped. The husband followed, but without the vines, and was killed. Today's young men may prove that they have learned the escape trick and will not be fooled again (Menz 1993).

Physics Background

The action of a bungee cord is similar to that of a spring. Thus, in modeling bungee jumping we employ a critical law of physics concerning springs, Hooke's Law.

Hooke's Law pertains to springs that are perfectly elastic, so that they can "spring back" fully. Thus, the law can be applied only as long as the spring has not been stretched too much. The law states that within the elastic limit of the spring, a restoring force (F) applied to a spring is in proportion to and in the opposite direction of the spring's displacement (s) from its equilibrium position. Thus, for a spring constant k, which varies depending on the spring, we have the following formula for the restoring force:

$$F = -ks$$

Figure 4.2.1 illustrates the situations for stretched and compressed springs. When a resting spring is pulled or pushed, a force is exerted to restore the spring. As long as we do not stretch it beyond its elastic limit, the further we pull or push the spring, the more force going in the opposite direction results. For example, as we tug farther on an exercise spring, we feel more resistance.

> **Hooke's Law**   Within the elastic limit of a spring, where F is the applied force, k is the spring constant, and s is the displacement (distance) from the spring's equilibrium position, the following formula holds:
>
> $$F = -ks$$

Quick Review Question 1

   a. If displacement is in meters, give the units of the spring constant.
   b. For a displacement of 0.1 m and a spring constant of 5 kg/sec^2, give the restoring force along with its units.

Quick Review Question 2

This question reflects on Step 2 of the modeling process—formulating a model—for a vertical spring's length. Suppose a spring has a weight on the end that we pull down or

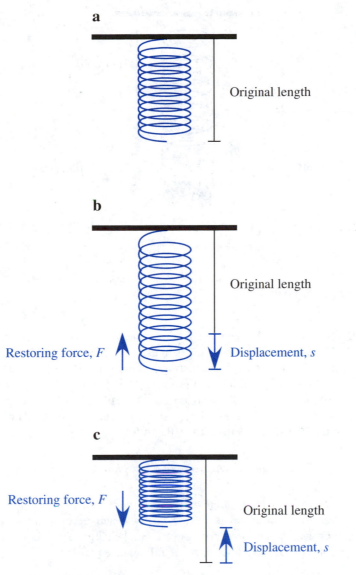

Figure 4.2.1   Original, stretched, and compressed spring

push up. Consider down as the positive direction. We simplify this first attempt at a model by ignoring friction and the weight of the spring. After completing this question and before continuing in the text, we suggest that you develop a model for the motion of the spring. The next section completes such a model.

a. Give the three lengths that sum to the total length of the spring.
b. Give the force(s).
c. Besides these lengths and forces, give other variables and constants for the model.

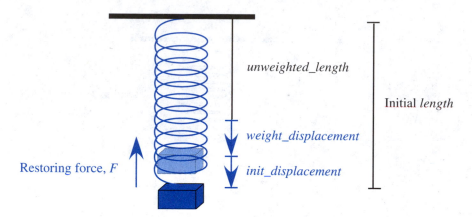

Figure 4.2.2   Vertical spring with attached weight

Vertical Springs

Before modeling bungee jumping, it is helpful to examine the action of a vertical spring, such as in Figure 4.2.2, hanging from a horizontal surface with an attached weight. Because we are considering lengths, having the positive direction be down is convenient. The initial length (*unweighted_length*) of the spring is augmented by a displacement due to the weight (*weight_displacement*) and an additional displacement due to stretching or compressing (*init_displacement*). Thus, we initialize the length of the spring (*length*) to be the following:

$$unweighted\_length + weight\_displacement + init\_displacement$$

We enter *unweighted_length* and *init_displacement* as parameters, but a system dynamics tool can calculate *weight_displacement* because the displacement due to weight, which is a force, conforms to Hooke's Law, $F = -ks$ or $s = -F/k$. Using the variables of the model, we have the following equation:

$$weight\_displacement = -weight/spring\_constant$$

A system dynamics diagram for the action of a vertical spring is similar to the diagram of the motion of a ball under the influence of air friction in Figure 4.1.4 of Module 4.1 on "Modeling Falling and Skydiving." We change the name of the stock *position* to *length* and replace the section concerning air friction with one involving the force due to Hooke's Law. Moreover, because down is positive, *acceleration_due_to_gravity* is +9.81 m/sec² instead −9.81 m/sec² as in the skydiving module. Figure 4.2.3 presents a model diagram of the action of a vertical spring experiencing no friction, or an **undamped vertical spring**, with changes to converters/variables from Figure 4.1.4 in color. With the total displacement at any instant being

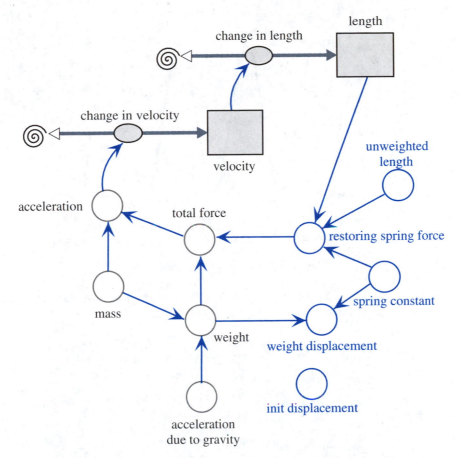

Figure 4.2.3   Model diagram of action of undamped vertical spring

length − *unweighted_length*, the following Hooke's Law equation yields the restoring force of the spring:

$$restoring\_spring\_force = -spring\_constant * (length - unweighted\_length)$$

For a simulation, suppose we consider hanging a 0.2 kg mass on the end of a 1 m spring that has a spring constant of 10 N/m. The 0.2 kg mass exerts a force of $F = mg = (0.2 \text{ kg})(9.81 \text{ m/sec}^2) = 1.962$ N, its weight. The resulting displacement because of the weight is 1.962 N/(10 N/m) = 0.1962 m = 19.62 cm. Thus, the length of the resting spring with an attached 0.2 kg mass is 1 m + 0.1962 m = 1.1962 m. If we then consider pulling the weight an additional 0.3 m = 30 cm, the simulation produces the graph of the length of the spring in Figure 4.2.4 with an initial length of 1 + 0.1962 + 0.3 = 1.4962 m. Because the spring is undamped, the simulation indicates perpetual oscillations. The length fluctuates from a maximum of 1.4962 m to a minimum of 0.8962 m. The equilibrium point, 1.1962 m, is the midway point between the two extremes and the length of the weighted motionless spring. The extremes of the oscillation are each *init_displacement* = 0.3 m from the equilibrium

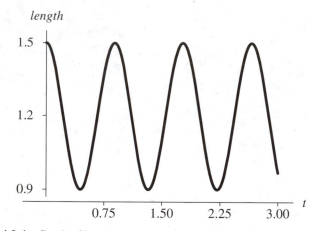

Figure 4.2.4   Graph of length (m) with respect to time (sec) in undamped spring

point. Such periodic oscillating motion is an example of **simple harmonic motion**, whose definition is below.

Definition   A **simple harmonic oscillator** satisfies the following
         conditions:

1. The system oscillates around an equilibrium position.
2. The equilibrium position is the point at which no net force exists.
3. The restoring force is proportional to the displacement.
4. The restoring force is in the opposite direction of the displacement.
5. The motion is periodic.
6. All damping effects are neglected.

Quick Review Question 3

Suppose a weight of 8 N hangs from a spring with unstretched length of 2 m and spring constant of 100 N/m. Then, the spring is compressed by 0.5 m. Using down as the positive direction and ignoring drag, give the following along with units:

    a. The displacement caused by the weight
    b. The equilibrium position
    c. The maximum length
    d. The sign of the initial restoring force
    e. The restoring force when the length of the spring is 3.15 m

A damped spring also is a simple harmonic oscillator. The diagram is as in Figure 4.2.3 with the addition of another force, drag, which becomes part of the total force. Exercises and projects refine the model to include friction. The resulting graph in Figure 4.2.5 shows a damped oscillation with the same period as for the corresponding spring with no friction.

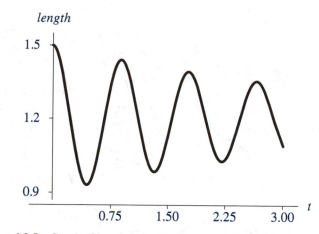

Figure 4.2.5   Graph of length (m) with respect to time (sec) in damped spring

Modeling a Bungee Jump

A model of a bungee jump is very similar to that of the weighted spring under the influence of air friction. As a simplifying assumption, we ignore the weight of the bungee cord. Thus, the forces are the weight of the jumper, the restoring force of the cord, and air resistance. The only difference between this model and that of a weighted damped spring is in the equation for *restoring_spring_force*. A vertical spring is fairly rigid, so that when the weight is above the equilibrium point, the spring exerts a restoring force in the opposite direction. However, when the bungee jumper is above the unweighted length, the cord is slack and, we can assume, exerts no downward force. This unweighted length is the length of the original, unstretched cord without the jumper. Thus, the value of *restoring_spring_force* is a conditional expression. If the length is stretched beyond *unweighted_length*, then the restoring spring force obeys Hooke's Law with the displacement being *length − unweighted_length*. When the jumper flies above *unweighted_length*, the bungee cord does not exert such a force. Thus, the value of *restoring_spring_force* has the following logic:

> if (*length > unweighted_length*) then
> > *restoring_spring_ force* ← −*spring_constant* ∗ (*length − unweighted_length*)
> else
> > *restoring_spring_ force* ← 0

Figure 4.2.6 presents a diagram for the motion of a bungee jumper with *total_force* summing *weight*, *air_friction*, and *restoring_spring_force* and with the additional converters/variables from Figure 4.2.3 in color. Figure 4.2.7 displays graphs for the bungee cord's length and the jumper's velocity for a cord with spring constant 6 N/m, unweighted length 30 m, and initial length, or distance from the top of the bridge, of 0 m. For this simulation, the jumper has mass 80 kg (equivalent to a weight of about 176.4 lb) and by jumping head first, has a small projected area of about 0.1 m². The graphs show the simple harmonic oscillation and the damping

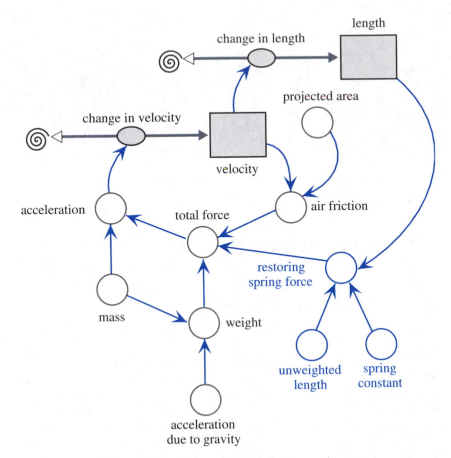

Figure 4.2.6   Diagram for motion of bungee jumper

motion due to drag. The rather pointed shape to the first hump of the velocity function is an artifact of the simulation in which time steps are discrete, in this case
0.01 sec. One of the projects explores choosing an appropriate bungee cord for a
jumper.

Quick Review Question 4

With down being the positive direction, give the formulas for each of the following
components of the diagram for a bungee jump (Figure 4.2.6).

    a. *total_force*
    b. *acceleration*
    c. *weight*
    d. *acceleration_due_to_gravity* to three significant digits
    e. *air_friction* assuming Newtonian friction, sea-level air density, and absolute
       value function *ABS*
    f. *restoring_spring_force* when *length* is greater than *unweighted_length*
    g. *restoring_spring_force* when *length* is less than *unweighted_length*

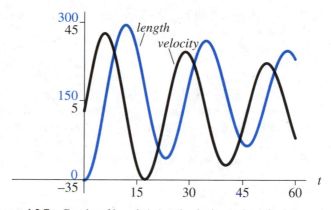

Figure 4.2.7 Graphs of length (m) and velocity (m/sec) for bungee jump

Exercises

1. Give the changes in the diagram and equations for the model of the damped spring to account for air friction.

2. **a.** Write a differential equation in terms of displacement s to model the motion of an undamped vertical spring with mass m. Recall that weight should equal the spring's restoring force. Use the following variables: s, displacement; t, time; m, mass; k, spring constant.

   b. Show that $s(t) = c_1 \cos (\sqrt{k/m}\ t) + c_2 \sin(\sqrt{k/m}\ t)$ is a solution to the differential equation.

   c. From Part b, determine the period of the vibrations.

   d. Using $k = 10$ N/m and $m = 0.2$ kg, determine the period of the vibrations. Does your answer agree with the graph in Figure 4.2.4?

3. Write a differential equation in terms of displacement s to model the motion of a damped vertical spring with mass assuming Newtonian friction.

4. Write a differential equation in terms of displacement s to model the motion of a damped vertical spring with mass assuming Stokes' friction.

5. Write differential equations in terms of displacement s to model the motion of a bungee jumper. The system should model two situations, $s \geq$ unweighted length and $s <$ unweighted length.

Projects

For additional projects, see Module 7.7 on "Electrical Circuits—A Complete Story." For all model development, use an appropriate system dynamics tool.

Download from the text's website a VerticalSpring *file with the model of the motion of an undamped vertical spring for Projects 1–7. Download a* Bungee *file with the model of the motion of a bungee jump for Project 8.*

1. Refine the model of the motion of an undamped vertical spring to account for air friction, and include a graph of *velocity*. For Parts a, b, and c, determine if

changing any of the following affects the period. If it does, determine a relationship between the parameter and the period.

    a. *init_displacement*

    b. *mass*

    c. *spring_constant*

    d. How does changing *spring_constant* affect the graph of *length*?

    e. Give the relationship between the length of the spring and its velocity. For example, when the velocity is zero, what is the position of the spring? At what stage(s) does the spring have a maximum velocity?

2. Refine the model of the motion of a spring to account for air friction. Run an experiment with a real spring and mass to determine the lengths at various times and the period. Estimate the spring constant using a system dynamics model.

3. Refine the model of the motion of a spring to account for drag using Stokes' friction. Suppose b is the constant of proportionality in Stokes' friction; define $u = b/(2m)$ for mass m. Also, define $w^2 = k/m$, where k is the spring constant. Show that when $u^2 > w^2$ with the damping coefficient large in comparison to the spring constant, the system is **overdamped** and displays no oscillation. However, when $u^2 < w^2$, the system is **underdamped** and does show oscillatory behavior.

4. The restoring force for a nonlinear hard spring is $ks(1 + a^2s^2)$, where k is the spring constant, s is the displacement, and a is a small constant. The restoring force for a nonlinear soft spring is $ks(1 - a^2s^2)$. Develop models for such springs. Discuss and compare their motions.

5. Model the motion of an aging spring by replacing the spring constant k with a decreasing function ke^{-at}, where a is a positive constant and t is time.

6. Adjust the equations of the spring model to compute the spring constant given a weight and a corresponding displacement.

7. Model a damped oscillator with parameters very close to those of the bungee jump. Discuss the results in comparison to those of the bungee jump.

8. A bungee jumper wants to have a "great ride," getting close to the ground without hitting it. Suppose the distance of the jumping bridge above a gorge is 80 m and the length of the cord is 30 m. Determine the most appropriate whole number spring constants for jumpers of masses 60 kg, 70 kg, and 80 kg. Employ comparison graphs. Discuss your results.

9. The **buoyancy** of a floating object is the restoring force to return the object to its normal floating layer after a vertical displacement. This force is equal to $-grA^s$, where g is the acceleration due to gravity, r is the fluid density, A is the cross-sectional area of the object, and s is the displacement from the normal floating layer. Design a model for the motion of a displaced object, and discuss the results of the simulation. Have the density of water be 1 g/cm^3.

Answers to Quick Review Questions

1. **a.** N/m or kg/sec^2 because $k = -F/s$, s is in m, and F is in N or kg m/sec^2

    b. $-(5 \text{ kg/sec}^2)(0.1 \text{ m}) = -0.5$ N

2. a. resting length of spring with no weight, displacement from that length due to the weight, initial displacement due to pulling down or pushing up the weight
   b. weight and restoring force of spring
   c. velocity, acceleration, mass, acceleration due to gravity, spring constant. A diagram of the model has the basic form of Module 4.1's Figure 4.1.4 for an object falling with friction. Before continuing, we suggest that you revise this figure and include equations to model the motion of a spring, using *length* instead of *position* and ignoring friction,
3. a. 8 N/(100 N/m) = 0.8 m
   b. 2 m + 0.8 m = 2.8 m
   c. 2.8 m + 0.5 m = 3.3 m
   d. Positive
   e. −(100 N/m)(3.15 m − 2.00 m) = −115 N
4. a. *weight + air_friction + restoring_spring_force*
   b. *total_force/mass*
   c. *mass * acceleration_due_to_gravity*
   d. 9.81
   e. *−0.65 * projected_area * velocity * ABS(velocity)*
   f. *−spring_constant * (length − unweighted_length)*
   g. 0

References

Danby, J. M. A. 1997. *Computer Modeling: From Sports to Spaceflight . . . From Order to Chaos*. Richmond, Va.: Willmann-Bell.

Higdon, Don, Bud Rorison, Allen Skinner, and Charlotte Trout. 2001. "Modeling Oscillatory Systems: Physics Component." *Building the Other Models*, Teacher Notes, National Computational Science Leadership Program, July 2001. http://www.ncsec.org/team13/science/buildext.htm

The MathWorks. 2005. "Problem: The Math Behind Bungee Jumping." The MathWorks. http://www.mathworks.com/academia/student_center/homework/matlab/physics_problem1.html

Menz, Paul G. 1993. "The Physics of Bungee Jumping." Bungee.com. http://www.bungee.com/bzapp/press/pt.html. Originally from *The Physics Teacher*, 31(8) (November).

Weisstein, Eric. 2003. "Eric Weisstein's World of Physics." Wolfram Research. http://scienceworld.wolfram.com/physics/

Zill, Dennis G. 2001. *A First Course in Differential Equations with Modeling Applications*. 7th ed. Pacific Grove, Calif.: Brooks/Cole Publishing Co.

MODULE 4.3

Tick Tock—The Pendulum Clock

Download

The text's website has available for download for various system dynamics tools the file *simplePendulum*, which contains a model in this module.

Introduction

When we think of a pendulum, many of us think about clocks—Grandfather clocks, mantel clocks, kitchen clocks, etc. We may even become nostalgic for the sound of such a clock in our grandparents' or parents' home. Such clocks had their origin in 1656, when a Dutch scientist, Christiaan Huygens, built a "pendulum" clock. Considering the early date, his first clock was incredibly accurate, losing less than a minute each day (Bellis 2003).

Galileo is usually credited with the invention of the pendulum, studying its motion during the sixteenth century. Although a popular myth is that he studied gravity by dropping objects from the top of the Leaning Tower of Pisa, Galileo actually used the motion of a pendulum (Boyd 2002).

Physicists use the pendulum as a classic example of energy conservation. A mass (**bob**) on a string is attached to a pivot point (i.e., a pendulum) and is acted upon by both gravity and tension. The mechanical energy of the mass is affected only by external forces and, therefore, is only influenced in this case by tension. Mechanical energy is equal to the total of kinetic and potential energy and, assuming no friction with the air, is always constant during the oscillation of the pendulum. If you pull up the mass, you increase the potential energy. When you release it, the potential energy decreases as the mass falls, but the kinetic energy increases as the speed of the mass increases (Jack and Kerr 2002).

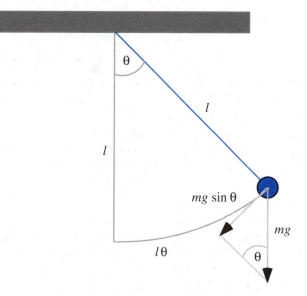

Figure 4.3.1   A simple pendulum

Simple Pendulum

We start our model of a pendulum's motion by considering a **simple pendulum**, which incorporates the following simplifying assumptions: All the mass for the bob is concentrated at a point; the stiff string has no mass; and friction does not exist. Figure 4.3.1 depicts such a simple pendulum with bob of mass m and string of length l being at an angle θ (in radians) off the vertical (Osinga 1999; Weisstein 2003). We consider the angle to be positive when to the right of the vertical and negative when to the left.

As noted in the diagram, the weight of the bob is mg, where $g = 9.81$ m/sec^2 is the acceleration due to gravity with down being the positive direction. The only force pulling the bob along an arc in this simple pendulum is the tangential component of the weight, whose magnitude is $mg\sin\theta$. Because the component points to the left when $\theta > 0$ and points to the right when $\theta < 0$, the force acting on the bob is $-mg\sin\theta$.

According to a geometric formula, the arc length from the bob to the vertical is the product of the string length and the angle in radians, $l\theta$. Thus, acceleration along the bob's path, the **angular acceleration**, is the second derivative of this arc length, or angular acceleration $= d^2(l\theta)/dt^2$. Because length is constant for a given pendulum, $d^2(l\theta)/dt^2 = l\, d^2(\theta)/dt^2$.

By Newton's Second Law of Motion, the force is equal to the mass times the acceleration. Thus, the force pulling the bob along the arc is as follows:

$$\text{force} = (\text{mass})(\text{angular acceleration}) = ml\, d^2(\theta)/dt^2$$

Equating this expression to the negative of the earlier component of weight along the arc, $-mg\sin\theta$, we have the following:

$$ml\, d^2(\theta)/dt^2 = -mg\sin\theta$$

Canceling m and dividing by l, we have the following formula for angular acceleration:

$$\text{angular acceleration} = d^2(\theta)/dt^2 = -g\,\sin\theta/l$$

Initial conditions include specifying that at time $t = 0$, $\theta = \pi/4$, and the initial angular velocity is $d(\theta)/dt = 0$.

Because we cancel m, we see that the mass of the bob is irrelevant for the angular acceleration. According to the formula, for a given acceleration due to gravity at a particular location, the angular acceleration of a simple pendulum depends only on the length of the string and the angle off the vertical.

Quick Review Question 1

This question reflects on Step 2 of the modeling process—formulating a model—for developing a model for a pendulum. We simplify this first attempt at a model by ignoring friction. After completing this question and before continuing in the text, we suggest that you develop a model for a pendulum.

    a. Determine variables for the model and their units in the metric system.
    b. Which of these variables is the rate of change of angular velocity?
    c. What is the flow into a stock/box variable of angular velocity?
    d. Give a differential equation relating angular acceleration ($d^2(\theta)/dt^2$), angle (θ), time (t), and pendulum length (l).
    e. What is the value of the flow into a stock/box variable of angle?

Angular acceleration is the rate of change of angular velocity, and angular velocity is the rate of change of the angle, or $d\theta/dt$. With this information, we are in a position to develop a model diagram, which appears in Figure 4.3.2. The stocks are

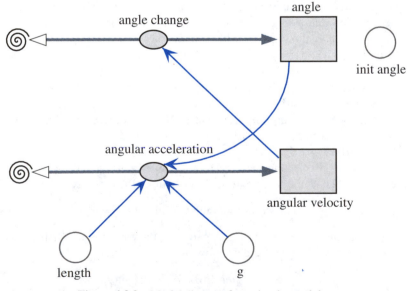

Figure 4.3.2   Model diagram for a simple pendulum

angle, which has an initial value of *init_angle*, and *angular_velocity* with an initial value of 0. Angular acceleration (*angular_acceleration*), whose formula developed above is −*g* sinθ/*l*, is in a flow going into *angular_velocity*. The flow into *angle*, *angle_change*, has a value equal to *angular_velocity*. Both *angular_acceleration* and *angle_change* are bidirectional to allow for increasing and decreasing stocks as the pendulum swings back and forth. Moreover, *angular_velocity* and *angle* should be allowed to accommodate negative and positive values. Various underlying equations that accompany this diagram appear in Equation Set 4.3.1.

Equation Set 4.3.1

Various underlying equations to accompany Figure 4.3.2

init_angle = 3.14159/4
length = 1
g = 9.81
angle_change = *angular_velocity*
angle(0) = *init_angle*
angular_acceleration = −*g* * sin(*angle*)/*length*
angular_velocity(0) = 0

For a string length of 1 m and initial angle of π/4, Figure 4.3.3 presents a plot of the angle θ, angular velocity, and angular acceleration versus time. With only the force involving weight, the pendulum moves back and forth forever.

Quick Review Question 2

The following questions relate to a simple pendulum with mass of 3 kg, length of 4 m, and initial angle of π/6.

   a. Give the magnitude of the initial tangential component of weight along with its units.
   b. Give the initial angular acceleration along with its units.

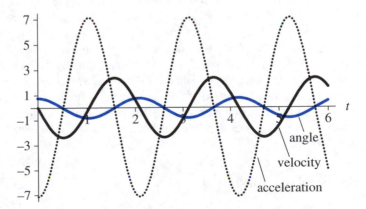

Figure 4.3.3   Plot of angle in radians, angular velocity in radians per sec, and angular acceleration in radians per sec² versus time in sec

Linear Damping

In a real pendulum, the motion is damped by friction. As indicated in Module 4.1, a simple model of such effect is Stokes' friction ($F = kv$). Thus, we assume that the damping force is proportional to the angular velocity, $d(\theta)/dt$. In this case of **linear damping**, the model for the forces is as follows, where k is a positive constant:

$$ml \, d^2(\theta)/dt^2 = -mg \, \sin\theta - kd(\theta)/dt$$

The initial conditions continue to be $\theta = \theta_0$ and $d(\theta)/dt = 0$ when $t = 0$. Projects explore various models of a pendulum's motion that incorporate friction (Danby 1997).

Quick Review Question 3

    a. Using linear damping, give the formula for angular acceleration.
    b. With linear damping, give the effect on the amplitude of the angular acceleration of increasing the mass: increases, decreases, no effect.

Pendulum Clock

As the exercises and projects explore, the angle θ of a pendulum does not affect its **period**, or the length of time to complete a full cycle. Thus, we can use the device in construction of a clock.

We consider the construction of a 60-second clock with only one hand as in Figure 4.3.4. Falling to the ground, a **weight** attached to a rope wound around a **drum** provides potential energy to run the clock. The drum also has an attached clock hand so that the hand moves as the drum does. The weight is prevented from falling to the floor immediately by the action of the **pendulum** and a toothed wheel, or **escapement gear**. The shaft of the pendulum bob attaches rigidly to a shaft with an **anchor** that has an associated lever arm. As the pendulum swings in one direction, this attachment moves the anchor and a tooth of the gear escapes the grasp of a **right** or **left stop** on the anchor, lowering the weight and producing a "tick" sound. Swinging in the opposite direction, the advancing gear hits the other stop with a "tock" sound. Besides regulating the lowering weight, the gear imparts enough of the falling weight's potential energy through the rigid shaft attachment to the pendulum for the latter to overcome friction that dampens its motion. Thus, the pendulum continues swinging for an extended period of time. Although we do not picture it here, through the interaction of gears, the mechanism also controls the hour and minute hands of the clock. A project models a pendulum clock with a second hand (Brain 2002; Britannica 2000).

Quick Review Question 4

For each of the following match the clock part: escapement gear with anchor and stops, pendulum, second hand, stop, weight.

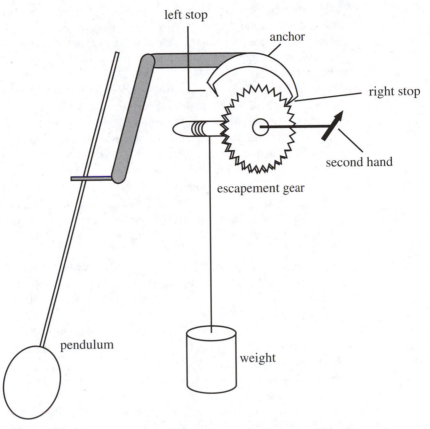

Figure 4.3.4 Pendulum clock with only a second hand

   a. Prevents weight from falling to floor
   b. Provides potential energy to run clock
   c. Regulates timing
   d. Transfers energy from weight to overcome friction

Exercises

1. a. Using a computational tool, evaluate $\sin \theta / \theta$ for values of θ getting closer and closer to 0.

   b. Using your answer from Part a, give an approximation for $\sin \theta$ for small angles θ.

   c. Using your answer from Part b, give a formula that is an approximation for the restoring force due to gravity for a simple pendulum.

   d. Using your answer from Part c and the formula for length of an arc, give a formula with independent variable of the arc length that is an approximation for the restoring force due to gravity for a simple pendulum.

e. Using your answer from Part a, give a differential equation that is an approximation for the angular acceleration for small angles θ.

f. Determine a, b, and c so that $\theta(t) = a\cos(ct) + b\sin(ct)$ is a solution to the differential equation of Part e with initial conditions $\theta(0) = \theta_0$ and $\theta'(0) = 0$.

2. **a.** Using the model for the simple pendulum in *simplePendulum* (see "Download") with length being 1, determine the period.

   b. Change the length of the string to 4 and determine the period.

   c. Change the length of the string to 9 and determine the period in relationship to your answer in Part a.

   d. Determine a relationship between the period and the length of string for a given acceleration due to gravity.

3. **a.** Using the model for the simple pendulum in *simplePendulum* (see "Download") with length being 1, determine the period.

   b. Change the acceleration due to gravity to 4×9.81 and determine the period.

   c. Change the acceleration due to gravity to 9×9.81 and determine the period in relationship to your answer in Part a.

   d. Determine a relationship between the period and the acceleration due to gravity for a given string length.

4. **a.** Using the model for the simple pendulum in *simplePendulum* (see "Download") with length being 1 and the acceleration due to gravity being 9.81 m/sec^2, determine the period.

   b. Determine a formula for period by using the answers from Part a, Exercise 2d and Exercise 3d.

5. Using the model for the simple pendulum in *simplePendulum* (see "Download") with length being 1, determine the periods for different initial angles. Does the angle have any effect on the period?

6. Write a description of what the three graphs in Figure 4.3.3 show. Describe the relative phases of the three curves. Is it reasonable that the magnitude of the angular velocity is greatest when the angle is zero? Is it reasonable that the angular acceleration is greatest when the angle is least, and vice versa?

Projects

For all model development, use an appropriate system dynamics tool.

1. According to the Conservation of Energy Principle, with only conservation forces in effect, the sum of a particle's potential and kinetic energies is constant throughout motion. The formula for potential energy is *mgh* and for kinetic energy is $0.5\ mv^2$, where h is height and v is velocity. Adjust the simple pendulum model in *simplePendulum* (see "Download") to illustrate the Conservation of Energy Principle. Note, h is the y-value of the bob.

2. Develop a model of the pendulum assuming damping as described in the section "Linear Damping." Determine the period. With the model and with mathematics, show the effect of increasing mass on the resistance. Does the period change as the amplitude diminishes?

3. Develop a model of the pendulum assuming constant-magnitude dry friction whose sign is opposite that of $d(\theta)/dt$.
4. Develop a model of the pendulum assuming dry friction whose magnitude is greater at angular velocities closer to zero.
5. Develop a model of a pendulum clock that completes a cycle in one second. Assume linear damping as modeled in Project 2. Have the impulse increase the angular velocity by an appropriate fixed amount at the bottom of a swing in one direction. If available in your system dynamics tool, a delay function that returns the value of an argument in the previous time step might be helpful. Approximately, how long does your model run before the clock "runs down"?
6. Develop a model of a pendulum clock. Assume dry friction as modeled in Project 4. Have the impulse increase the kinetic energy or equivalently, the angular acceleration, by an appropriate fixed amount (see Project 1). If available in your system dynamics tool, a delay function that returns the value of an argument in the previous time step might be helpful. Approximately, how long does your model run before the clock "runs down"?

Answers to Quick Review Questions

1. **a.** time in seconds; length of pendulum in meters; angle in radians; angular velocity, perhaps in radians/sec; angular acceleration in radians/sec^2
   b. angular acceleration
   c. angular acceleration
   d. angular acceleration $= d^2(\theta)/dt^2 = -g \sin\theta/l$
   e. angular velocity
2. **a.** $|mg\sin(\theta)| = (3)(9.81)\sin(\pi/6)$ kg m/sec$^2 = (3)(9.81)(0.5)$ N $= 14.7$ N
   b. $-g \sin \theta/l = -(9.81)\sin(\pi/6)/4$ m/sec$^2 = -1.23$ m/sec^2
3. **a.** $d^2(\theta)/dt^2 = -g \sin \theta/l - (k\, d(\theta)/dt)/(ml)$
   b. increases
4. **a.** escapement gear with anchor and stops
   b. weight
   c. pendulum
   d. escapement gear with anchor and stops

References

Bellis, Mary. 2003. "The Invention of Clocks Part 3: Mechanical Pendulum Clocks and Quartz Clocks." About.com. http://inventors.about.com/library/weekly/aa072801a.htm

Boyd, Thomas M. 2002. "Pendulum Measurements." Colorado School of Mines. http://www.mines.edu/fs_home/tboyd/GP311/MODULES/GRAV/NOTES/pend.html (accessed May 15, 2003).

Brain, Marshall. 2002. "How Pendulum Clocks Work." HowStuffWorks. http://science.howstuffworks.com/clock.htm

Danby, J.M.A. 1997. *Computer Modeling: From Sports to Spaceflight . . . From Order to Chaos*. Richmond, Va.: Willmann-Bell: 408.

Encyclopedia Britannica. 2000. "Pendulum Clock." *Britannica Online*. http://www.britannica.com/clockworks/pendulum.html (accessed May 18, 2003).

Jack, Trevor, and Stephanie Kerr. 2002. "Pendulum Motion." Albertson College of Idaho. http://www.albertson.edu/physics/PHY271_F01/Projects-2002/Pendulum/Pendulum Motion Page.htm (accessed May 24, 2003)

Osinga, Hinke. 1999. "The classical pendulum." California Institute of Technology. http://www.cds.caltech.edu/~hinke/courses/CDS280/pendulum.html

Weisstein, Eric. "Simple Pendulum." "Eric Weisstein's World of Physics," Wolfram Research. http://scienceworld.wolfram.com/physics/SimplePendulum.html

MODULE 4.4

Up, Up, and Away—Rocket Motion

Download

The text's website has available for download for various system dynamics tools the following file containing a framework for the model in this module: *Rocket*.

Introduction

"Of course Peter had been trifling with them, for no one can fly unless the fairy dust has been blown on him."

The Adventures of Peter Pan
J. M. Barrie

Human beings have long looked to the sky with a yearning to fly and have long experimented with various methods and contrivances to accomplish this goal. Around 1500 A.D., a mandarin named Wan-Hu attempted to fly to the moon in a wicker chair to which were attached 47 "rockets"—actually 47 bamboo tubes filled with black powder (Dvir 2003; Lethbridge 2000). Unfortunately, the innovator was unable to conduct other experiments with the potential of rockets to propel human beings into the sky. The successful launching of human beings into space would have to wait several centuries. Now, the launching of rockets with or without human beings is quite an ordinary event. The space above earth is littered with various types of satellites placed into orbit by rockets.

The Chinese generated a form of black powder or "gun powder" during the first century from charcoal, saltpeter, and sulfur (Lethbridge 2000; Schombert 2003). Initially, they used this powder for fireworks; but sometime around the year 1000 A.D., they adapted this powder for use in fire arrows. These fire arrows were made by attaching powder-filled bamboo tubes to arrows and launching them with a bow. By 1232, they had modified these arrows by attaching tubes, open at one end and capped at the other end, to long sticks. They lit the powder, and the first true rockets were launched toward their Mongol attackers. The tips of these rockets were coated with either flammable materials or poison. How effective these rockets were as weapons is questionable, but the Chinese successfully warded off these invaders. Furthermore, the Mongols developed their own rockets and helped to spread their use to Europe. In fact, the word "rocket" probably originated from an Italian word "rochetta," coined by Muratori in his description of fire arrows used in medieval times (Lethbridge 2000; Schombert 2003).

From the time of the Chinese "fire arrows," rockets have continued to play important military roles. During the last half of the twentieth century, however, rockets have taken on new roles in the exploration of the universe. Currently, satellites carried by rockets are providing us with three-dimensional views of polar ice sheets to give us insight into climate and its effects on life. Rockets have launched space telescopes and propelled probes to Mars, to the edge of our solar system, and beyond. More than 400 people have traveled into space borne by rockets since 1961 (NASA).

Physics Background

Before embarking on our development of a model of rocket motion, we need to consider some of the physics fundamentals. We have already worked extensively with Newton's Second Law, $F = ma$, where F is a force acting on an object of mass m and imparting an acceleration, a (see the section on "Physics Background" from Module 4.1, "Modeling Falling and Skydiving"). In that same section, we discussed drag on an object.

With rockets, we consider another mechanical force, **thrust**. A rocket's engine generates thrust through the acceleration of a mass of gas through the bottom, propelling the rocket in the opposite direction. Thus, the forces obey **Newton's Third Law of Motion**: "For every action, there is an equal and opposite reaction." The concept for a rocket is the same as that of a filled and released balloon, where expelled gas under pressure causes the balloon to fly around the room.

> **Definition**   **Thrust** is a mechanical force caused by the acceleration of a mass of fluid and in the opposite direction to flow.

Suppose c is the velocity of the gas relative to the rocket and v is the velocity of the rocket, so that $c + v$ is the velocity of the gas in space. If m is the mass and up is positive, then the **thrust** of the engine (T) is as follows:

$$T = c\frac{dm}{dt}$$

Over a period of time Δt, we have the following discrete version:

$$T = c\frac{\Delta m}{\Delta t}$$

or

$$T \, \Delta t = c \, \Delta m$$

Quick Review Question 1

    a. Select all appropriate units of measure for thrust: kg, kg m/sec^2, kg/sec^2, m/sec^2, mph, N, N sec, lb, lb/sec^2, sec.

    b. With up being positive, suppose a rocket is traveling up at a speed of 500 m/sec, and the speed of the downward gas is 800 m/sec. Give the value of c.

    c. Suppose over a period of 0.1 sec, 2 kg of propellant burns. Give the engine thrust.

As rocket fuel burns, the mass of the fueled rocket decreases. From time t to time $t + \Delta t$, the **impulse** is the product of the thrust and Δt, as follows:

$$\mathbf{\mathit{I = T\,\Delta t}}$$

During that period, the **specific impulse** (I_{sp}) is the impulse per newton (or pound) of fuel, or the quotient of impulse and the weight of the burned fuel, Δw.

$$I_{sp} = \frac{I}{\Delta w} = \frac{T\Delta t}{(\Delta m)g}$$

Solving for T, we have the following value of thrust from time t to time $t + \Delta t$:

$$T = I_{sp}g\,\frac{\Delta m}{\Delta t}$$

Letting Δt approach 0, we have the equivalent derivative form:

$$T = I_{sp}g\,\frac{dm}{dt}$$

Definitions   An **impulse** is the product of the thrust and the length of time. **Specific impulse** is the impulse per unit weight of burned fuel, or the quotient of impulse and the change in the fuel's weight.

As with earlier models of motion, our model of the motion of a rocket incorporates acceleration. In this case, we wish to have a formula for acceleration due to thrust. Because thrust is a force, for acceleration a we have the following equation by Newton's Second Law:

$$T = ma$$

Substituting for T and solving for acceleration, the following is true:

$$I_{sp}g\,\frac{dm}{dt} = ma$$

or

$$a = \frac{I_{sp}g\,\dfrac{dm}{dt}}{m}$$

Quick Review Question 2

    a. Select all appropriate units of measure for impulse: kg, kg m/sec^2, kg/sec^2, m/sec^2, mph, N, N sec, lb, lb/sec^2, sec.

    b. Suppose a fuel burning for 2 seconds imparts a thrust of 75 newtons to a rocket. Give the impulse.

    c. Select all appropriate units of measure for specific impulse: kg, kg m/sec^2, kg/sec^2, m/sec^2, mph, N, N sec, lb, lb/sec^2, sec.

    d. Suppose 0.5 kg of the fuel of Part b burns during 2 seconds. Give the specific impulse.

System Dynamics Model

The model of the motion of a ball tossed into the air in Figure 4.1.1 of Module 4.1 on "Modeling Falling and Skydiving" serves as a basis for the rocket-motion model. For the extension, we make several assumptions:

- The only forces acting on the rocket are gravitation and thrust derived from burning fuel.
- Acceleration due to gravity is constant.
- The earth is flat.
- The rocket is vertical.
- The rocket has only one stage.

Quick Review Question 3

This question reflects on Step 2 of the modeling process—formulating a model—for developing a model for rocket motion. We employ the above simplifying assumptions. After completing this question and before continuing in the text, we suggest that you develop a model for rocket motion.

    a. Building on the model in Example 1 of Module 4.1 on "Modeling Falling and Skydiving," determine additional variables for the rocket-motion model and their units in the metric system.

    b. Give a differential equation for change in total mass (dm/dt) as a function of the constants mass of initial unburned fuel (f) and time to burn (b). Use the simplifying assumption that dm/dt is constant.

    c. Give a differential equation for acceleration, or change in velocity (dv/dt), in terms of total mass (m), change in total mass (dm/dt), specific impulse (I_{sp}), and acceleration due to gravity (g).

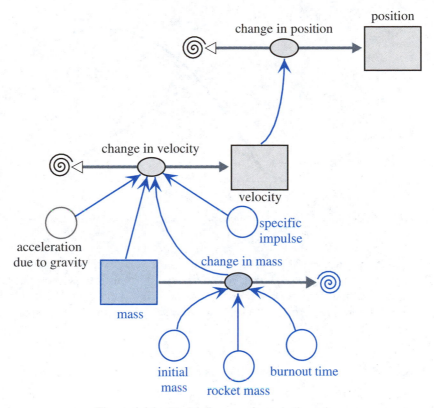

Figure 4.4.1   Model diagram of a rocket's motion

Extending the model in "Modeling Falling and Skydiving," thrust from burning fuel also impacts the motion of a rocket. The *change_in_velocity*, or acceleration, now involves acceleration due to this thrust as well as acceleration due to gravity. The extended model has an additional stock (box variable), *mass*, that contains the total mass of the fuel and rocket, which has mass *rocket_mass*. This stock has an initial value of *initial_mass*. We assume that while fuel is present, the flow out, *change_in_mass*, is constant and consists of the mass of the initial unburned fuel divided by the time for it to burn (*burnout_time*). After burnout, *change_in_mass* becomes zero. Figure 4.4.1, which is similar to Figure 4.1.1, contains a model diagram of a rocket's motion.

Quick Review Question 4

Refer to Figure 4.4.1 to give the formulas for the following:

    a. The initial mass of the unburned fuel
    b. The change in mass per unit of time of rocket with fuel

Figure 4.4.2 displays a graph of position (in color) and velocity versus time when *initial_mass* = 5000 kg, *rocket_mass* = 1000 kg, *burnout_time* = 60 sec, and

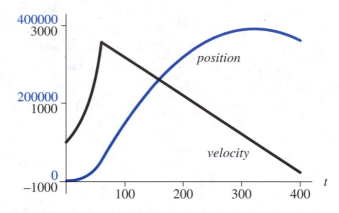

Figure 4.4.2   Graph of position (m) and velocity (m/sec) versus time for a rocket

specific_impulse = 200 sec. The graph shows the velocity increasing more and more until the fuel completely burns. With a velocity of about 2567 m/sec = 2.567 km/sec at that instant, the rocket continues to rise to a height of about 388,500 m = 388.5 km before starting its descent. Various projects complete and expand upon the model in Figure 4.4.1.

Exercises

1. Write the differential equations in the model developed in this module.
2. Revise the model's differential equation for acceleration (dv/dt) to accommodate Newtonian friction (see section "Physics Background" from Module 4.1, "Modeling Falling and Skydiving"). The following formula approximates the density of the Earth's atmosphere:

$$D = 1.225e^{-0.1385y} \text{ kg/m}^3 \text{ for altitudes } y < 100 \text{ km}$$

   The coefficient of drag for a very shiny rocket could be as low as 0.6, while rougher surfaces command higher values closer to 1.
3. Revise the model's differential equation for acceleration (dv/dt) where acceleration due to gravity is not constant but decreases with altitude according to the formula $gR^2/(R + y)^2$, where R is the radius of the earth, which is approximately 6.378×10^6 m, g is the acceleration due to gravity at sea level, and y is the distance of the rocket above the earth's surface. Continue to use $c = I_{sp} g$. The next exercise develops the formula for acceleration due to gravity at altitude y.
4. This exercise develops the formula for acceleration due to gravity at altitude y from the previous exercise.
   a. Newton's Gravitation Law states that the gravitation force between two objects of masses m_1 and m_2 is as follows:

   $$F = G\frac{m_1 m_2}{d^2}$$

where G is the universal gravitational constant $(6.672 \times 10^{-11}$ N m^2 kg$^{-2})$ and d is the distance between the centers of mass of the objects. Let R be the radius of the earth, m_e its mass, and m the mass of an object. Write Newton's Gravitation Law for the weight of the object at the earth's surface using these masses.

b. Write the weight of the object at the earth's surface using Newton's Second Law and g.

c. Setting Parts a and b equal, solve for g, and simplify.

d. Consider the object at an altitude y above the surface of the earth. Write Newton's Gravitation Law for the weight of the object at height y.

e. Write the weight of the object using Newton's Second Law and g_y, the acceleration due to gravity at altitude y.

f. Setting Parts d and e equal, solve for g_y and simplify.

g. Evaluate the ratio g_y/g and simplify.

h. Solve for g_y.

5. National Association of Rocketry (NAR) codes, such as C6-3, appear on hobby rocket motors, as follows:

- The letter, which can be from 1/2A to K, specifies the total impulse with C indicating 5.01 to 10.00 N sec. A letter's range is double that of its predecessor, so that an impulse in the range of 1/2A is the smallest, and that in the range of K is the largest. Thus, the total impulse for rockets with letter B is from 2.51 to 5.00 N sec.
- The subsequent number, such as 6, indicates **average thrust** (in newtons). The average thrust with total impulse indicates the length of time over which the motor releases its total energy.
- The number after the dash, such as 3, gives **time delay** (in seconds), or the time from motor burnout until activation of a recovery parachute. During that time, the rocket coasts to a higher altitude and slows.

For the following questions involving total impulse, use the higher range value, such as 10.00 N sec for type-C motors.

a. Approximate the length of time for which a C6-5 motor delivers its energy.

b. Repeat Part a for a C4-5 motor.

c. Repeat Part a for a C10-5 motor.

d. Repeat Part a for a C5-3 motor.

e. Which of the above three engines is most powerful?

f. The C5-3 burns 12.7 g of propellant. Calculate its specific impulse.

Projects

For all model development, use an appropriate system dynamics tool.

1. Complete the model of rocket motion described in this module and begun in *Rocket* (see "Download"). Plot position (altitude) and velocity with respect to time. Obtain the maximum altitude and velocity. Try various parameter values, such as those for a hypothetical rocket with initial mass of 5000 kg,

Table 4.4.1

Rocket Engine Specifications (see Exercise 5 also) (Culp)

Engine Type	Maximum Lift (g)	Initial Mass (g)	Propellant Mass (g)
A10-3T	141.5	7.9	3.78
C5-3	226.4	25.5	12.7
C6-3	113.2	24.9	12.48
D12-5	283.0	43.1	24.93

Table 4.4.2

Comparison of Chemical and Solar Electric Engines, Each with Total Impulse $= 6 \times 10^7$ N sec (Wallace)

Engine	c (m/sec)	I_{sp} (sec)	T (N)	Burn Time (sec)	Propellant Mass (kg)
Centaur (chemical)	−4300	440	66,000	880	13,600
Electrostatic Thruster	−29,400	3000	1	5×10^7	2000

rocket mass of 1000 kg, burnout time of 60 sec, and specific impulse of 200. Also, consider values of real engines, such as those in Table 4.4.1 with code information (see Exercise 5). Write a discussion of the results. Augment your work by having a comparative plot of altitude and velocity versus time for various rocket masses. Discuss the impact of various rocket masses on the altitude and velocity at burnout. Similarly, investigate the impact of various specific impulses.

2. Complete the model of rocket motion described in this module and begun in *Rocket* (see "Download"). Use your model to compare a Centaur Upper Stage System, which is a chemical system, and the Electrostatic Thruster System, which utilizes the sun's nuclear energy as well as a propellant (see Table 4.4.2). Your comparison should include maximum velocity, propellant mass, thrust time, and the types of missions advisable for each (Wallace).

3. The first model assumption above was to ignore the effects of drag. In this project, refine the model to accommodate Newtonian friction (see Exercise 2). Investigate the impact of drag on altitude and speed at burnout. Discuss the results, including the reasonableness of the assumption to ignore drag.

4. The first model assumption above was that acceleration due to gravity is constant. In this project, refine the model to consider decreasing acceleration due to gravity as the rocket gains altitude according to the formula in Exercise 3. Investigate the impact on altitude and speed at burnout. Discuss the results, including the range of altitudes at which the assumption that acceleration due to gravity is constant seems reasonable.

5. Develop a model of a single-stage rocket in which after burnout and a time delay a parachute deploys so that the rocket falls safely to earth (see Module 4.1, "Modeling Falling and Skydiving").

6. Develop a model for a two-stage rocket. Each stage has an engine with propellant. After the initial burn, the rocket coasts for a few seconds before second-stage ignition occurs. Discuss the advantages and disadvantages of such an arrangement.

Answers to Quick Review Questions

1. a. kg m/sec^2, N, lb
   b. -300 m/sec
   c. $c \, \Delta m/\Delta t = (-300)(2)/(0.1) = -6000$ N
2. a. N sec
   b. $I = T\Delta t = (75)(2) = 150$ N sec
   c. sec
   d. $I_{sp} = I/\Delta w = I/(\Delta mg) = (150$ N sec$)/((0.5$ kg$)(9.81$ m/sec$^2)) = 30.6$ sec
3. a. rocket mass in kg, total mass of rocket and fuel in kg, change in total mass in kg, time for fuel to burn in sec, specific impulse in sec
   b. $dm/dt = f/b$
   c. Acceleration is the sum of acceleration due to gravity and acceleration due to thrust. Thus, $a = dv/dt = \dfrac{I_{sp} g (dm/dt)}{m} + g.$
4. a. *initial_mass – rocket_mass*
   b. (*initial_mass – rocket_mass*)/*burnout_time*

References

Culp, Randy. "Rocket Simulations." http://my.execpc.com/~culp/rockets/rckt_sim.html

———. "East Coast Model Center Engine Specifications." *Estes Catalog*. http://my.execpc.com/~culp/rockets/estes_spec.html

Danby, J.M.A. 1997. *Computer Modeling: From Sports to Spaceflight . . . From Order to Chaos*. Richmond, Va.: Willmann-Bell.

Dvir, Tal. 2003. "Four decades after Gagarin, China finally reaches for the stars." *Telegraph Newspaper Online*, May 10. http://www.telegraph.co.uk/news/main.jhtml?xml=/news/2003/10/05/wspace05.xml&sSheet=/portal/2003/10/05/ixportal.html

Lethbridge, Cliff. 2000. "History of Rocketry Chapter 1: Ancient Times Through the 17th Century." Spaceline. http://www.spaceline.org/history/1.html

NASA Aeronautics Resources. "Brief History of Rockets." NASA. http://www.grc.nasa.gov/WWW/K-12/TRC/Rockets/history_of_rockets.html

National Association of Rocketry. "Standard Motor Codes." http://www.nar.org/NARmotors.html

Ratliff, Lisa, Larry Storm, James Blattman, and Gae McGibbon. "Rocketry: Modeling and Models." Computational Science in Education, National Computational Science Education Consortium. http://www.ncsec.org/cadre2/webview.cfm?menu=1.1&team=14&cadre=2

Schombert, James. 2003. "Roots Of Rocketry." University of Oregon. http://zebu.
uoregon.edu/~js/space/lectures/lec01.html

Wallace, Sherry. "Solar Electric Propulsion Fundamental Physics." NASA In-Space
Propulsion Technologies. http://www.inspacepropulsion.com/tech/solar_elec_
physics.pdf

Weisstein, Eric. 2003. "Eric Weisstein's World of Physics." Wolfram Research.
http://scienceworld.wolfram.com/physics/

5

SIMULATION TECHNIQUES

MODULE 5.1

Computational Toolbox—Tools of the Trade: Tutorial 2

Prerequisite: Module 2.1 on "Computational Toolbox—Tools of the Trade: Tutorial 1"

Download

From the textbook's website, download Tutorial 2 in the format of your computational tool or in pdf format. We recommend that you work through the tutorial and answer all Quick Review Questions using the corresponding software.

Introduction

Various computer software tools are useful for graphing, numeric computation, and symbolic manipulation. This second computational toolbox tutorial, which is available from the textbook's website in your system of choice, prepares you to use the tool to complete projects for this and subsequent chapters. The tutorial introduces the following functions and concepts:

- Lists
- Plotting data
- Comments
- Appending

The module gives computational examples and Quick Review Questions for you to complete and execute in the desired software system.

MODULE 5.2

Euler's Method

Download

The text's website has available for download for various system dynamics tools the file *unconstrainedError* file, which contains a model for the "Error" section below.

Introduction

With system dynamics tools, we often have the choice of simulation techniques, such as Euler's Method, Runge-Kutta 2, Runge-Kutta 4, and others. These numerical methods are for solving ordinary differential equations and estimating definite integrals for which the indefinite integral does not exist. In this module, we discuss the most straightforward of these, Euler's Method.

Reasoning behind Euler's Method

In Module 3.2, "Unconstrained Growth and Decay," to simulate $dP/dt = 0.10P$ with $P_0 = 100$, we employ the following underlying equations with *INIT* meaning "initial" and *dt* representing a small change in time, Δt:

```
growth_rate = 0.10
INIT population = 100
growth = growth_rate * population
population = population + growth * dt
```

These equations, which we enter explicitly or implicitly with a model diagram, represent the following finite difference equations using **Euler's Method**:

```
growth_rate = 0.10
population(0) = 100
growth(t) = growth_rate * population(t - Δt)
population(t) = population(t - Δt) + growth(t) * Δt
```

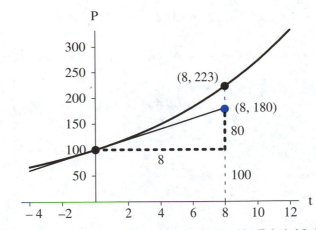

Figure 5.2.1   Actual point, (8, 223), and point obtained by Euler's Method, (8, 180)

The population at one time step, *population(t)*, is the population at the previous time step, *population(t − Δt)*, plus the estimated change in population, *growth(t)* ∗ Δt. The derivative of population with respect to time is *growth*, and

$$growth(t) = growth\_rate * population(t - \Delta t)$$
$$= 0.10 * population(t - \Delta t)$$

or $dP/dt = 0.10P$. The change in the population is the flow (in this case, *growth*) times the change in time, Δt; and the flow (*growth*) is the derivative of the function at the previous time step, $t - \Delta t$.

Figure 5.2.1 illustrates Euler's Method to estimate $P_1 = P(8)$ for the above differential equation by starting with $P_0 = P(0) = 100$ and using $\Delta t = 8$. In this situation, $t = 8$, $t - \Delta t = 0$, and the derivative at that time is $P'(0) = 0.1(100) = 10$. We multiply Δt, 8, by this derivative at the previous time step, 10, to obtain the estimated change in P, 80. Consequently, the estimate for P_1 is as follows:

$$\text{Estimate for } P_1 = \text{previous value of } P + \text{estimated change in } P$$
$$= P_0 + P'(0)\Delta t$$
$$= 100 + 10(8)$$
$$= 180$$

In Module 3.2, "Unconstrained Growth and Decay," we discovered analytically that the solution to the above differential equation is $P = 100\, e^{0.1t}$. Because the graph of the actual function is concave up, this estimated value, 180, is lower than the actual value at $t = 8$, $100e^{0.1(8)} \approx 223$.

Quick Review Question 1

For $dP/dt = 10 + P/5$ with $P_0 = 500$ and $\Delta t = 0.1$, calculate the following:

   a. dP/dt at $t = 0$
   b. Euler's estimate of P_1

Algorithm for Euler's Method

Following the description above, Algorithm 1 presents Euler's Method.

Algorithm 1   **Euler's Method**

$t \leftarrow t_0$

$P(t_0) \leftarrow P_0$

Initialize *SimulationLength*
while $t < SimulationLength$ do the following:

$\quad t \leftarrow t + \Delta t$

$\quad P(t) = P(t - \Delta t) + P'(t - \Delta t)\Delta t$

Thus, simulation uses a sequence of times—t_0, t_1, t_2, . . . —and calculates a corresponding sequence of estimated populations—P_0, P_1, P_2, ... In Algorithm 1, $t_n = t_{n-1} + \Delta t$ or $t_{n-1} = t_n - \Delta t$, and $P_n = P_{n-1} + P'(t_{n-1})\Delta t$.

However, as illustrated in Module 2.2 on "Errors," repeatedly accumulating Δt into t usually produces an accumulation error. To minimize error, we calculate the time as the sum of the initial time and an integer multiple of Δt. Using the functional notation $f(t_{n-1}, P_{n-1})$ to indicate the derivative dP/dt at step $n - 1$, Algorithm 2 presents a revised Euler's Method that minimizes accumulation of error.

Algorithm 2   **Revised Euler's Method** to minimize error accumulation
           of time with $f(t_{n-1}, P_{n-1})$ indicating the derivative dP/dt at step
           $n - 1$

Initialize t_0 and P_0

Initialize *NumberOfSteps*
for n going from 1 to *NumberOfSteps* do the following:

$\quad t_n = t_0 + n\,\Delta t$

$\quad P_n = P_{n-1} + f(t_{n-1}, P_{n-1})\,\Delta t$

Quick Review Question 2

Match each of the symbols below to its the meaning in Algorithm 2 for Euler's Method. Here, "previous" means "immediately previous."

   A. Change in time between time steps
   B. Derivative of function at estimated value of function for current time step
   C. Derivative of function at estimated point for previous time step

D. Estimated value of function at current time step
E. Estimated value of function at previous time step
F. Initial time
G. Initial value of function
H. Number of current time step
I. Time at current time step
J. Time at previous time step

a. t_n
b. t_0
c. n
d. Δt
e. P_n
f. P_{n-1}
g. $f(t_{n-1}, P_{n-1})$
h. t_{n-1}

Error

As we saw in Module 3.2, "Unconstrained Growth and Decay" the analytical solution to $dP/dt = 0.10P$ with $P_0 = 100$ is $P = 100e^{0.10t}$. Even with Algorithm 2, comparison of the results of Euler's Method with the analytical solution reveals an accumulation error. As Figure 5.2.2 and an *unconstrainedError* file illustrate (see "Download"), we can adjust the unconstrained growth model to demonstrate the variation.

The converter/variable *actual_population* evaluates $P_0 e^{rt}$ or $100e^{0.10t}$. The formula does not use the changing value of *population*, but the initial population, *initial_population*. With *Time* as the current value of time and *EXP* as the exponential function, the equation in the converter *actual_population* is as follows:

```
initial_population * EXP(growth_rate * Time)
```

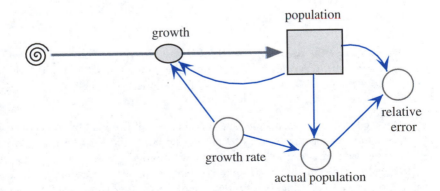

Figure 5.2.2   Unconstrained growth model with monitoring

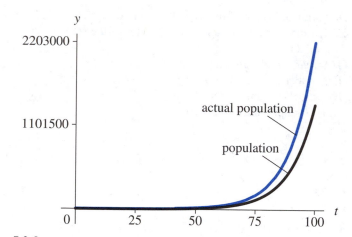

Figure 5.2.3   Graphs of analytical solution and Euler's Method solution with $\Delta t = 1$

Similarly, the converter *relative_error* computes the relative error as the absolute value (*ABS*) of the difference in Euler's estimate and the analytical population with the result divided by the latter, as follows:

```
ABS(population - actual_population)/actual_population
```

Figure 5.2.3 presents graphs of the analytical solution and the solution using Euler's Method with Δt being 1. As demonstrated, the simulated solution is below the analytical one. At the end of the run, at time 100, the analytical value for the population is 2,202,647, while the simulated solution using Euler's Method produces 1,378,061, so that the relative error is over 37.4%. For Δt being cut in half, the relative error is almost cut in half to 21.5% at time 100. The new simulated population is 1,729,258, which is considerably closer to the analytical solution of 2,202,647. If we cut the time step in half again so that Δt is 0.25, the relative also reduces by about half to 11.6% at time 100. Thus, the relative error is proportional to Δt. We say that the relative error is **on the order of Δt, O(Δt)**.

Quick Review Question 3

The analytical solution to $dP/dt = 10 + P/5$ with $P_0 = 500$ of Quick Review Question 1 is $P = 550e^{t/5} - 50$. Part b of that question showed that for $\Delta t = 0.1$ the Euler's Method estimate of P_1 is 511. Calculate the relative error as a percentage with four decimal places.

Of the three simulation techniques in this chapter, Euler's Method is the easiest to understand and has the fastest execution time but is the least accurate. We usually can reduce the error of the Euler Method by employing a smaller Δt, which unfortunately slows the simulation. Despite its shortcomings, the reasoning behind Euler's Method serves as an excellent introduction to the other techniques, Runge-Kutta 2 and Runge-Kutta 4, because each of these has Euler's Method embedded as the first step in its simulation.

Exercises

1. Use $dP/dt = 0.10P$ with $P_0 = 100$ and Euler's Method to calculate P_2 starting with $P_1 = 180$ at $t = 8$ in Figure 5.2.1.

   In Exercises 2–5, for each differential equation with initial condition and Δt, calculate the following using Euler's Method and any other requested computation:
   a. *The first estimated point, such as P_1 where the differential equation is in terms of dP/dt*
   b. *The second estimated point, such as P_2*

2. $dP/dt = 0.10P$ With $P_0 = 100$ and $\Delta t = 2$
   c. The relative error for P_1

3. The logistic equation $dP/dt = 0.5(1 - P/1000)P$ with $P_0 = 20$ and $\Delta t = 2$
   c. The relative error for P_1, where the exact solution is $P(t) = \dfrac{10}{0.01+0.49e^{-0.5t}}$

4. The rate of change of the number of particles of radioactive carbon-14 in a dead tree $dA/dt = -2.783 \, e^{-0.000121t}$ with $A_0 = 23{,}000$ particles and $\Delta t = 0.2$yr
   c. The relative error for A_2, where the exact solution is $A(t) = 23000 \, e^{-0.000121t}$

5. The Gompertz differential equation $dN/dt = kN\ln(M/N)$ with $N(0) = 200$, $k = 0.1$, $M = 1000$, and $\Delta t = 0.5$
   c. The relative error for N_2, where the exact solution is $N(t) = Me^{\ln(N_0/M)e^{-kt}}$

Projects

Using Algorithm 2 for Euler's Method, develop a computational tool file to perform the simulations of Projects 1–7. Run the simulation for the indicated length of time and perform any other requested tasks.

1. Calculate P from $t = 0$ through $t = 100$, where $dP/dt = 0.10P$ with $P_0 = 100$ and $\Delta t = 2$. Calculate the relative error at each time step using the solution $P = 100e^{0.1t}$. Repeat the computation with $\Delta t = 0.25$. Check your results using a system dynamics tool. Use your results in a discussion of relative error.

2. Calculate P from $t = 0$ through $t = 100$ for logistic equation $dP/dt = 1.05(1 - P/1000)P$ with $P_0 = 500$ and $\Delta t = 2$. Calculate the relative error at each time step using the solution $P(t) = \dfrac{10}{0.01+0.49e^{-0.5t}}$. Repeat the computation with $\Delta t = 0.25$. Check your results using a system dynamics tool. Use your results in a discussion of relative error.

3. Suppose the instantaneous rate of change of the number of particles (A) of radioactive carbon-14 in a gram of a dead tree is $dA/dt = -2.783 \, e^{-0.000121t}$ particles/year from the time t the tree dies with $A_0 = 23{,}000$ particles. Use Euler's Method to estimate of the total change in the number of particles of carbon-14 between years 10 and 20. Calculate the exact value of the definite integral with calculus or an appropriate computational tool and compute the relative error.

4. The Gompertz differential equation, which is one of the best models for predicting the growth of cancer tumors, follows:

$$\frac{dn}{dt} = KN \ln\left(\frac{M}{N}\right)$$

where N is the number of cancer cells. Calculate N from $t = 0$ through $t = 20$, where $k = 0.1$, $M = 1000$, and $\Delta t = 0.5$. Using the solution $N(t) = Me^{\ln(N_0/M)e^{-kt}}$ calculate the relative error at each time step. Repeat the computation with $\Delta t = 0.25$. Check your results using a system dynamics tool. Use your results in a discussion of relative error.

5. Estimate $\int_1^5 (-t^2 + 10t + 24)\, dt$ using $\Delta t = 0.25$. Calculate the exact value using calculus or an appropriate computational tool and compute the relative error of the simulated result.

6. Estimate $\frac{1}{\sqrt{2\pi}} \int_0^2 e^{-t^2}\, dt$ using $\Delta t = 0.1$. The corresponding indefinite integral does not exist. The function being integrated is the normal distribution with mean 0 and standard deviation 1.

7. Calculate $h(t)$ and $s(t)$ from $t = 0$ through $t = 50$ using $\Delta t = 0.25$ for the following system of differential equations:

$$\begin{cases} \dfrac{ds}{dt} = 2\,s - 0.02\,hs, & s_0 = 100 \\[2mm] \dfrac{dh}{dt} = 0.01\,sh - 1.06\,h, & h_0 = 15 \end{cases}$$

As Module 6.4, "Predator-Prey Model," discusses, this system is a model for predator (h) and prey (s) populations.

Answers to Quick Review Questions

1. **a.** 110 because $10 + (500)/5 = 110$
   b. 511 because $500 + 0.1(110) = 511$
2. **a.** t_n **I.** Time at current time step
   b. t_0 **F.** Initial time
   c. n **H.** Number of current time step
   d. Δt **A.** Change in time between time steps
   e. P_n **D.** Estimated value of function at current time step
   f. P_{n-1} **E.** Estimated value of function at previous time step
   g. $f(t_{n-1}, P_{n-1})$ **C.** Derivative of function at estimated point for previous time step
   h. t_{n-1} **J.** Time at previous time step
3. 0.0217% because

$$550e^{0.1/5} - 50 = 511.111 \quad \text{and}$$
$$|(511 - 511.111)|/511.111 = 0.000217 = 0.0217\%$$

References

Burden, Richard L., and J. Douglas Faires. 2001. *Numerical Analysis.* 7th ed. Pacific Grove, Calif.: Brooks/Cole Publishing Co.

Danby, J.M.A. 1997. *Computer Modeling: From Sports to Spaceflight . . . From Order to Chaos*. Richmond, Va.: Willmann-Bell.

Woolfson, M. M., and G. J. Pert. 1999. *An Introduction to Computer Simulation.* Oxford, U.K.: Oxford University Press.

Zill, Dennis G. 2001. *A First Course in Differential Equations with Modeling Applications*. 7th ed. Pacific Grove, Calif.: Brooks/Cole Publishing Co.

MODULE 5.3

Runge-Kutta 2 Method

Introduction

In Module 5.2, "Euler's Method," which is a prerequisite to the current module, we discuss the simplest of this chapter's simulation techniques for solving differential equations and computing definite integrals numerically. In this section, we consider a second and better simulation technique, **Euler's Predictor-Corrector (EPC) Method**, also called **Runge-Kutta 2**.

Euler's Estimate as a Predictor

In the current module, we consider the example of Module 5.2 on "Euler's Method," $dP/dt = 0.10P$ with $P_0 = 100$. As in that section, $f(t_n, P_n)$ is sometimes a more convenient notation for the derivative dP/dt at step n. Thus, at $(t, P) = (0, 100)$, the derivative is $f(0, 100) = 0.1(100) = 10$. According to that technique, using the slope of the tangent line at (t_{n-1}, P_{n-1}), we have the following computation for t_n and estimation of P_n:

$$t_n = t_0 + n\,\Delta t$$
$$P_n = P_{n-1} + f(t_{n-1}, P_{n-1})\,\Delta t$$

As Figure 5.2.1 of "Euler's Method" and Fig. 5.3.1 illustrates for $t_0 = 0$ and $\Delta t = 8$, the estimate at $t_1 = 8$ is the vertical coordinate of the point on the tangent line, $100 + 8(10) = 180$.

Corrector

To estimate (t_n, P_n), we would really like to use the slope of the chord from (t_{n-1}, P_{n-1}) to (t_n, P_n) instead of the slope of the tangent line at (t_{n-1}, P_{n-1}). As in Figure 5.3.2, if we know the slope of the chord between $(0, P(0)) = (0, 100)$ and

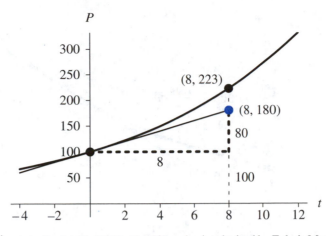

Figure 5.3.1   Actual point, $(8, P(8)) \approx (8, 223)$, and point obtained by Euler's Method, $(8, 180)$

$(8, P(8)) = (8, 100e^{0.10(8)}) \approx (8, 223)$ is approximately $\frac{223 - 100}{8 - 0} = \frac{123}{8}$, we can evaluate $P(8) = P(0) + slope\_of\_chord * \Delta t = 100 + \left(\frac{123}{8}\right)8 = 223$. However, to evaluate the slope of the chord, we must know $P(8) \approx 223$, which is the value we are trying to estimate. If we know the actual value, there is no need to estimate.

Although we do not know the slope of the chord between $(0, P(0))$ and $(8, P(8))$, we can estimate it as approximately the average of the slopes of the tangent lines at $P(0)$ and $P(8)$:

$$\begin{pmatrix} \text{slope of the chord} \\ \text{between } (0,\ P(0)) \text{ and } (8,\ P(8)) \end{pmatrix} \approx \frac{(\text{slope of tan at } P(0)) + (\text{slope of tan at } P(8))}{2}$$

Figure 5.3.3 depicts these two tangent lines.

How can we find the slope of the tangent line at $P(8)$ when we do not know $P(8)$? Instead of using the exact value, which we do not know, we predict $P(8)$ as in

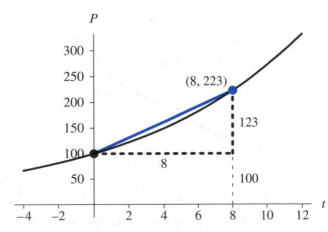

Figure 5.3.2   Actual point, $(8, P(8)) \approx (8, 223)$, along the chord between $(0, 100)$ and $(8, 223)$

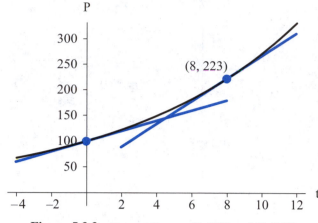

Figure 5.3.3   Tangent lines at $(0, P(0))$ and $(8, P(8))$

Euler's Method. As the computation in the first section on "Euler's Estimate as a Predictor" shows, in this case, the prediction is $Y = 180$. We use the point $(8, 180)$ in derivative formula to obtain an estimate of slope at $t = 8$. In this case, the slope of the tangent line at $(8, 180)$, or the derivative, is $f(8, 180) = 0.1(180) = 18$. Using 18 as the approximate slope of the tangent line at $(8, P(8))$, we estimate the slope of chord between $(0, P(0))$ and $(8, P(8))$ as the following average of tangent line slopes:

$$\text{slope of chord} \approx (10 + 18)/2 = 0.5(10 + 18) = 14$$

As Figure 5.3.4 shows, using 14, the corrected estimate is $P_1 = 100 + 14(8) = 212$, which is closer to the actual value of 223.

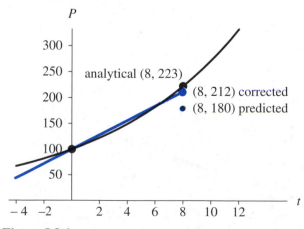

Figure 5.3.4   Predicted and corrected estimation of $(8, P(8))$

Quick Review Question 1

Quick Review Question 1 of Module 5.2 on "Euler's Method" considered $dP/dt = 10 + P/5$ with $P_0 = 500$ and $\Delta t = 0.1$. Part a calculated the derivative at $t = 0$ to be 110, and Part b evaluated Euler's estimate of P_1 to be 511. Calculate the corrected estimate for P_1 using the technique of this section and decimal notation for your answer.

Runge-Kutta 2 Algorithm

Computations of the previous section illustrate Euler's Predictor-Corrector Method for estimating P numerically given a differential equation involving dP/dt. The algorithm for **Euler's Predictor-Corrector (EPC) Method**, or **Runge-Kutta 2**, is the same as Euler's with only one more statement in the loop to obtain the corrected value.

> **Algorithm for Euler's Predictor-Corrector (EPC) Method**, or **Runge-Kutta 2**, with $f(t_{n-1}, P_{n-1})$ indicating the derivative dP/dt at step $n - 1$
>
> Initialize t_0 and P_0
> Initialize *NumberOfSteps*
> for n going from 1 to *NumberOfSteps* do the following:
> $\quad t_n = t_0 + n \, \Delta t$
> $\quad Y_n = P_{n-1} + f(t_{n-1}, P_{n-1}) \, \Delta t$, which is the Euler's Method estimate
> $\quad P_n = P_{n-1} + 0.5 \, (f(t_{n-1}, P_{n-1}) + f(t_n, Y_n)) \, \Delta t$

Quick Review Question 2

Match each of the symbols below to its meaning in the Algorithm for Euler's Predictor-Corrector (EPC) Method. Here, "previous" means "immediately previous"; "EPC estimate" means "estimated value of function using the EPC Method"; and "Euler estimate" means "estimated value of function using Euler's Method."

A. Average of derivatives of function at previous EPC estimate and current Euler estimate

B. Average of derivatives of function at previous Euler estimate and current EPC estimate

C. Derivative of function at EPC estimate for current time step

D. Derivative of function at EPC estimate for previous time step

E. Derivative of function at Euler estimate for current time step

F. Derivative of function at Euler estimate for previous time step

G. EPC estimate at current time step

H. EPC estimate at previous time step

I. Euler estimate at current time step
J. Euler estimate at previous time step
a. Y_n
b. P_{n-1}
c. $f(t_{n-1}, P_{n-1})$
d. P_n
e. $f(t_n, Y_n)$
f. $0.5 \left(f(t_{n-1}, P_{n-1}) + f(t_n, Y_n) \right)$

Error

As noted above, the actual slope of the chord is $(P(8) - 100)/8 \approx (222.6 - 100)/8 \approx$ 15.3, but 14 is certainly a better slope to use than 10 from Euler's Method. With Euler's Method, $P_1 = 180$, giving a relative error of $(180 - P(8))/P(8) \approx |180 - 222.6|/222.6 \approx 0.191 \approx 19.1\%$. We get a much better estimate with Euler's Predicator-Corrector (Runge-Kutta 2) Method, $P_1 = 212$, which has a relative error of $|212 - P(8)|/P(8) \approx |212 - 222.6|/222.6 \approx 0.047 \approx 4.7\%$.

As we saw in the "Error" section of Module 5.2, "Euler's Method," the relative error of Euler's method is on the order of Δt, $O(\Delta t)$. If we cut the time interval Δt in half, the relative error is halved as well. Using the same model with the Runge-Kutta 2 simulation method, Table 5.3.1 shows estimates of $P(100)$, whose actual value to 0 decimal places is 2,202,647, for $\Delta t = 1$, 0.5, and 0.25. As the time interval is cut by $1/2$, the relative error is cut by about $(1/2)^2 = (1/4)$. Thus, the relative error of the EPC method is $O((\Delta t)^2)$, or on the order of $(\Delta t)^2$. Thus, although in each EPC algorithm iteration we must compute a correction, the EPC method is more accurate than Euler's Method.

Quick Review Question 3

The analytical solution to $dP/dt = 10 + P/5$ with $P_0 = 500$ of Quick Review Question 1 is $P = 550e^{t/5} - 50$. That question showed that the Euler's Predictor-Corrector Method estimate of P_1 is 511.11 for $\Delta t = 0.1$. Calculate the relative error as a percentage rounded to four decimal places.

Table 5.3.1
Estimates of $P(100)$ and Relative Errors for Various Changes in Time Using Runge-Kutta 2 Simulation Method, where $dP/dt = 0.10P$ with $P_0 = 100$

	EPC Estimates at Time 100	
Δt	*Estimated P*(100)	*Relative Error*
1.00	2,168,841	1.53%
0.50	2,193,824	0.40%
0.25	2,200,396	0.10%

Exercises

Repeat the exercises of Module 5.2 on "Euler's Method" using the Runge-Kutta 2 Method. Compare the relative errors with those of the corresponding exercises from Module 5.2.

Projects

Repeat the projects of Module 5.2 on "Euler's Method" using the Runge-Kutta 2 Method.

Answers to Quick Review Questions

1. 511.11   because $Y_1 = 511$; $f(0.1, \ 511) = 10 + 511/5 = 112.2$; $500 + (0.5)$ $(110 + 112.2) (0.1) = 511.11$

2. a. Y_n **I.** Euler estimate at current time step

 b. P_{n-1} **H.** EPC estimate at previous time step

 c. $f(t_{n-1}, P_{n-1})$ **D.** Derivative of function at EPC estimate for previous time step

 d. P_n **G.** EPC estimate at current time step

 e. $f(t_n, Y_n)$ **E.** Derivative of function at Euler estimate for current time step

 f. $0.5 \ (f(t_{n-1}, P_{n-1}) + f(t_n, Y_n))$ **A.** Average of derivatives of function at previous EPC estimate and current Euler estimate

3. 0.0002% because $550e^{0.1/5} - 50 = 511.111$ and $|511.11 - 511.111|/511.111 = 0.000002 = 0.0002\%$

References

Burden, Richard L., and J. Douglas Faires. 2001. *Numerical Analysis*. 7th ed. Pacific Grove, Calif.: Brooks/Cole Publishing Co.

Danby, J.M.A. 1997. *Computer Modeling: From Sports to Spaceflight . . . From Order to Chaos*. Richmond, Va.: Willmann-Bell.

Woolfson, M. M., and G. J. Pert. 1999. *An Introduction to Computer Simulation*. Oxford, U.K.: Oxford University Press.

Zill, Dennis G. 2001. *A First Course in Differential Equations with Modeling Applications*. 7th ed. Pacific Grove, Calif.: Brooks/Cole Publishing Co.

MODULE 5.4

Runge-Kutta 4 Method

Introduction

Of the three integration techniques of this chapter—Euler's, Runge-Kutta 2, and Runge-Kutta 4 Methods—the last is the most involved but the most accurate. The relative errors of the techniques are $O(\Delta t)$, $O(\Delta t^2)$, and $O(\Delta t^4)$, respectively, with the names Runge-Kutta 2 and 4 indicating the exponents of Δt. Thus, the latter technique improves the most as Δt gets smaller.

To illustrate Runge-Kutta 4 Method, we again use the example $f(t, P) = dP/dt = 0.10P$ with $P_0 = 100$ and $\Delta t = 8$ to show the derivation of P_1 from P_0. To estimate P_n, the technique adds to P_{n-1} a weighted average of four estimates—∂_1, ∂_2, ∂_3, and ∂_4—of the change in P.

First Estimate ∂_1 Using Euler's Method

As with the Runge-Kutta 2 Method, in the Runge-Kutta 4 Method we employ the estimated value of the function from Euler's Method for the first predicted change in P. As the section "Reasoning behind Euler's Method" from Module 5.2 explains, we multiply the derivative of the function at (t_0, P_0) times Δt for the change in the value of the function from the initial value P_0 to the new estimate P_1. In our example, $f(0, 100) = 0.1(100) = 10$, so the first estimate of the change in P is $\partial_1 = f(0, 100) \Delta t = 10 (8) = 80$. Figure 5.4.1 illustrates this change with a boldface dashed line in color to the estimated point that is also in color.

In general, the **first estimate** of $\Delta P = P_n - P_{n-1}$ is as follows:

$$\partial_1 = f(t_{n-1}, P_{n-1}) \Delta t$$

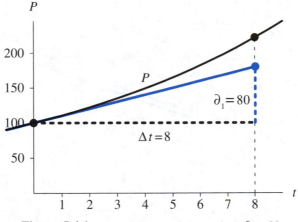

Figure 5.4.1   First estimate of change in P, $\partial_1 = 80$

Quick Review Question 1

Suppose $dP/dt = -P^2/1000$, $t_{30} = 1$, and $P_{30} = 500$. Evaluate ∂_1 for $\Delta t = 6$.

Second Estimate ∂_2

To calculate the second estimate of ΔP for the above example, we use the point halfway between the initial point (t_0, P_0), and point from Euler's estimate, $(t_0 + \Delta t, P_0 + \partial_1)$, in Figure 5.4.1. The midpoint is on the tangent line to the graph of the function P at $(t_0, P_0) = (0, 100)$. Its first coordinate is $t_0 + 0.5\Delta t = 0 + 0.5(8) = 4$, and its second coordinate is $P_0 + 0.5\partial_1 = 100 + 0.5(80) = 140$. Figure 5.4.2 depicts this point, $(4, 140)$, in color.

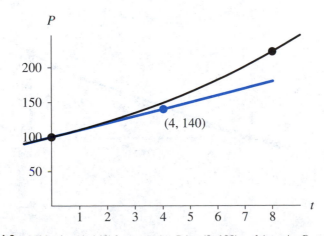

Figure 5.4.2   Midpoint $(4, 140)$ between $(t_0, P_0) = (0, 100)$ and $(t_0 + \Delta t, P_0 + \partial_1) = (8, 180)$

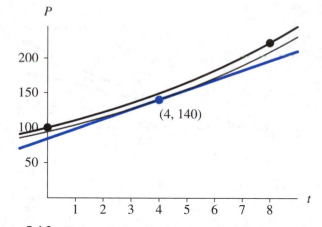

Figure 5.4.3   Estimated slope at midpoint between (0, 100) and (8, 180)

We calculate the derivative f for this midpoint using the derivative formula $f(t, P) = 0.1\,P$, as follows:

$$f(4, 140) = 0.1(140) = 14$$

Figure 5.4.3 shows with less thickness the exponential function through (4, 140) that has derivative 14 at $t = 4$. Thus, at $t = 4$ the curve's tangent line, which is in color, has slope 14.

For the second estimate of the change in P, ∂_2, we determine the change in the vertical direction for this line for $\Delta t = 8$, as follows:

$$\partial_2 = ((0.1)(140))\,(8) = 14\,(8) = 112$$

Figure 5.4.4 pictures a line of the same slope (14) that passes through the initial point (0, 100). After a change in t of 8 units, P increases by 112. Thus, the second

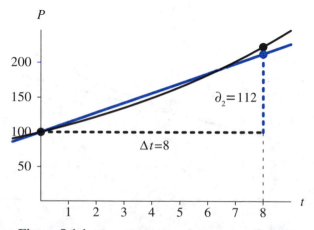

Figure 5.4.4   Second estimate of change in P, $\partial_2 = 112$

estimate of P_1 is $100 + 112 = 212$. The actual value of P at $t = 8$ is about 222.6, so 212 is an improvement over the estimate using Euler's Method, $100 + 80 = 180$. Figure 5.4.4 depicts the new estimated point in color as being significantly closer to the actual point than Euler's estimate in Figure 5.4.1. The improvement comes from making a midway correction.

The second estimate for the change in P employs the estimated slope at the point $(t_{n-1} + 0.5\Delta t, P_{n-1} + 0.5\partial_1)$, as follows:

$$\partial_2 = f(t_{n-1} + 0.5\Delta t, P_{n-1} + 0.5\partial_1)\Delta t$$

Quick Review Question 2

Suppose $dP/dt = -P^2/1000$, $t_{30} = 1$, and $P_{30} = 500$. Quick Review Question 1 showed that $\partial_1 = -1500$ for $\Delta t = 6$.

    a. Give the t coordinate of the point at which to calculate the derivative for ∂_2.
    b. Give the P coordinate of the point at which to calculate the derivative for ∂_2.
    c. Evaluate ∂_2

Third Estimate ∂_3

For the third estimate ∂_3, we use the same process as for the second estimate on the line in Figure 5.4.4 that passes through the initial point $(0, 100)$ and the second estimate point, $(t + \Delta t, P_0 + \partial_2) = (8, 212)$. First, we find the midpoint, $(4, 156)$, between the endpoints (see Figure 5.4.5).

Using the derivative formula, $f(t, P) = 0.1\,P$, we estimate the slope of the curve at $t = 4$ as $f(4, 156) = 0.1(156) = 15.6$. The line through $(4, 156)$ with slope 15.6 appears in color in Figure 5.4.6.

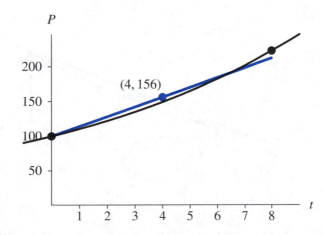

Figure 5.4.5   Midpoint $(4, 156)$ between $(t_0, P_0) = (0, 100)$ and $(t_0 + \Delta t, P_0 + \partial_2) = (8, 212)$

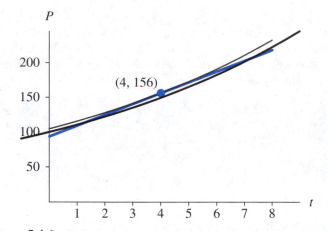

Figure 5.4.6   Estimated slope at midpoint between (0, 100) and (8, 212)

Using the slope of this line, we determine the third estimated change in P over $\Delta t = 8$, ∂_3, as follows:

$$\partial_3 = ((0.1)(156))\,(8) = 15.6\,(8) = 124.8$$

Figure 5.4.7 displays this third estimate of ΔP, ∂_3, as the length of the boldface vertical dashed line to the point, (8, 224.8), both of which are in color.

> The **third estimate** for the change in P employs the estimated slope at the point $(t_{n-1} + 0.5\Delta t,\ P_{n-1} + 0.5\partial_2)$, as follows:
>
> $$\partial_3 = f(t_{n-1} + 0.5\Delta t,\ P_{n-1} + 0.5\partial_2)\Delta t$$

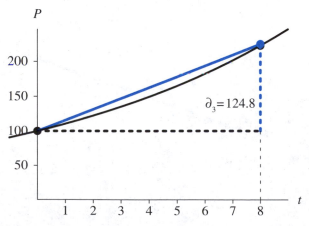

Figure 5.4.7   Third estimate of change in P, $\partial_3 = 124.8$

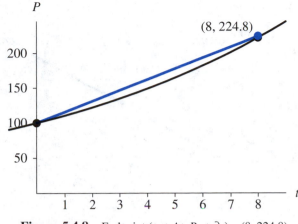

Figure 5.4.8   Endpoint $(t_0 + \Delta t, P_0 + \partial_3) = (8, 224.8)$

Quick Review Question 3

Suppose $dP/dt = -P^2/1000$, $t_{30} = 1$, and $P_{30} = 500$. Quick Review Question 2 showed that $\partial_2 = -375$ for $\Delta t = 6$. Express your answers to one decimal place.

    a. Give the t coordinate of the point at which to calculate the derivative for ∂_3.
    b. Give the P coordinate of the point at which to calculate the derivative for ∂_3.
    c. Evaluate ∂_3.

Fourth Estimate ∂_4

The fourth estimate ∂_4 of the change in P over the interval of length Δt occurs at the end of the interval. As Figure 5.4.8 illustrates, using the third estimate ∂_3, the endpoint is $(t_0 + \Delta t, P_0 + \partial_3) = (8, 224.8)$ for the example under discussion.
    With $dP/dt = f(t, P) = 0.1P$, The following computation estimates the slope at the endpoint:

$$f(8, 224.8) = 0.1(224.8) = 22.48$$

Figure 5.4.9 shows the endpoint along with the exponential function and tangent line of slope 22.48 through that point.
    With this slope, we estimate ∂_4, the increase in P as t increases by $\Delta t = 8$, as follows:

$$\partial_4 = ((0.1)(224.8))\,(8) = 22.48\,(8) = 179.84$$

This fourth estimate of ΔP is the length of the boldface vertical dashed line to the point, (8, 279.84), both of which are in color in Figure 5.4.10. Using ∂_4, 279.84 is the new estimate of P_1.

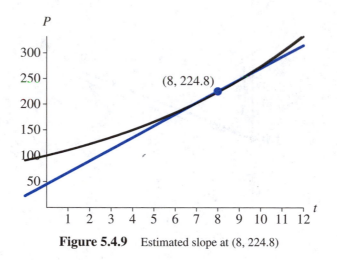

Figure 5.4.9   Estimated slope at (8, 224.8)

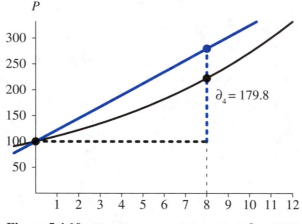

Figure 5.4.10   Fourth estimate of change in P, $\partial_4 = 179.84$

> The **fourth estimate** for the change in P employs the estimated slope at the point $(t_{n-1} + \Delta t, P_{n-1} + \partial_3)$, as follows:
>
> $$\partial_4 = f(t_{n-1} + \Delta t, P_{n-1} + \partial_3)\Delta t$$

Quick Review Question 4

Suppose $dP/dt = -P^2/1000$, $t_{30} = 1$, and $P_{30} = 500$. Quick Review Question 3 showed that $\partial_3 = -585.9$ for $\Delta t = 6$. Express your answers to one decimal place.

a. Give the t coordinate of the point at which to calculate the derivative for ∂_4.

b. Give the P coordinate of the point at which to calculate the derivative for ∂_4.
c. Evaluate ∂_4.

Using the Four Estimates

We have obtained estimates of the rate of change of P with respect to t, $dP/dt = f(t, P)$, at four places on the interval of length Δt, at each end and at the midpoint twice. Using these computations, we derived four estimates ($\partial_1, \partial_2, \partial_3,$ and ∂_4) of the change in P over the interval from $t_0 = 0$ to $t_0 + \Delta t = 8$. Figure 5.4.11 shows, corresponding to the black points, the estimates at the left and right endpoints, $\partial_1 = 80$ and $\partial_4 = 179.84$, respectively, and, corresponding to the points in color, the two estimates at the midpoint, $\partial_2 = 112$ and $\partial_3 = 124.8$. Each value indicates a length of vertical dashed line in color from a height of $P_0 = 100$ to a point whose second coordinate is the corresponding estimate of P_1.

To determine the Runge-Kutta 4 estimate of P_1, we add to $P_0 = 100$ a weighted average of $\partial_1, \partial_2, \partial_3,$ and ∂_4. Giving twice the weight to the estimates at the midpoint, the computation is as follows:

$$P_1 = P_0 + (\partial_1 + 2\partial_2 + 2\partial_3 + \partial_4)/6$$
$$= 100 + (80 + 2 \cdot 112 + 2 \cdot 124.8 + 179.84)/6$$
$$= 100 + 122.24$$
$$= 222.24$$

The **Runge-Kutta 4 estimate** of P_n is as follows:

$$P_n = P_{n-1} + (\partial_1 + 2\partial_2 + 2\partial_3 + \partial_4)/6$$

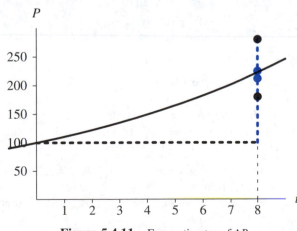

Figure 5.4.11   Four estimates of ΔP

Quick Review Question 5

Suppose $dP/dt = -P^2/1000$, $t_{30} = 1$, and $P_{30} = 500$. For $\Delta t = 6$, we found the following estimations of ΔP:

∂_i	Quick Review Question
$\partial_1 = -1500$	1
$\partial_2 = -375$	2
$\partial_3 = -585.9$	3
$\partial_4 = -44.3$	4

Evaluate Runge-Kutta 4's estimate of P_{31} to one decimal place.

Runge-Kutta 4 Algorithm

The Runge-Kutta 4 Algorithm below combines the computation of the four estimates of ΔP, the weighted average, and the final estimate of P_n into a loop.

Runge-Kutta 4 Algorithm with $f(t_{n-1}, P_{n-1})$ indicating the derivative dP/dt at step $n-1$

    Initialize t_0 and P_0
    Initialize *NumberOfSteps*
    for n going from 1 to *NumberOfSteps* do the following:
        $t_n = t_0 + n\,\Delta t$
        $\partial_1 = f(t_{n-1}, P_{n-1})\,\Delta t$
        $\partial_2 = f(t_{n-1} + 0.5\Delta t, P_{n-1} + 0.5\partial_1)\Delta t$
        $\partial_3 = f(t_{n-1} + 0.5\Delta t, P_{n-1} + 0.5\partial_2)\Delta t$
        $\partial_4 = f(t_{n-1} + \Delta t, P_{n-1} + \partial_3)\Delta t$
        $P_n = P_{n-1} + (\partial_1 + 2\partial_2 + 2\partial_3 + \partial_4)/6$

Quick Review Question 6

Match each of the symbols below to its meaning in the Runge-Kutta 4 Algorithm. Here, "previous" means "immediately previous"; and "estimate" means "estimated value of function using the indicated method."

    A. Derivative at midpoint between (t_{n-1}, P_{n-1}) and point with Euler estimate as second coordinate
    B. Derivative at midpoint between (t_{n-1}, P_{n-1}) and point with EPC estimate as second coordinate
    C. Estimate of ΔP using a midpoint

    D. Estimate of ΔP at right endpoint
    E. Euler estimate at current time step
    F. Euler estimate at previous time step
    G. Runge-Kutta 2 estimate
    H. Runge-Kutta 4 estimate
    I.   Weighted average of intermediate estimates of ΔP
    a. ∂_1
    b. $f(t_{n-1} + 0.5\Delta t, P_{n-1} + 0.5\partial_1)$
    c. ∂_3
    d. ∂_4
    e. $(\partial_1 + 2\partial_2 + 2\partial_3 + \partial_4)/6$
    f. P_n

Error

For the above example, the analytical solution of P_1 to two decimal places is 222.55, while the Runge-Kutta 4 estimate is 222.24. Thus, as the following shows, the relative error is small:

$$|222.24 - 222.55|/222.55 = 0.0014 = 0.14\%$$

An even more dramatic illustration of the improvement in accuracy of the Runge-Kutta 4 Method over Euler's and Runge-Kutta 2 (Euler's Predictor-Corrector, EPC) Methods occurs at the estimate of $P(100)$. The analytical solution of $P(100)$ to no decimal places is 2,202,647. Table 5.4.1 lists the relative errors for the three techniques using $\Delta t = 1$, 0.5, and 0.25. With the Runge-Kutta 4 method at $\Delta t = 0.25$, the relative error is extremely small, and the rounded estimate and analytical solutions are identical.

    Producing such small errors, simulations can usually have larger step sizes, or Δt values, with the Runge-Kutta 4 Method than with the other two techniques. However, the computation is slower because on each step, the Runge-Kutta 4 algorithm computes the derivative f four times instead of one or two times. Thus, a trade-off of time for accuracy exists.

Table 5.4.1
Relative Errors of $P(100)$ for Various Time Changes and
Simulation Methods, where $dP/dt = 0.10P$ with $P_0 = 100$

	Relative Errors at Time 100		
Δt	*Euler's*	*EPC*	*Runge-Kutta 4*
1.00	37.4%	1.53%	0.000767%
0.50	21.5%	0.40%	0.000050%
0.25	11.6%	0.10%	0.000003%

Exercises

Repeat the exercises of Module 5.2 "Euler's Method" using the Runge-Kutta 4 Method. Compare the relative errors with those of the corresponding exercises from Module 5.2, "Euler's Method," and Module 5.3, "Runge-Kutta 2 Method."

6. Download from the text's website the file *simplePendulum* in one of the system dynamics tools. Figure 4.3.3 of Module 4.3 on "Tick Tock—The Pendulum Clock" shows a plot of a simple pendulum's angle, angular velocity, and angular acceleration versus time that is the result of the simulation with $\Delta t = 0.01$ and Runga-Kutta 4 integration. Run the simulation with $\Delta t = 0.01$ using in turn Runga-Kutta 4, Runga-Kutta 2, and Euler's Methods or whatever methods are available with your system dynamics tool. Describe any anomalies in the graphs. Repeat the simulations and description for $\Delta t = 0.1$. Discuss the implications of your findings.

Projects

Repeat the projects of Module 5.2, "Euler's Method," using the Runge-Kutta 4 Method.

Answers to Quick Review Questions

1. -1500 because the product of the derivative and Δt at (t_{30}, P_{30}) is $f(1, 500)\Delta t = (-(500^2)/1000)(6) = -1500$
2. **a.** 4 because $t_{30} + 0.5\Delta t = 1 + 6/2 = 4$
   b. -250 because $P_{30} + 0.5\partial_1 = 500 + (-1500/2) = 500 - 750 = -250$
   c. -375 because $f(t_{30} + 0.5\Delta t, P_{30} + 0.5\partial_1)\Delta t = f(4, -250)\Delta t = (-(-250)^2/1000)6 = -375$.
3. **a.** 4 because $t_{30} + 0.5\Delta t = 1 + 6/2 = 4$
   b. 312.5 because $P_{30} + 0.5\partial_2 = 500 + (-375/2) = 500 - 187.5 = 312.5$
   c. -585.9 because $f(t_{30} + 0.5\Delta t, P_{30} + 0.5\partial_2)\Delta t = f(4, 312.5)\Delta t = (-(312.5)^2/1000) 6 = -585.9$
4. **a.** 7 because $t_{30} + \Delta t = 1 + 6 = 7$
   b. -85.9 because $P_{30} + \partial_3 = 500 - 585.9 = -85.9$
   c. -44.3 because $f(t_{30} + \Delta t, P_{30} + \partial_3)\Delta t = f(7, -85.9)\Delta t = (-(-85.9)^2/1000) 6 = -44.3$
5. -77.7 because $500 + (-1500 + 2(-375) + 2(-585.9) + -44.3)/6 = -77.7$
6. **a.** ∂_1 **F.** Euler estimate at previous time step
   b. $f(t_{n-1} + 0.5\Delta t, P_{n-1} + 0.5\partial_1)$ **A.** Derivative at midpoint between (t_{n-1}, P_{n-1}) and point with Euler estimate as second coordinate

   c. ∂_3 **C.** Estimate of ΔP using a midpoint
   d. ∂_4 **D.** Estimate of ΔP at right endpoint

e. $(\partial_1 + 2\partial_2 + 2\partial_3 + \partial_4)/6$ I. Weighted average of intermediate estimates of ΔP

f. P_n H. Runge-Kutta 4 estimate

References

Burden, Richard L., and J. Douglas Faires. 2001. *Numerical Analysis.* 7th ed. Pacific Grove, Calif.: Brooks/Cole Publishing Co.

Danby, J.M.A. 1997. *Computer Modeling: From Sports to Spaceflight . . . From Order to Chaos.* Richmond, Virginia: Willmann-Bell.

Woolfson, M. M., and G. J. Pert. 1999. *An Introduction to Computer Simulation.* Oxford, U.K.: Oxford University Press.

Zill, Dennis G. 2001. *A First Course in Differential Equations with Modeling Applications.* 7th ed. Pacific Grove, Calif.: Brooks/Cole Publishing Co.

6

SYSTEM DYNAMICS MODELS WITH INTERACTIONS

MODULE 6.1

Competition

Download

The text's website has available for download for various system dynamics tools the file *sharkCompetition*, which contains a submodel for this module.

Community Relations

In any population of organisms, the individuals are interacting with each other and with their environment. Populations, which are made up of only one species, are also interacting with other species in a particular area in what we term a **community**. These interactions influence the composition and dynamics of the community through time. Some of these interactions are robust, while others are not so robust, or even very weak. The magnitude of these interactions is dependent on the extent of their niche overlap. An **ecological niche** can be defined as the complete role that a species plays in an ecosystem. The more overlap two species have, the stronger the interaction will be. Two of these interactions between species are competition and predator-prey relationships.

Competition Introduction

Everyone is familiar with competition. We compete for attention in families, for grades in school, for jobs and promotions, for parking spaces, and on and on. Competition is integral to most economic activity. Through competition in human societies, wages and prices are set; quantities and types of products manufactured are selected; businesses succeed or fail; and resources are distributed. Economic and social competition may occur even in noncapitalist systems.

More broadly, competition is a basic characteristic of all communities, human and nonhuman. It may occur within a population of the same species (**intraspecific**), like the human species, or it may occur between populations of different species (**interspecific**). Competitive interactions affect species distribution, community organization, and species evolution.

Simply speaking, **competition** is the struggle between individuals of a population or between species for the same limiting resource. If one individual (species) reduces the availability of the resource to the other, we term that type of competition **exploitative** or **resource depletion**. This interaction is indirect and may involve removal of the resource or denial of living space. If there is direct interaction between individuals (species), where one interferes with or denies access to a resource, we term that competition **interference**. In this form, there may be physical contests for territory or resource. Interference may also, as in some plants, involve the production of toxic chemicals.

Modeling Competition

Sometimes two species are not eating each other but are competing for the same limited food source. For example, white tip sharks (WTS) and black tip sharks (BTS) in an area might feed on the same kinds of fish in a year when the fish supply is low. We anticipate that a large increase in one species, such as BTS, might have a detrimental effect on the ability of the other species, such as WTS, to obtain an adequate amount of food and, therefore, to thrive. Also, we expect that superior hunting skills of one species would diminish the food supply for the other species. As one species grows, the other shrinks and vice versa.

In an unconstrained growth model (see Module 3.2), which ignores competition and limiting factors, we consider a population's (P) births to be proportional to the number of individuals in the population (r_1P) and its deaths to follow a similar proportionality (r_2P). Thus, in this model, the rate of change of the population is $dP/dt = r_1P - r_2P = (r_1 - r_2)P$, so that the solution is an exponential function, $P = P_0e^{(r_1 - r_2)t}$.

However, with competition, a competing species has a negative impact on the rate of change of a population. In this situation, we can model the number of deaths of each species as being proportional to its population size and the population size of the other species. Thus, for B the population of black tip sharks and W the population of white tip sharks, the number of deaths of each species is proportional to the product BW. Moreover, the constant of proportionality associated with this proportionality for one species reflects the hunting skills of the other species. Consequently, we have the following equations for the change in the number of deaths of each species:

Δ(deaths of WTS) = wBW,　　　　　where w is a WTS death proportionality constant

Δ(deaths of BTS) = $bWB = bBW$,　　where b is a BTS death proportionality constant

Figure 6.1.1 illustrates the interaction with the number of each species of shark affecting the deaths of the other species. Equation Set 6.1.1 gives some of the corresponding equations and constants, which in this case model births as being unconstrained. The set of numbers serves as an example and, although realistic, does not

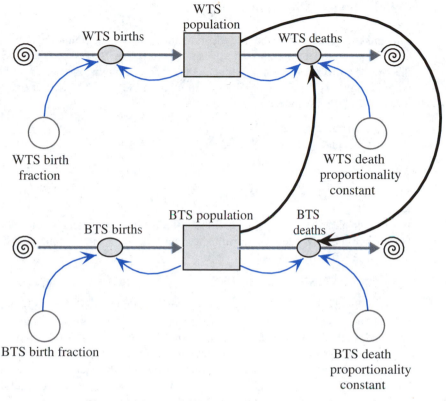

Figure 6.1.1   Model diagram of competition of species

represent any actual population. Typically, a computational scientist uses actual field data to establish reasonable parameters for a model.

Equation Set 6.1.1

Some equations to accompany Figure 6.1.1

BTS_population(0) = 15
BTS_birth_ fraction = 1
BTS_births = *BTS_birth_ fraction* * *BTS_population*
BTS_death_proportionality_constant = 0.20
BTS_deaths = (*BTS_death_ proportionality_constant* * *WTS_population*) * *BTS_population*
WTS_population(0) = 20
WTS_birth_ fraction = 1
WTS_births = *WTS_population* * *WTS_birth_ fraction*
WTS_death_proportionality_constant = 0.27
WTS_deaths = (*WTS_death_ proportionality_constant* * *BTS_population*) * *WTS_population*

Quick Review Question 1

This question reflects on Step 2 of the modeling process—formulating a model—for developing a model for competition. As above, let W be the number of WTS and B, the number of BTS. We simplify this model by assuming unconstrained births. After completing this question and before continuing in the text, we suggest that you develop a model for competition.

a. Give an equation for WTS births.
b. Give an equation for WTS deaths.

Quick Review Question 2

If all other parameters are equal and the WTS death proportionality constant (w) is larger than the BTS death proportionality constant (b), which population should be larger after a few time steps?

      A. WTS   **B.** BTS   **C.** Impossible to determine

With populations only inhibited by the competition for food, we might have a situation, such as illustrated in Figure 6.1.2 and Table 6.1.1. In this case, the white tip sharks initially outnumber the black tip sharks. However, the WTS death proportionality constant ($w = 0.27$) is larger than the BTS death proportionality constant ($b = 0.20$). Early in the simulation, the population of both species decreases. Eventually, the white tip sharks die out and the black tip sharks thrive. The projects and exercises explore situations that have different initial populations and constants of proportionality and, consequently, different results.

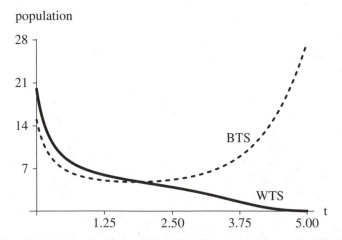

Figure 6.1.2 Graph of results of simulation from Figure 6.1.1, where the WTS death proportionality constant (w) is 0.27, the BTS death proportionality constant (b) is 0.20, and time (t) is in months

Table 6.1.1
Table of Results of Simulation from Figures 6.1.1 and 6.1.2 where $w = 0.27$ and $b = 0.20$

Time (months)	WTS	BTS
0	20.00	15.00
1	6.57	5.37
2	4.69	4.84
3	3.08	6.00
4	0.99	10.83
5	0.02	27.43

Exercises

1. **a.** Write the differential equations for modeling competition with unconstrained growth for both populations.
   b. Find all equilibrium solutions to these equations.
2. **a.** Write the differential equations for modeling competition with constrained growth for both populations.
   b. Find all equilibrium solutions to these equations.
3. What would be the effect on each of the following of increased intraspecific competition? *Hint:* Increased competition would be reflected in higher population densities.
   a. Mortality in terms of number of pines/acre
   b. Fertility in terms of number of seeds/plant/m^2
   c. Average adult weight in terms of average adult bluegill wt/liter of water
   d. Rate of growth in terms of increase in mallard duckling wt/time

Projects

For additional projects, see Module 7.8 on "Fueling Our Cells—Carbohydrate Metabolism."

For all model development, use an appropriate system dynamics tool.

1. **a.** Using your system dynamics tool's *sharkCompetition* file, which contains a model for competing species, find values for the initial populations and the constants of proportionality in which one population becomes extinct.
   b. Find values for which the two populations reach equilibrium.
   c. Discuss the results.
   d. Adjust the model to have the populations constrained by carrying capacities (see Module 3.3 on "Constrained Growth").
   e. Adjust the parameters several times obtaining different results.
   f. Explain the models and discuss the results.
2. Argentine ants (*Linepithema humile*) are native to South America, but have been invading the temperate zone of North America from the turn of the

twentieth century. With its large and aggressive workers, Argentine ants are generally able competitively to exclude many native ant species. This success comes from the ant's ability to use exploitive, as well as interference competitive mechanisms (Holway 1999).

    a. Develop a model of **exploitive competition** for the Argentine ant versus a native ant. The competitive factors include discovery time and rate of recruitment. The Argentine ant might discover a food source faster and attract other workers to the food source more quickly than the native ant.

    b. Develop a model of **interference competition** for the Argentine ant versus a native ant. The competitive factors include physical inhibition/removal and chemical repellents. Argentine ants might fight off or remove native ants from the food source, or they might use chemicals to repel them.

3. Model intraspecific competition. See Exercise 3 for examples. Discuss mortality and rate of growth in response to increasing intraspecific competition.

4. Plants can produce chemicals that, when released to the soil, inhibit the growth of other plants. These chemicals can act by inhibiting respiration, photosynthesis, cell division, protein synthesis, mineral uptake, or altering the function of membranes. For instance, sandhill rosemary (*Ceratiola ericoides*), an evergreen shrub found along the coastal plain of the southeastern United States, produces ceratiolin. This chemical washes from the leaves and degrades to hydrocinnamic acid, a compound that effectively inhibits seed germination of many competing species (Hunter and Menges 2002).

    Assume that this chemical is increasingly effective at germination inhibition with increasing concentrations. Assume the highest concentration released to be 60 ppm and that concentration decreases linearly from the tips of the outermost leaves (for periods without rain).

    a. Model inhibition of a competing plant species, where the effective concentrations of the toxin are between 20 and 60 ppm.

    b. Model inhibition for this species with 2 cm rain per day. Set your own decrease in concentration per cm of rain for your model.

5. Model the interferance competition of titmice versus other birds at feeders.

6. Model an environment with two competing species of plants—Species A and Species B—and two essential resources—phosphorus and nitrogen. The renewal rate for each resource is 0.4 units/month. The availabilities of phosphorus and nitrogen are 12 units and 28 units, respectively. The two species have starting populations of 12 plants. The maximum growth rates for Species A and B are 1.2 plants/month and 1.0 plants/month, respectively; while their mortality rates are 0.5 plants/month. Species A requires 10 units of phosphorus and 5 units of nitrogen to achieve half its maximum growth rate, while Species B requires 5 units of phosphorus and 10 units of nitrogen for this rate. The phosphorus consumption rates for Species A and B are 0.5/month and 0.2/month, respectively; and the nitrogen consumption rates are 0.2/month and 0.5/month, respectively. Explain the model and discuss the results. Will this scenario result in equilibrium (Tilman 1980)?

7. Do one of the previous projects in this module as a simulation written in a programming language or with another computational tool.

Answers to Quick Review Questions

1. a. cW, where c is a birth rate
   b. wBW or wWB, where w is a death proportionality constant
2. B. BTS, because a larger portion of the white tip sharks are dying

References

Holway, David A. 1999. "Competitive Mechanisms Underlying the Displacement of Native Ants by the Invasive Argentine Ant." *Ecology*, Vol. 80, No. 1: 238–251.

Hunter, Molly E., and Eric S. Menges. 2002. "Allelopathic Effects and Root Distribution of *Ceratiola ericoides* (Empetraceae) on Seven Rosemary Scrub Species." *American Journal of Botany*, Vol. 89, no. 7: 1113–1118.

Smith, Robert L., and Thomas M. Smith. 2001. *Ecology and Field Biology.* 6th ed. San Francisco, Calif.: Benjamin Cummings Publishing Co.

Tilman, D. 1980. "Resources: A Graphical-Mechanistic Approach to Competition and Predation." *American Naturalist*, 116: 362–393.

MODULE 6.2

Spread of SARS

Downloads

The text's website has available for download for various system dynamics tools *SIR* and *SARSRelationships* files, which contain models for the examples of this module.

Introduction

Imagine being a college student in New York City and being told not to leave the city. That's what happened in 2003 in Beijing, when thousands of people were ordered to stay home and college students told to stay in Beijing. Quarantine procedures were instituted for those who were thought to have had "intimate contact" with others who showed signs of a new rapidly spreading respiratory disease. More than 40 had died in the capital, and thousands of people in China were displaying symptoms of this pneumonia. Imagine the feelings of fear and panic that Beijing residents must have had—people in masks, disinfecting their homes, and hoarding of food and other necessities.

This new disease was called **SARS, Severe Acute Respiratory Syndrome**, with the first case occurring on November 16, 2002 in southern China. Chinese health officials reported the outbreak to the World Health Organization (WHO) on February 11, 2003. By April 2, the total reported cases of SARS was 2000; and by July, the count was over 8400 with more than 800 dead. In response to the initial report, WHO coordinated the investigation into the cause and implemented procedures to control the spread of this disease. The control measures were extremely effective, and the last new case was reported on June 12, 2003.

By the third week in March several laboratories worldwide had identified the probable causative agent—*SARS-CoV*, the SARS coronavirus. Coronaviruses represent a large group of +-stranded RNA-containing viruses associated with various respiratory and gastrointestinal illnesses. Although the human diseases associated with these viruses have been mild previously, this coronavirus is quite different. Like many respiratory pathogens, SARS is spread by close personal contact and perhaps by airborne transmission.

The Centers for Disease Control and Prevention (CDC) in the United States uses clinical epidemiological and laboratory criteria to diagnose SARS. Severe cases exhibit a fever higher than 38°C and one or more respiratory symptoms—difficulty breathing, cough, shortness of breath. Additionally, the person must show radiographic evidence (lung infiltrates) of pneumonia, or **respiratory distress syndrome (RDS)**. RDS is an inflammatory disease of the lung, characterized by a sudden onset of edema and respiratory failure. A few others qualified if they exhibited an unexplained respiratory illness that resulted in death and an autopsy confirmed RDS with no identifiable cause. Epidemiological evidence might include close contact with a known SARS patient or travel to a region with documented transmission within ten days of onset of symptoms. Laboratory tests confirm SARS if they reveal one of the following (CDC SARS):

- antibody to SARS virus in specimens obtained during acute illness or more than 28 days after onset of illness
- SARS viral RNA detected by RT-PCR
- SARS virus

On July 5, 2003, the World Health Organization declared that SARS had been contained. It resulted in 812 deaths, but the toll might have been much higher, if WHO and other health agencies had not acted so quickly and effectively (WHO). Besides the direct effect on the victims and their families, SARS became a major drag on the economies of China, Taiwan, and Canada. Hong Kong's unemployment rate climbed to an unprecedented 8.3%, and travel warnings for Toronto cost Canada an estimated $30 million per day. One can only imagine the impact of this disease being spread into Africa, where there are poor healthcare systems and the astronomical HIV infections rates generate immunologically compromised populations.

SARS is an interesting disease for modeling, particularly because there is so much epidemiological information. We still have much to learn about SARS, and we still have no available, effective treatment.

SIR Model

Before developing a model for the spread of SARS, we consider the simpler situation of a disease in a closed environment in which there are no births, deaths, immigration, or emigration. A 1978 *British Medical Journal* article reported on such a situation—influenza at a boys' boarding school. On January 22, only one boy had the flu, which none of the other boys had ever had. By the end of the epidemic on February 4, 512 of the 763 boys in the school had contracted the disease (Murray 1989; NCSLIP; SUCCEED).

To model this spread of influenza, we employ the **SIR Model**, which W. O. Kermack and A. G. McKendrick developed in 1927 (Kermack and McKendrick 1927). Many systems models of the spread of disease, including the SARS model later in this

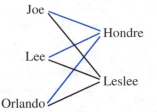

Figure 6.2.1   Possible contacts between S and I

module, are extensions of the SIR Model. The name derives from the following three populations considered:

>   **Susceptibles** (S) have no immunity from the disease.
>   **Infecteds** (I) have the disease and can spread it to others.
>   **Recovereds** (R) have recovered from the disease and are immune to further infection.

The model gives the differential equations for the rate of change for each of these populations. We assume that after a certain amount of time, an individual with the flu recovers. Thus, the rate of change of the number of recovereds is proportional to the number of infecteds.

Quick Review Question 1

With the constant of proportionality being the recovery rate (a), give the differential equation for the rate of change of the number of recovereds.

As the answer to Quick Review Question 1 states, the differential equation for the rate of change of the number of recovereds is $dR/dt = aI$ for recovery rate a. If the time unit is in days and d is the number of days that someone remains infected, we can consider a to be $1/d$. For example, if a boy is usually sick with the flu for two days, then $d = 2$ and $a = 0.5$/day; so that approximately half the infected boys get well in a day.

A susceptible boy at the boarding school becomes infected with influenza by having contact with an infected boy. The number of such possible contacts is the product of the sizes of the two populations, SI. For example, suppose the set of susceptibles is $S = \{$Joe, Lee, Orlando$\}$ and the set of infecteds is $I = \{$Hondre, Leslee$\}$. As Figure 6.2.1 pictures, $(3)(2) = 6$ possible interactions exist between pairs of boys in different sets. The virus in Hondre can spread through contact to Joe, Lee, and Orlando. Similarly, Joe can become infected with the virus from Hondre or Leslee. With no new students entering the school, the number of susceptibles can only decrease, and the rate of change of the number of boys in this set is directly proportional to the number of possible contacts, SI, between susceptibles and infecteds. In Module 6.1 on "Competition," we saw that competitors' death rates of change exhibit the same proportionality to a product of population sizes because of interactions.

Quick Review Question 2

a. Is the rate of change of S positive, zero, or negative?

b. With $r > 0$ being the constant of proportionality, give a differential equation for the rate of change of S.

As the answers to Quick Review Question 2 reveal, for positive constant of proportionality r, $dS/dt = -rSI$. The constant r, called the **transmission constant**, represents the infection rate and indicates the infectiousness of the disease and the interactions among the students. In the case of the boys' school, we use 0.00218 per day. Thus, $0.00218 = 0.218\%$ of the possible contacts result in the disease being spread from one child to another.

Only susceptibles become infected, and infecteds eventually recover. What I gains comes from what S has lost; and what I loses, R gains. Thus, the differential equation for the rate of change of the number of infecteds is the sum of the negatives of the other two rates of change:

$$dI/dt = -dS/dt - dR/dt$$

Quick Review Question 3

Give the differential equation for the rate of change of the number of infecteds in terms of S, I, R, the infection rate (r), and the recovery rate (a).

Figure 6.2.2 presents a diagram for the SIR model with *susceptibles, infecteds*, and *recovereds* replacing the symbols S, I, and R, respectively, and with *infection_rate* and *recovery_rate* representing the constants of proportionality r and a, respectively. Some of the corresponding equations and constants for a particular simulation appear in Equation Set 6.2.1.

Equation Set 6.2.1

Some equations and constants for SIR model in Figure 6.2.2

susceptibles(0) = 762
infection_rate = 0.00218

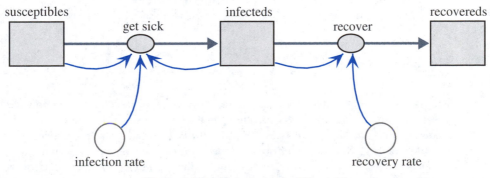

Figure 6.2.2   Diagram for the SIR model

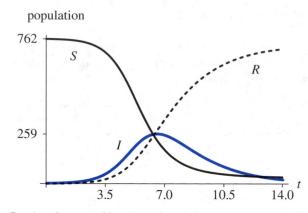

Figure 6.2.3 Graphs of *susceptibles* (*S*), *infecteds* (*I*), and *recovereds* (*R*) versus time (*t*) in days

$get\_sick = infection\_rate * susceptibles * infecteds$
$infecteds(0) = 1$
$recovery\_rate = 0.5$
$recover = recovery\_rate * infecteds$
$recovereds(0) = 0$

The graphs of the three populations that result from running the simulation are in Figure 6.2.3. The number of *susceptibles* decreases slowly at first before experiencing a rapid decline and subsequent leveling. In contrast, the number of *recovereds*, which is initially 0, has a graph that appears similar to the logistic curve. When the number of *susceptibles* decreases sharply, the *infecteds* increase to their maximum. Afterwards, as the number of *infecteds* decreases, the number of *recovereds* rises. Although not mimicking the final numbers exactly, this model does capture the trend of the data along with the epidemic increase and decrease.

Quick Review Question 4

Answer the following questions referring to Figure 6.2.3:

    a. On what day was the number of cases the largest?
    b. On what day were most of the boys sick or recovered?
    c. On what day were most of the boys recovered?

SARS Model

Marc Lipsitch in collaboration with others developed a model for the spread of **Severe Acute Respiratory Syndrome (SARS)** and used the model to make predictions on the impact of public health efforts to reduce disease transmission (Lipsitch et al. 2003). Such efforts included **quarantine** of exposed individuals to separate them, perhaps by confinement to their homes, from the susceptible population and **isolation** of those who had SARS to remove them to strictly supervised hospital areas with no contacts other than by healthcare personnel. The Lipsitch model is an extension of the SEIR model, which is an extension of the SIR model. Besides the populations considered by SIR, the **SEIR Model (Susceptible-Exposeds-Infecteds-Recovereds)** has an intermediate **Exposed (*E*)** population of individuals who have the disease but are not yet infectious. The Lipsitch model modifies SEIR to allow for quarantine, isolation, and death. The modelers make the following simplifying assumptions:

1. There are no births.
2. The only deaths are because of SARS.
3. The number of contacts of an infected individual with a susceptible person is constant and does not depend on the population density.
4. For susceptible individuals with exposure to the disease, the quarantine proportion (q) is the same for non-infected as for infected people.
5. Quarantine and isolation are completely effective. Someone who has the disease and is in quarantine or isolation cannot spread the disease.

The populations considered are as follows:

susceptible (*S*)—do not have but can catch SARS from infectious individuals

susceptible_quarantined (S_Q)—do not have SARS, quarantined because of exposure, so cannot catch SARS

exposed (*E*)—have SARS, no symptoms, not yet infectious

exposed_quarantined (E_Q)—have SARS, no symptoms, not yet infectious, quarantined because of exposure

infectious_undetected (I_U)—have undetected SARS, infectious

infectious_quarantined (I_Q)—have SARS, infectious, quarantined

infectious_isolated (I_D)—have SARS, infectious, isolated

SARS_death (*D*)—dead due to SARS

recovered_immune—recovered from SARS, immune to further infection

Because we are assuming that quarantine is completely effective, only someone in the *susceptible* (*S*) category can catch SARS. Transmission to a susceptible can only occur through exposure to an individual in the *infectious_undetected* (I_U) category. Those with SARS in other categories are under quarantine or isolation, or are not yet infectious.

Quick Review Question 5

After completing this question and before continuing in the text, we suggest that you make a diagram with stocks (box variables) and flows only to represent possible transitions between categories. For each of the following, give the possible category(ies):

a. Flows out of *S* into what categories?
b. Flows into *S* from what categories?
c. Flows into *D* from what categories?

Without inclusion of converters and connectors, Figure 6.2.4 displays a diagram with the stocks that represent these populations along with the flows between them. As illustrated, a susceptible individual who is exposed to SARS and moved from the *susceptible* group can be quarantined with or without the disease (to *exposed_quarantined*

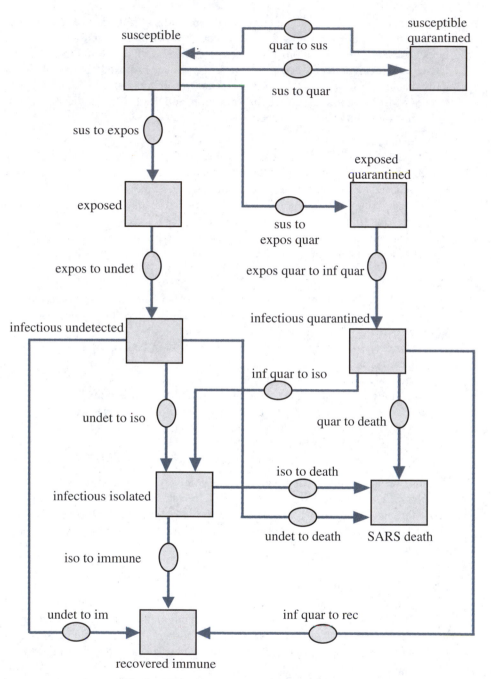

Figure 6.2.4   Initial diagram of relationships for SARS

or *susceptible_quarantined*, respectively) or can be infected and not quarantined (to *exposed*). A susceptible, quarantined person who does not have SARS (in *susceptible_quarantined*) eventually is released from quarantine (to *susceptible*). An exposed but not yet infectious individual who does have SARS, whether quarantined or not (in *exposed_quarantined* or *exposed*, respectively), eventually becomes infectious (to *infectious_quarantined* or *infectious_undetected*, respectively). Regardless of quarantine status, an infectious individual can recover (to *recovered_immune*), go into isolation after discovery (to *infectious_isolated*), or die (to *SARS_death*). Isolated patients who are sick with SARS can recover or die.

Quick Review Question 6

Using this model, indicate if each of the following situations is possible or not:

   a. A susceptible person dies of SARS.
   b. A person who has undetected SARS in the early stages recovers without ever becoming infectious.
   c. Someone in quarantine diagnosed with SARS recovers without going into isolation.
   d. Someone who has recovered from SARS becomes infected with the disease again.
   e. Someone is transferred from isolation to quarantine.

The model employs the following parameters:

   b—probability that a contact between person in *infectious_undetected* (I_U) and someone in *susceptible* (S) results in transmission of SARS
   k—mean number of contacts per day someone from *infectious_undetected* (I_U) has with someone in *susceptible* (S). By assumption, the value does not depend on population density
   m—per capita death rate
   N_0—initial number of people in the population
   p—fraction per day of exposed people who become infectious; this fraction applies to the transitions from *exposed* (E) to *infectious_undetected* (I_U) and from *exposed_quarantined* (E_Q) to *infectious_quarantined* (I_Q). Thus, $1/p$ is the number of days in the early stages of SARS for a person to be infected but not infectious.
   q—fraction per day of individuals in *susceptible* (S) who have had exposure to SARS that go into quarantine, either to category *susceptible_quarantined* (S_Q) or to *exposed_quarantined* (E_Q)
   u—fraction per day of those in *susceptible_quarantined* (S_Q) who are allowed to leave quarantine, returning to the *susceptible* (S) category. Thus, $1/u$ is the number of days for a susceptible person to be in quarantine
   v—per capita recovery rate. This rate is the same for the transition from category *infectious_undetected* (I_U), *infectious_isolated* (I_D), or *infectious_quarantined* (I_Q) to category *recovered_immune*
   w—fraction per day of those in *infectious_undetected* (I_U) who are detected and isolated and thus transferred to category *infectious_isolated* (I_D)

Quick Review Question 7

a. Suppose it takes an average of 5 days for someone who has SARS but is not infectious to progress to the infectious stage. Give the value of p along with its units.

b. Give the formula for the rate of change of exposed individuals who are not quarantined to move into the phase of being infectious and undetected, from E to I_U.

c. Give the formula for the rate of change of exposed individuals who are quarantined to move into the phase of being infectious and quarantined, from E_Q to I_Q.

d. Suppose 10% of the people who have been in quarantine but who do not have SARS are allowed to leave quarantine each day. Give u and the average number of days for a susceptible person to be in quarantine.

e. Suppose the duration of quarantine is 16 days. If someone has not developed symptoms of SARS during that time period, he or she may leave quarantine. Give the corresponding parameter and its value.

f. Give the formula for the rate of change of susceptible, quarantined individuals leaving quarantine, from S_Q to S.

As illustrated in Figure 6.2.4, three paths exist for someone to leave *infectious_undetected* (I_U)—to *recovered_immune* at a rate of v, to *SARS_death* at a rate of m, or to *infectious_isolated* (I_D) at a rate of w. Thus, the total rate of change to leave *infectious_undetected* (I_U) is $(v + m + w)$/day. For example, if $v = 0.04$, $m = 0.0975$, and $w = 0.0625$, $v + m + w = 0.2$/day. In this case, $1/(v + m + w) = 5$ days is the average duration of the potential to infect susceptibles.

By assumption, k is the number of contacts an undetected infectious person has with a susceptible each day, regardless of population density. Thus, with N_0 being the initial population size, k/N_0 is the fraction per day of such contacts. Because b is the probability of such a contact infecting the susceptible, the product $(k/N_0)b$ is the transmission constant. As in the SIR model, the product $I_U S$ gives the total number of possible interactions. Thus, $(k/N_0)b\, I_U S = kbI_U S/N_0$ is the number of new cases of SARS each day. Of these new cases, a fraction (q) go into category *exposed_quarantined* (E_Q), while the remainder, fraction ($1 - q$), go into *exposed* (E).

Quick Review Question 8

a. Suppose $k = 10$ contacts/day, and $N_0 = 10,000,000$ people. Give the percentage of contacts per day.

b. Suppose 6% of contacts between an infectious and a susceptible person result in transmission of the disease. Give the corresponding parameter and its value.

c. Using your answers to Parts a and b, what percentage of all possible contacts results in transmission of SARS each day?

d. If the sizes of *infectious_undetected* (I_U) and *susceptible* (S) are 5000 and 9,000,000, respectively, give the total number of possible contacts.

e. Using your answers to Parts c and d, give the number of contacts that result in transmission of SARS.

   f. Suppose $q = 0.1 = 10\%$ of the individuals who have had contact with an infectious person go into quarantine. Give the number of those from Part e who go into *exposed_quarantined* (E_Q).

   g. Give the formula for the rate of change from *susceptible* (S) to *exposed_quarantined* (E_Q).

   h. Assuming $q = 0.1$, give the number of those from Part e who go into *exposed* (E).

   i. Give the formula for the rate of change from *susceptible* (S) to *exposed* (E).

For those transferring from *susceptible* (S) to *susceptible_quarantined* (S_Q), although they have been exposed to an infectious person, the disease was not transmitted to them. The fraction of total possible contacts $I_U S$ is (k/N_0), and the probability of non-transmittal is $(1 - b)$. Thus, the total number of non-transmission contacts is $(k/N_0)(1 - b)I_U S = k(1 - b)I_U S/N_0$. However, only a fraction (q) of those go into quarantine. Thus, the rate of change of those going from *susceptible* (S) to *susceptible_quarantined* (S_Q) is $qk(1 - b)I_U S/N_0$.

Quick Review Question 9

Using the values from Quick Review Question 8, determine the rate of change of those going from *susceptible* (S) to *susceptible_quarantined* (S_Q).

Reproductive Number

Several exercises deal with the differential equations for this SARS model, and a project completes the model. In this model, an important value in evaluating the effectiveness of quarantine and isolation is the **reproductive number R**, which is the expected number of secondary infectious cases resulting from an average infectious case once the epidemic is in progress. For example, if on the average an infectious individual transmits SARS to three other people who eventually become infectious, then the reproductive number is $R = 3$. Such a number results in the dangerous situation of exponential growth of the disease. One person transmits infectiousness to three other people, who each cause three other people to become infectious, and so forth. In such a situation, at stage n of transmission, 3^n new people eventually become infectious. For example, at stage $n = 13$, 3^{13} or over 1.5 million new people would get sick. Because of such exponential growth, it very important that R be below 1. With $R < 1$, there is no epidemic. For $R > 1$, there is an epidemic. The larger the reproductive number, the more virulent the epidemic.

> **Definition** The **reproductive number R** is the expected number of secondary infectious cases resulting from an average infectious case once the disease has started to spread.

For this SARS model, an undetected infectious person has k contacts per day with a susceptible person. Of these, with a probability b of transmission, kb new cases of SARS result each day. Because the average duration of infectiousness is $1/(v + m + w)$ days (see explanation after Quick Review Question 7), without quarantine being a factor, one infectious person eventually gives rise to $R = kb/(v + m + w)$ secondary infectious cases of SARS. However, when a fraction, q, go into quarantine so that a fraction $(1 - q)$ do not, the reproductive number is $R = \frac{kb}{v+m+w}(1 - q)$. The larger q is, the smaller R is, and the less severe the impact of the disease is.

Quick Review Question 10

Evaluate the reproduction number R, using the values of Quick Review Question 8 and text material: $k = 10$ contacts/day, $b = 0.06$, $v = 0.04$, $m = 0.0975$, $w = 0.0625$, and $q = 0.1$.

Examining the initial value of R (R_0), the death rate, and other factors, members of WHO and other health organizations realized that they must act quickly with bold measures involving quarantine and isolation to avoid a major, worldwide epidemic of SARS. Computer simulations with scenario analyses verified the seriousness of the disease. Thanks to aggressive actions, a terrible catastrophe was averted.

Exercises

1. Write the system of differential equations for the SIR model using a transmission constant of 0.0058 and a recovery rate of 0.04.

In the SARS model, give the differential equation for each rate of change in Exercises 2–10.

2. dS_Q/dt 3. dE/dt 4. dE_Q/dt 5. dS/dt 6. dI_U/dt
7. dI_D/dt 8. dI_Q/dt 9. $d(recovered\_immune)/dt$ 10. dD/dt

11. a. For a reproductive number of $R = 3$, give the number of new people that will eventually become infectious at stage $n = 10$ of transmission of the disease.
    b. Give the total number of people who will eventually become infectious.
    c. Repeat Part a for $n = 15$.
    d. Repeat Part b for $n = 15$.

Projects

For additional projects, see Module 7.8 on "Fueling Our Cells—Carbohydrate Metabolism."
For all model development, use an appropriate system dynamics tool.

1. Adjust the SIR model to allow for vaccination of susceptible boys. Assume that 15% are vaccinated each day, and make a simplifying assumption that

immunization begins immediately. Discuss the effect on the duration and intensity of the epidemic. Consider the impact of other vaccination rates.

2. Adjust the SIR model to allow for vaccination of susceptible boys. Assume that 15% are vaccinated each day and that immunization begins after three days. Discuss the effect on the duration and intensity of the epidemic. Consider the impact of other vaccination rates.

3. Adjust the SIR model to allow for vaccination of susceptible boys. Assume that all children are vaccinated two days before a boy comes down with the flu and that immunization begins after four days. Discuss the effect on the duration and intensity of the epidemic. Consider the impact of other vaccination rates.

4. Develop an SEIR model of disease.

5. Complete the Lipsitch SARS model introduced in the text. Have the model evaluate R. Produce graphs and a table of appropriate populations, including *susceptible*, *recovered_immune*, *SARS_death*, and the total of the five categories of infecteds. Employ the following parameters: $k = 10$/day; $b = 0.06$; $1/p = 5$ days; $v = 0.04$, $m = 0.0975$, and $w = 0.0625$, so that $v + m + w = 0.2$/day and $1/(v + m + w) = 5$ days; $1/u = 10$ days; $N_0 = 10,000,000$ people. Vary q from 0 upward. Discuss the results.

6. After developing the model of Project 5, with a fixed value of q, test other ranges of k from 5 to 20 per day. Discuss the results.

7. After developing the model of Project 5, with a fixed value of q, test other ranges of $1/(v + m + w)$ from 1 to 5 days.

8. Adjust the model of Project 5 so that the simulation is allowed to run for a while before quarantine and isolation measures that reduce R to below 1 are instituted. Discuss the implications on the number of people quarantined and on the health care system of not taking aggressive measures initially.

9. Develop a model of strep throat. Bacterium Group A *Streptococcus* causes strep throat, which occurs most frequently in school-aged children. The bacterium spreads through direct or airborne contact with the mucus from an infected person. Symptoms start from one to five days after exposure and include fever, sore throat, and tender and swollen neck glands. If untreated, people with strep throat are infectious for 10 to 21 days. Usually, 24 hours after antibiotic treatment, those who are ill are no longer contagious. The spread of strep throat can be minimized by infectious people covering their mouths when sneezing or coughing and by washing their hands frequently (EDCP).

10. Develop a model of the viral infection mumps. Symptoms include painful and swollen salivary glands, painful swallowing, fever, weakness, fatigue, and a tender, swollen testicle. Infection is spread through breathing of infected saliva droplets. About one-third of those with mumps experience no symptoms. If present, symptoms usually start two to three weeks after infection. The person is contagious from approximately one day before salivary gland swelling occurs and remains contagious for at least another three days. As the swelling diminishes, so does the degree of the contagion. Before licensing of the mumps vaccine in 1967, the United States had more than 200,000 cases per year. Since then, the country has had fewer than 1000 cases per year (Mayo Clinic 2002).

11. Diphtheria has been virtually eradicated in the United States because of a vaccine, which was introduced in the 1920s. Before that time, the United States had 100 to 200 cases per 100,000 people. The disease is still a problem in developing countries. Two types of diphtheria exist, respiratory and cutaneous. The former is more serious, and death results in 5% to 10% of those cases. The disease is spread through respiratory and physical contact. The incubation period for the disease is usually 2 to 5 days. Develop a model for respiratory diphtheria (CDC-diptheria).

12. Using data and mathematical models implemented in spreadsheets, the Dutch Ministry of Health, Welfare and Sports developed "a national plan to minimize effects of pandemic influenza." Through scenario analysis, scientists examined various intervention options and estimated the number of hospitalizations and deaths. In the base case, in which no intervention was possible, they assumed 30% of the population would become ill with influenza. In the Influenza Vaccination Scenario, they considered two strategies:

    1. Vaccinate two risk groups, persons 65 years of age or older ($N = 2.78$ million (M)) and healthcare workers ($N = 0.80$ M)
    2. Vaccinate the total population ($N = 15.6$ M)

    They assumed the vaccine to be 56% effective in preventing hospitalizations and deaths for the older at-risk group and 80% effective for those younger than 65. Develop a model for the first strategy. With no intervention, assume a hospitalization rate (per 100,000) for influenza and influenza-related illnesses of 125 (per 100,000) for persons 65 years of age or older and a rate (per 100,000) of 50 for the younger age group; and assume death rates (per 100,000) of 56 and 15, respectively, for the two age groups. (In the actual study, scientists considered three age groups and a more involved set of input variables.) (van Genugten et al. 2003)

13. Develop a model for the second strategy in Project 12.

14. Develop models for the two strategies in Project 12, discuss the results, and make recommendations.

Answers to Quick Review Questions

1. $dR/dt = aI$
2. **a.** Negative while people are getting sick because the number of susceptibles is decreasing
   b. $dS/dt = -rSI$
3. $dI/dt = rSI - aI$
4. **a.** Day 7
   b. Day 6
   c. Day 8
5. **a.** E, E_Q, S_Q
   b. S_Q
   c. I_U, I_Q, I_D

6. **a.** no
   b. no
   c. yes
   d. no
   e. no
7. **a.** 0.2/day
   b. pE
   c. pE_Q
   d. 0.1/day, 10 days
   e. $u = 1/16$ per day $= 0.0625$/day
   f. $u\,S_Q$
8. **a.** $k/N_0 = 10/10,000,000 = 0.000001 = 0.0001\%$ per day
   b. $b = 0.06$
   c. $(0.000001)(0.06) = 0.00000006 = 0.000006\%$/day
   d. $(5000)(9,000,000) = 45,000,000,000$
   e. $(0.00000006)(45,000,000,000) = 2700$
   f. $(0.1)(2700) = 270$ people
   g. $qkbI_US/N_0$
   h. $(1 - 0.1)(2700) = (0.9)(2700) = 2430$ people; or $2700 - 270 = 2430$ people
   i. $(1 - q)kbI_US/N_0$
9. $qk(1 - b)I_US/N_0 = (0.1)(10)(1 - 0.06)(5000)(9,000,000)/(10,000,000) = 4230$ people
10. $R = (1 - q)kb/(v + m + w) = (1 - 0.1)(10)(0.06)/(0.04 + 0.0975 + 0.0625) = 2.7$

References

CDC (Centers for Disease Control and Prevention). "Diphtheria." Department of Bacterial and Mycotic Diseases. http://www.cdc.gov/

———. "Updated Interim U.S. Case Definition for Severe Acute Respiratory Syndrome (SARS)." Department of Health and Human Services. http://www.cdc.gov/ncidod/sars

Epidemiology and Disease Control Program. "Strep Throat Fact Sheet." Community Health Administration, Maryland Department of Health & Mental Hygiene. http://www.edcp.org/factsheets/strepthr.html

Kermack, W. O., and A. G. McKendrick. 1927. "A Contribution to the Mathematical Theory of Epidemics." *Proceedings of the Royal Society of London*, Series A, Vol. 115, No. 772: 700–721.

Lipsitch, Marc, et al. 2003. "Transmission Dynamics and Control of Severe Acute Respiratory Syndrome." *Sciencexpress Report*, May 23. http://www.sciencexpress.org/23 May 2003/Page 1/10.1126/science.1086616

Mayo Clinic Staff. 2002. "Mumps." Mayo Foundation for Medical Education and Research, August 5. http://www.mayoclinic.com/invoke.cfm?id=DS00125

Murray, J. D. 1989. *Mathematical Biology*. New York: Springer-Verlag.

National Computational Science Leadership Program. "Influenza Epidemic in a Boarding School." The Shodor Educational Foundation, Inc. http://www.shodor.org/ncslp/talks/basicstella/sld009.htm

Project SUCCEED. "CASE STUDY: Influenza Epidemic in a Boarding School."
    The Shodor Educational Foundation, Inc. http://www.shodor.org/succeed/models/
    flu/case/index.html. Original from Murray 1989.

van Genugten, Marianne L. L., Marie-Louise A. Heijnen, and Johannes C. Jager.
    2003. "Pandemic Influenza and Healthcare Demand in the Netherlands: Scenario
    Analysis." *Emerging Infectious Diseases* 9 (5), Centers for Disease Control and
    Prevention (CDC). http://www.medscape.com/viewarticle/453679

WHO (World Health Organization). "Severe Acute Respiratory Syndrome." The
    United Nations. http://www.who.int/topics/sars/en/

MODULE 6.3

Enzyme Kinetics

Download

The text's website has available for download for various system dynamics tools the file *substrate*, which contains a submodel for this module.

Introduction

Enzymes catalyze, or hasten, chemical reactions for biological systems. Although most enzymes are proteins, there are some RNA enzymes as well. They are remarkably adapted for this role, because in minute quantities they are very specific. Enzymes do not influence the direction of the reaction. Unchanged by the reaction, they can be used over and over again. Without them, even spontaneous reactions would not proceed fast enough to support living cells. Enzymes can increase the rate of reaction by a factor of up to 10^{20}. Additionally, they are "regulatable" by both physical and chemical factors. Many biochemists study the activities of enzymes, the factors that regulate and determine the speed of the catalyzed reaction.

Archibald Garrod, studying the disease **alkaptonuria**, proposed in 1902 that the instructions for producing specific enzymes in the cell were inherited. He elaborated on his work in his *Inborn Errors of Metabolism*, published in 1923. Essentially his hypothesis was that diseased individuals lacked a normal enzyme in the catabolism of proteins. This enzyme deficiency resulted from receiving one recessive gene from each parent. Later investigation confirmed that alkaptonurics lack the activity of homogentisate dioxygenase. This enzyme normally converts homogentisate, one of the intermediate compounds in the breakdown of the amino acid tyrosine, into maleylacetoacetate. Hence, homogentisate accumulates in and darkens various body tissues (e.g., bone, skin, prostate), causing symptoms of arthritis. Some is eliminated in the urine, which will turn dark, if allowed to stand. We now understand the bases of many metabolic diseases caused by defective enzymes.

All the attention to metabolic diseases has catalyzed great interest in enzymes and how they work. One area of focus has been on the rate of enzyme activity and its control. The quantitative study of enzyme activity is **enzyme kinetics**. What factors influence the rate of an enzymatic reaction? Commonly, we consider things like substrate concentration, enzyme concentrations, cofactors, inhibitors, *pH*, and temperature.

Michaelis-Menten Equation

For most enzymes, if for substrate S and enzyme E you increase **substrate concentration** (usually denoted **[S]**) and hold **enzyme concentration ([E])** constant, the resulting **initial velocities**, or **reaction rates, of the reaction** (v) produce an asymptotic curve. In other words, v increases rapidly at first as you increase [S]. Then, the rate of increase in v decreases, and v approaches a **limit of the reaction rate**, called v_{max}. No further increases in [S] will increase velocity. With some assumptions, the **Michaelis-Menten equation** describes this relationship between [S] and v, as follows:

$$v = \frac{v_{max}[S]}{K_m + [S]}$$

The graph of this model appears in Figure 6.3.1. K_m is the **Michaelis-Menten constant** and is equal to the [S] where $v = v_{max}/2$. K_m is an indicator of the enzyme's affinity for the substrate. The lower the K_m value, the higher the affinity, so it takes less substrate to reach half of v_{max} and the enzyme is a better catalyst for the reaction. Table 6.3.1 provides K_m values for several enzyme-substrate combinations.

We begin by considering a simple reaction in which one substrate S in the presence of an enzyme E converts to one product P. Michaelis and Menten hypothesized that the enzyme catalyzes the reaction by reacting with the substrate to form an intermediate

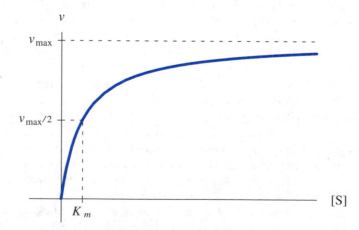

Figure 6.3.1   Graph of initial reaction velocity versus substrate concentration for Michaelis-Menten equation

Table 6.3.1
Michaelis-Menton Constants (Becker et al. 2003; Kimball 2003)

Enzyme	Substrate	K_m (mMoles/liter)
Acetylcholinesterase	Acetylcholine	0.09
Carbonic anhydrase	CO_2	12
Catalase	H_2O_2	1100
Chymotrypsin	Gly-Tyr-Gly	108
Fumarase	Fumarate	0.005
Triose phosphate isomerase	Glyceraldehyde-3-phosphate	0.5
Beta-lactamase	Benzylpenicillin	0.02

enzyme-substrate complex ES. This complex experiences a catalytic reaction to form the enzyme E and the product P. Figure 6.3.2 depicts the enzyme reaction, and the following diagram represents the situation where k_1, k_2, k_3, and k_4 are rate constants:

$$E + S \underset{k_2}{\overset{k_1}{\rightleftarrows}} ES \underset{k_4}{\overset{k_3}{\rightleftarrows}} E + P \tag{1}$$

Thus, Diagram 1 for the chemical reactions indicates the following about the rate constants:

- k_1—reaction rate of substrate S in the presence of enzyme E to intermediate enzyme-substrate complex ES
- k_2—reverse reaction rate of intermediate enzyme-substrate complex ES back to substrate S and enzyme E
- k_3—reaction rate of intermediate enzyme-substrate complex ES to product P and enzyme E
- k_4—reverse reaction rate of product P and enzyme E back to intermediate enzyme-substrate complex ES

To derive their model, Michaelis and Menten made the following simplifying assumptions:

1. The reaction rate is determined before very much product is formed. Consequently, the reverse reaction from E + P to ES is negligible.
2. k_3 is small in comparison to k_1 and k_2; that is, the rate of product formation is slow in comparison to the rate of ES formation and the rate of ES dissociation to E + S.

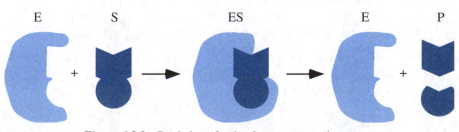

E   S   ES   E   P

Figure 6.3.2   Depiction of a simple enzyme reaction

3. [S] is much greater than [E], so that [S] is virtually constant.
4. [E] + [ES] is constant.

Under these assumptions, the Michaelis-Menten equation models reaction 1 as follows (Danby 1997):

$$v = \frac{v_{max}[S]}{K_m + [S]}$$

Exercise 3 derives this equation.

Quick Review Question 1

a. Using Assumption 1 above, give an approximate value for k_4.
b. Using Assumption 3 above, give an approximate value for $d[S]/dt$.
c. Using assumptions above, give the approximate relationship between the initial velocity of the reaction, v, and the rate of change of the product, $d[P]/dt$.

Although the Michelis-Menten equation captures the relationship of reaction velocity to substrate concentration, K_m and v_{max} are difficult to ascertain from its graph, such as in Figure 6.3.1. Hans Lineweaver and Dean Burk reorganized the equation into a form that is more helpful for determination of these constants (Danby 1997). As Exercise 5 develops, taking the reciprocal of both sides, they solved for $1/v$ in terms of $1/[S]$, as follows:

$$\frac{1}{v} = \frac{K_m}{v_{max}}\left(\frac{1}{[S]}\right) + \frac{1}{v_{max}}$$

K_m/v_{max} and $1/v_{max}$ are constants. Moreover, considering $x = 1/[S]$ to be an independent variable and $y = 1/v$ to be a dependent variable, the equation has the form of a line, $y = mx + b$. Thus, in the graph of $1/v$ versus $1/[S]$, the slope is K_m/v_{max}, and the vertical intercept is $1/v_{max}$. Setting $1/v$ equal to zero, we find that the horizontal intercept is $-1/K_m$. Figure 6.3.3 presents the graph of $1/v$ versus $1/[S]$ for the Michaelis-Menten equation in Figure 6.3.1. The following Quick Review Question determines v_{max} and K_m from this graph.

Quick Review Question 2

Determine the following for Figure 6.3.3, where $1/v_{max} = 0.2$ and $-1/K_m = -4$:

a. The vertical intercept
b. v_{max}
c. The horizontal intercept
d. K_m
e. K_m/v_{max}
f. The slope

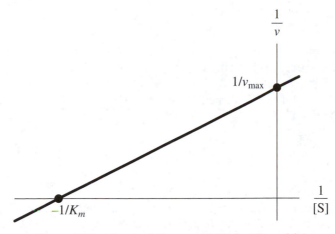

Figure 6.3.3   Graph of $1/v$ versus $1/[S]$ for Figure 6.3.1

Differential Equations

In the simplest case of reaction 1, the removal of S's concentration depends on the interaction of one substrate molecule with one enzyme molecule. As with the interaction of competing species (see Module 6.1 on "Competition") and the interaction of susceptibles and infecteds in the SIR model of disease spreading (see Module 6.2 on "Spread of SARS"), the rate of change of removal of [S] is proportional to the product of the concentrations of E and S, $k_1[E][S]$. Moreover, some of the enzyme-substrate complex ES reverts to form a molecule of E and a molecule of S. Thus, the rate of change of the formation of [S] is proportional to the concentration of ES, $k_2[ES]$. Because the rate of change of the concentration of S is equal to the rate of change of formation minus the rate of change of removal, we have the following differential equation:

$$d[S]/dt = k_2[ES] - k_1[E][S]$$

Because in this simplified reaction a molecule of enzyme reacts with a molecule of substrate, the rate of change of removal of [E] is the same as the rate of change of removal of [S], namely, $k_1[E][S]$. However, [E] can be formed from [ES] by forward and backward reactions. The rate of change of each of these reactions is proportional to [ES]. With rate constants of k_3 and k_2 for the forward and backward reactions, respectively, the rate of change of the forward reaction for the formation of [E] is $k_3[ES]$, while the rate of change of the backward reaction equals $k_2[ES]$. Thus, the total rate of change of the formation of [E] is the sum of these two values.

Quick Review Question 3

Considering the simple reaction of (1), give the formula for each of the following quantities:

   a. The rate of change of the formation of [E]
   b. $d[E]/dt$

    c. $d[P]/dt$
    d. $d[ES]/dt$

Quick Review Question 4

In the model diagram for the simple reaction of (1), what do the stocks (box variables) represent?

Model

The model mimics the differential equations of the last section. A stock (box variable) exists for each of the four concentrations, [E], [S], [ES], and [P]. Recall that the rate of change of [S] is as follows:

$$d[S]/dt = k_2[ES] - k_1[E][S]$$

Thus, the flow into [S] has connectors/arrows from the stock for [ES] and from a converter storing the value of the rate constant k_2, and its equation is the product of these two values. With connectors/arrows from the converter for k_1 and the stocks for [E] and [S], the flow out of [S] is the product of these values. Figure 6.3.4 depicts a submodel for [S] with $d[S]/dt = k_2[ES] - k_1[E][S]$. The figure does not include some diagram elements, such as flows, associated with [E], [ES], and [P].

Quick Review Question 5

In a submodel for [E] with differential equation $d[E]/dt = (k_2 + k_3)[ES] - k_1[E][S] - k_4[E][P]$, give the number of connectors/arrows that

    a. go to the flow into [E]'s stock
    b. go to the flow that leaves [E]'s stock
    c. come out of [E]'s stock

    The velocity of the reaction is the rate of change of [P], $d[P]/dt$. This derivative (*rate_of_P*) in a model is the rate of change of formation minus the rate of change of removal of [P], $k_3[ES] - k_4[E][P]$, or the flow into [P] minus the flow out of [P]. For use in graphing, we have a converter/variable evaluating this difference. In an actual chemical reaction, we consider the initial velocity of the reaction because of restrictive factors, such as the amount of substrate. In this model, for a particular substrate concentration, the rate of change of [P] approaches a limit. To observe the impact of substrate concentration on velocity, we have a system dynamics tool generate a comparative graph of *rate_of_P* versus *S_concentration* with *S_concentration* taking on values 0.0, 1.5, 3.0, ... , 30.0 millimoles (mM). For each value, we run the simulation using the Runge-Kutta 4 method for 3 seconds with $\Delta t = 0.05$ seconds. Figure 6.3.5 displays a resulting graph for parameters of $k_1 = 0.05$, $k_2 = 0.1$, $k_3 = 0.02$, and $k_4 = 0.0$ sec^{-1}. The tops of the columns reveal a graph similar to that of Figure 6.3.1. Tabular output gives the top of the last column as about 0.0001837 mM/sec.

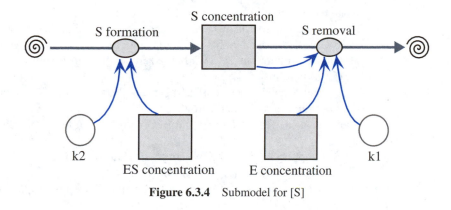

Figure 6.3.4   Submodel for [S]

Quick Review Question 6

The following questions refer to Figure 6.3.5.

  a. Select from the tick values the best estimate of v_{max}.
  b. Using your answer from Part a, estimate K_m to the nearest whole number.

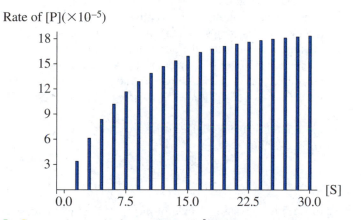

Figure 6.3.5   Comparative graph of *rate_of_P* ($\times 10^{-5}$ mM/sec) versus *S_concentration* (mM)

Exercises

  1. For the simple model of enzyme kinetics in (1), give the relationship between $d[E]/dt$ and $d[ES]/dt$.
  2. In the Michaelis-Menten equation, by considering the instant when $[S] = K_m$, show that K_m is equal to the substrate concentration when $v = v_{max}/2$.
  3. In this exercise, we derive the Michaelis-Menten equation (Danby 1997).
    a. The first assumption is that the reverse reaction from E + P to ES is negligible, so that k_4 is zero or very small. Write the differential equation $d[E]/dt$ for assuming $k_4 = 0$.
    b. Write the differential equation $d[ES]/dt$ for assuming $k_4 = 0$.

c. Write the differential equation $d[P]/dt$ for assuming $k_4 = 0$.

d. The first assumption also states that the reaction rate is determined very early. Give the derivative that the initial velocity v of the reaction equals.

e. The first assumption has $d[S]/dt$ being zero. Using this assumption, solve for $[E][S]$ in terms of $[ES]$.

f. Using your answer to Part e, $[E][S]$ is proportional to $[ES]$. Give the constant of proportionality and call it K_m.

g. Using your answers to Parts e and f, solve for K_m in terms of $[ES]$, $[E]$, and $[S]$.

h. By Assumption 4, $[E] + [ES]$ is constant. Call this constant $[E_0]$. Using this equation, solve for $[E]$ in terms of $[ES]$ and $[E_0]$.

i. Substitute your solution of $[E]$ from Part h into the answer to Part g, and solve for $[ES]$.

j. Substitute your solution of $[ES]$ from Part i into the differential equation for $d[P]/dt$ in Part c.

k. As $[S]$ increases, the initial velocity of the reaction, $v = d[P]/dt$, approaches its maximum, v_{max}. Moreover, as $[S]$ increases, K_m is small in comparison to $[S]$, so that $[S]$ and $K_m + [S]$ are approximately the same. Using your answer to Part j, we can also say that $d[P]/dt$ approaches what value?

l. Using Parts j and k, solve for $v = d[P]/dt$, and obtain the Michaelis-Menten equation.

4. The **Briggs-Haldane model**, assumes that very soon the rate of change of $[ES]$ is small in comparison to $[E]$ and $[S]$, so that $d[ES]/dt$ is almost zero. Using this assumption and the differential equation from your answer to Part b of Exercise 3, solve for $[ES]$ in terms of $[E][S]$. Define K as an appropriate proportionality constant, and in a similar fashion to Exercise 3 compute the solution of $v = d[P]/dt$ as $k_3[E_0][S]/(K + [S])$, where $[E_0]$ is the constant $[E] + [ES]$ (Danby 1997).

5. From the Michaelis-Menten equation, derive the solution of $1/v$ in terms of $1/[S]$, as follows:

$$\frac{1}{v} = \frac{K_m}{v_{max}}\left(\frac{1}{[S]}\right) + \frac{1}{v_{max}}$$

6. In Reaction 1, one molecule of the enzyme reacts with one molecule of the substrate. In the following reaction, one molecule of the enzyme reacts with n molecules of the substrate, and we assume that the reverse reaction from $E + P$ to ES is negligible:

$$E + nS \underset{k_2}{\overset{k_1}{\rightleftarrows}} ES \xrightarrow{k_3} E + P \qquad (2)$$

Write a differential equation for each of the parts below. Because the reaction involves one molecule of enzyme for n molecules of substrate, we take the product of $[E]$ and n copies of $[S]$, that is $[E][S]^n$, where appropriate (Danby 1997).

a. $d[S]/dt$

b. $d[E]/dt$

c. $d[P]/dt$

d. $d[ES]/dt$

7. Give the Michaelis-Menten approximation for the reaction in Exercise 6.

Projects

For additional projects, see Module 7.8 on "Fueling Our Cells—Carbohydrate Metabolism."

For all model development, use an appropriate system dynamics tool.

1. Model the following simplified reaction:

Plot substrate concentration [S] and product concentration [P] versus time. Describe the growth of [P] and decline of [S]. With your simulator, determine experimentally $[P]_{eq}$, $[S]_{eq}$, when equilibrium occurs, and the equilibrium constant K_{eq} for several values of k_1 and k_2. Do the experimental results agree with the analytical ones?

2. Model the complete Reaction 1. Generate tables with time, the values of all concentrations, and the rate of change of [P]. Generate the following graphs: [S] and [P] versus time, the rate of change of [P] versus time, and the rate of change of [P] versus [S]. Generate a comparative graph as in Figure 6.3.5. Estimate v_{max} and K_m from this graph. Using the generated data, plot $1/v$ versus $1/[S]$ with a computational tool, such as a spreadsheet. From the graph, estimate v_{max} and K_m.

3. Model the complete Reaction 1. Also, compute the Briggs-Haldane and Michaelis-Menten approximations. The Briggs-Haldane approximation assumes that after an initial period, the rate of formation of [ES] is small in comparison with the change in [S] and change in [P] (see Exercise 4). This approximation for the rate of change of the concentration of [P] is as follows:

$$\frac{d[P]}{dt} = \frac{k_3([E]+[ES])[S]}{\dfrac{(k_2+k_3)}{k_1}+[S]}$$

The Michaelis-Menten approximation is as follows:

$$\frac{d[P]}{dt} = \frac{k_3([E]+[ES])[S]}{\dfrac{k_2}{k_1}+[S]}$$

Find parameter values that satisfy the assumptions of the approximations. Graph and compare the rate of change of [P] for your simulation and the approximations versus time (Danby 1997).

4. Continue Project 2. Explore the situation of having a very low substrate concentration, where $[S] \ll K_m$. In this case, [S] is much lower than K_m, so that $K_m + [S]$ is approximately K_m. Simplify the Michaelis-Menten equation in this situation. Have your model compute this v. Compare your model with the computed value. How does v vary with [S] (Danby 1997)?

5. Continue Project 2. Explore the situation of having very high substrate concentration, where $[S] \gg K_m$. In this case, [S] is much higher than K_m, so that $K_m + [S]$ is approximately [S]. Simplify the Michaelis-Menten equation in this situation. Have your model compute this v. Compare your model with the computed value. How does v vary with [S]? How does v_{max} vary with [E] (Danby 1997)?

6. Continue Project 2. Explore the situation of $[S] = K_m$. Simplify the Michaelis-Menten equation in this situation. Have your model compute this v. Compare your model with the computed value. What is the meaning of K_m (Danby 1997)?

7. Model the complete Reaction 2 of Exercise 6. Generate tables with time, the values of all concentrations, and the rate of change of [P]. Generate the following graphs: [S] and [P] versus time, the rate of change of [P] versus time, and the rate of change of [P] versus [S]. Generate a comparative graph as in Figure 6.3.5. Estimate v_{max} and K_m from this graph. Using the generated data, plot $1/v$ versus $1/[S]^n$ with an appropriate computational tool, such as a spreadsheet. From the graph, estimate v_{max} and K_m.

Answers to Quick Review Questions

1. **a.** $k_4 \approx 0$ because the reverse reaction from $E + P$ to ES is negligible
   b. $d[S]/dt \approx 0$ because [S] is virtually constant, giving an almost 0 value for the rate of change of [S] with respect to time
   c. $v = d[P]/dt$. By Assumption 1, the reverse reaction from $E + P$ to ES is negligible, so the rate of change in [P] comes from formation, not from deformation of P. By Assumption 4, $[E] + [ES]$ is constant, so that the total amount of the enzyme, which is in the free form of E and ES, is constant. Moreover, $d[S]/dt \approx 0$ (see Part b). Thus, the initial velocity of the reaction is the rate of change of [P].

2. **a.** $1/v_{max} = 0.2$
   b. $1/0.2 = 5$
   c. $-1/K_m = -4$
   d. $K_m = -1/(-4) = 0.25$
   e. $K_m/v_{max} = 0.25/5 = 0.05$
   f. 0.05

3. **a.** Rate of change of formation of $[E] = k_2[ES] + k_3[ES] = (k_2 + k_3)[ES]$
   b. $d[E]/dt = (k_2 + k_3)[ES] - k_1[E][S] - k_4[E][P]$
   c. $d[P]/dt = k_3[ES] - k_4[E][P]$
   d. $d[ES]/dt = k_1[E][S] + k_4[E][P] - (k_2 + k_3)[ES]$

4. The four concentrations, [E], [S], [ES], and [P]

5. **a.** 3, from k_2, k_3, and [ES]
   b. 5, from k_1, [E], [S], k_4, and [P]
   c. 4, one to each flow

6. **a.** $v_{max} \approx 18 \times 10^{-5}$ mM/sec $= 0.00018$ mM/sec

b. $K_m \approx 5.0$ mM. If $v_{\max} = 0.00018$ mM/sec, $v_{\max}/2 = 0.00009$ mM/sec, which occurs at approximately $S\_concentration = 5.0$ mM.

References

Becker, Wayne M., Lewis J. Kleinsmith, and Jeff Hardin. 2003. *The World of the Cell*. 5th ed. San Francisco, Calif.: Benjamin Cummings: 140–145.

Danby, J.M.A. 1997. *Computer Modeling: From Sports to Spaceflight . . . From Order to Chaos*. Richmond, Va.: Willmann-Bell: 351–367.

Garrod, Archibald E. 1902. "The incidence of alkaptonuria: A study in chemical individuality." *Lancet*, ii:1616–1620.

———. 1923. *Inborn Errors of Metabolism*. 2nd ed. London: Henry Frowde and Hodder & Stoughton.

Kimball, John W. 2003. "Enzyme Kinetics." Kimball's Biology Pages. http://users.rcn.com/jkimball.ma.ultranet/BiologyPages/E/EnzymeKinetics.html

Paul, Carol Ann. 1999. "Enzyme Kinetics." Wellesley College. http://www.wellesley.edu/Biology/Concepts/IntroEnzymeKinetics.html

MODULE 6.4

Predator-Prey Model

Download

The text's website has available for download for various system dynamics tools a *Predator-Prey* file, which contains the model of this module.

Introduction

One of the interspecific interactions (see Module 6.1 on "Competition") common to biological communities is the **predator-prey** relationship. When one species (**predator**) consumes another species (**prey**), while the latter is still living, the action is **predation**. Predation might involve the consumption of a young squirrel by a hawk, but examples also include tomato hornworms consuming tomato plant leaves and a tapeworm feeding off its mammalian host. Predator-prey interactions are important influences on population levels and ecosystem energy flow.

One of the most interesting characteristics of this type of relationship is that both predators and prey develop fascinating adaptations, which normally develop over long periods of time. Predator adaptations usually involve better prey detection and capture, whereas prey adaptations normally involve improved abilities to escape and avoid detection.

So, let's consider a 3/4 inch frog, commonly called a poison dart frog. We might expect that such a small animal would, to avoid predation, come out only at night or adopt some camouflaged coloration. However, this brazen creature forages for small invertebrates during the day (prey may also be predators) and is brilliantly colored (bright red, yellow, etc.). How might it manage then to avoid predation? The answer lies in the skin of the frog, which contains toxic, alkaloid chemicals that cause paralysis and/or death in the predator. Over time, predators associate the coloration with the toxic nature of the prey, and hence, avoid that prey. So the bright coloration is termed warning or **aposematic coloration**.

Lotka-Volterra Model

In the 1920s, mathematicians Vito Volterra and Alfred Lotka proposed independently a model for populations of a predator species and its prey, such as hawk and squirrel populations in a certain area. For simplicity, we assume that a hawk hunts only squirrels and that no other animal eats squirrels. If the hawk's only food source is squirrel and the number of squirrels diminishes significantly, then scarcity of food will result in starvation for some of the hawks. With reduced numbers of hawks, the squirrel population should increase.

Quick Review Question 1

This question reflects on the predator-prey situation before we begin the discussion.

    a. Do predator-prey interactions have a direct impact on the births or deaths of the prey?

    b. Based on other interaction models of this chapter, we can model the prey deaths as being directly proportional to what?

    c. If we consider prey births as being unconstrained, we can model prey births as being directly proportional to what?

    d. Are predator-prey interactions advantageous or disadvantageous for predators?

    e. Based on other interaction models of this chapter, we can model predator births as being directly proportional to what?

    f. If we consider predator deaths as being unconstrained, we can model the predator deaths as being directly proportional to what?

Let s be the number of squirrels in the area and h be the number of hawks. If no hawks are present, the change in s is as in the unconstrained model (see Module 3.2 on "Unconstrained Growth and Decay"):

$$\Delta s = s(t) - s(t - \Delta t) = (k_s * s(t - \Delta t)) * \Delta t \quad \text{for constant } k_s$$

However, this prey's population is reduced by an amount proportional to the product of the number of hawks and the number of squirrels, $h(t - \Delta t) * s(t - \Delta t)$. Thus, with a proportionality constant k_{hs} for this reduction, the change in the number of squirrels is as follows:

$$\Delta s = s(t) - s(t - \Delta t)$$
$$= (k_s * s(t - \Delta t) - k_{hs} * h(t - \Delta t) * s(t - \Delta t)) * \Delta t$$

We can interpret the term $k_{hs} * h(t - \Delta t) * s(t - \Delta t)$ in a couple of ways. First, $h(t - \Delta t) * s(t - \Delta t)$ is the maximum number of distinct interactions of hawks with squirrels. For example, for $h(t - \Delta t) = 3$ hawks and $s(t - \Delta t) = 2$ squirrels, $(3)(2) = 6$ possible pairings exist. The decrease in the number of squirrels is proportional to this product, where the constant of proportionality, k_{hs}, is related to the hunting ability of the hawks and the survival ability of the squirrels. A second interpretation of

$k_{hs} * h(t - \Delta t) * s(t - \Delta t) = (k_{hs} * h(t - \Delta t)) * s(t - \Delta t)$ is that the size of the squirrel population decreases in proportion to the size of the hawk population.

While the squirrel population decreases with more contacts between the predator and prey, the hawk population increases. Moreover, the death rate of hawks is proportional to the number of hawks. Thus, the change in the hawk population is as follows:

$$\Delta h = h(t) - h(t - \Delta t)$$
$$= (k_{sh} * s(t - \Delta t) * h(t - \Delta t) - k_h * h(t - \Delta t)) * \Delta t$$

for constants k_{sh} and k_h.

Thus, we can express the predator-prey model, known as the **Lotka-Volterra model**, as the following pair of difference equations for the change in prey (here, change in the squirrel population, Δs) and change in predator (here, change in the hawk population, Δh):

$$\Delta s = (k_s * s(t - \Delta t) - k_{hs} * h_t * s(t - \Delta t)) * \Delta t \tag{1}$$
$$\Delta h = (k_{sh} * s(t - \Delta t) * h(t - \Delta t) - k_h * h(t - \Delta t)) * \Delta t$$

or as the following pair of differential equations:

$$\frac{ds}{dt} = k_s s - k_{hs} hs$$
$$\tag{2}$$
$$\frac{dh}{dt} = k_{sh} sh - k_h h$$

Figure 6.4.1 contains a diagram for the predator-prey model with the prey population affecting the number of predator births and the predator population influencing the number of prey deaths.

Quick Review Question 2

Consider the following Lotka-Volterra difference equations:

$$\Delta x = (2 * x(t - \Delta t) - 0.02 * y(t - \Delta t) * x(t - \Delta t)) * \Delta t \quad \text{with } x(0) = 100$$
$$\Delta y = (0.01 * x(t - \Delta t) * y(t - \Delta t) - 1.06 * y(t - \Delta t)) * \Delta t \quad \text{with } y(0) = 15$$

a. Which equation (Δx, Δy, both, or neither) models the change in predator population?

For each of the following questions, indicate the appropriate answer from the following choices:

A. 2	**B.** 0.02	**C.** −0.02	**D.** 0.01
E. 1.06	**F.** −1.06	**G.** 100	**H.** 15

b. Which number represents the predator birth fraction?
c. Which number represents the prey birth fraction?

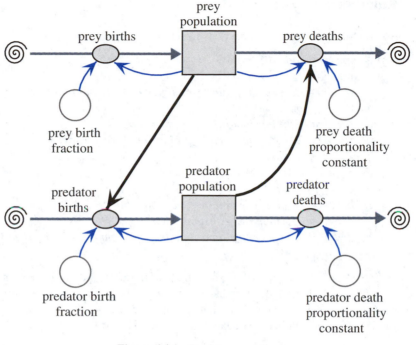

Figure 6.4.1   Predator-prey diagram

d. Which number represents the predator death proportionality constant?
e. Which number represents the prey death proportionality constant?
f. What is the initial number of predators?
g. What is the initial number of prey?

Particular Situations

Historical Note   During the Cultural Revolution in China in 1958–1960, Chairman Mao Zedong decreed that all sparrows be killed because they ate too much of the crops and they seemed to be only for pleasure anyway. With reduction in its main predator, the insect population increased dramatically. The insects destroyed much more of the crops than the birds ever did. Consequently, the Chinese reversed the decision that caused the imbalance (Heights 2002).

Returning to the hawks and squirrels example, some of the model's equations and constants appear in Equation Set 6.4.1. In that example, *prey_birth_fraction* $(k_s) = 2$, *prey_death_proportionality_constant* $(k_{hs}) = 0.01$, *predator_birth_fraction* $(k_{sh}) = 0.01$, *predator_death_proportionality_constant* $(k_h) = 1.06$, the initial *prey_population* $(s_0) = 100$, and the initial *predator_population* $(h_0) = 15$.

Equation Set 6.4.1

Some of the equations and constants for model in Figure 6.4.1

$predator\_population(0) = 15$
$predator\_birth\_fraction = 0.01$
$predator\_births = (predator\_birth\_fraction * prey\_population) *$
    $predator\_population$
$predator\_death\_proportionality\_constant = 1.06$
$predator\_deaths = predator\_death\_proportionality\_constant *$
    $predator\_population$
$prey\_population(0) = 100$
$prey\_birth\_fraction = 2$
$prey\_births = prey\_birth\_fraction * prey\_population$
$prey\_death\_proportionality\_constant = 0.02$
$prey\_deaths = (prey\_death\_proportionality\_constant * predator\_population) *$
    $prey\_population$

Table 6.4.1 and Figure 6.4.2 show the varying prey and predator populations as time advances through 12 months. Shortly after the squirrel or prey population increases, the hawk or predator population does likewise. As the predators kill off their food supply, the number of predators decreases. Then, the cyclic process starts over.

Quick Review Question 3

To the nearest whole number, what is the period (in months) of the cyclic functions for population in Figure 6.4.2?

Figure 6.4.3 shows the graph of a solution to the difference or differential equations with the prey population along the horizontal axis and the predator population along the vertical axis. With the initial predator population being 15 and prey

Table 6.4.1
Table of Prey and Predator Populations over 12-month period

Months	Prey Population	Predator Population
0.000	100.00	15.00
1.000	449.58	62.00
2.000	30.43	280.24
3.000	5.63	108.55
4.000	10.54	40.32
5.000	45.61	17.59
6.000	244.25	19.97
7.000	215.76	298.60
8.000	7.91	173.18
9.000	6.52	63.69
10.000	21.30	24.81
11.000	109.68	14.61
Final	470.44	74.28

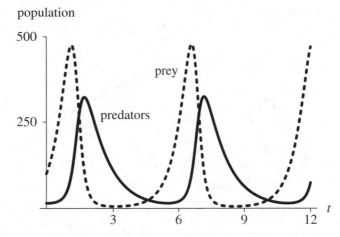

Figure 6.4.2   Graph of populations versus time in months

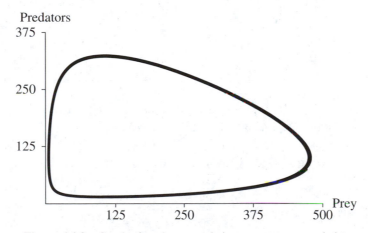

Figure 6.4.3   Graph of predator population versus prey population

population being 100, the plot starts at the bottom toward the left and proceeds counterclockwise as time progresses. Initially, with few predators endangering them, the prey population reaches a maximum of about 475 when the predator population is about 100. Then, with the graph developing to the left and up, we see that the prey population starts decreasing as the predator population continues to increase with the abundant supply of its food, the prey. At the graph's high point, about (107, 322), with approximately 107 prey, the predator population achieves a maximum of 322 individuals. That same number of predators, about 107, occurs toward the bottom of the graph when the prey only number about 15. After a maximum, the number of predators falls off rapidly because of the limited food supply, and the number of prey decreases as well. Eventually, on the bottom part of the graph, with the diminished number of predators, the prey are able to stage a

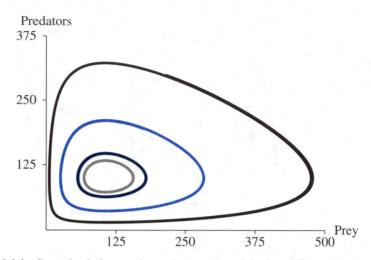

Figure 6.4.4   Several solutions to the predator-prey model using different initial conditions and the following coloration:

Predator	Prey	Color of Graph
15	100	black
75	125	gray
135	150	dark color
195	175	light color

come-back, and the cyclical process begins again. Figure 6.4.4 illustrates several such solutions employing different initial conditions.

Quick Review Question 4

The following are the Lotka–Volterra differential equations for the particular model we have been considering:

$$ds/dt = 2s - 0.02hs$$
$$dh/dt = 0.01sh - 1.06h$$
with $s(0) = 100$ and $h(0) = 15$

a. Indicate all that must be true for the system to be in equilibrium: $ds/dt = 0$; $s = 0$; $dh/dt = 0$; $h = 0$; all of these; none of these.
b. A trivial solution for equilibrium is $s = 0$ and $h = 0$. Find a nontrivial solution, where $s \neq 0$ and $h \neq 0$.

Exercises

1. Give two sets of Lotka-Volterra equations with all coefficients being different that represent a system in equilibrium, such that the number of prey is always 3000 and the number of predators is always 500.

2. Write the differential or difference equations for a predator-prey model where there is a carrying capacity M for the predator. See differential equation 1 or difference equation 2 in Module 3.3 on "Constrained Growth."

3. The blue whale, which can grow to 30 meters in length, is a baleen whale whose favorite food is Antarctic krill, a small shrimp that is about 5 centimeters long. The difference equation for the change in the krill population is similar to that for Δs in (1), except the birth term must be logistic (see Equation 2 in Module 3.3 on "Constrained Growth"). The difference equation for the change in the number of blue whales is a logistic equation, except that the carrying capacity is not a constant but is proportional to the krill population. Write the difference equations to model this system (Greenwood 1983).

Projects

For additional projects, see Module 7.8 on "Fueling Our Cells—Carbohydrate Metabolism," Module 7.9 on "Mercury Pollution—Getting on Our Nerves," and Module 7.10 on "Managing to Eat—What's the Catch?"

For all model development, use an appropriate system dynamics tool.

1. Develop a model where the prey birth fraction (k_s) is periodic, such as follows:

$$k_s = f + a \cos(p \cdot t), \quad \text{where } a < f$$

Have a table of population numbers, a graph of populations versus time, and a graph of one population versus the other. Determine values for the parameters so that the system is periodic, and then determine values where the system is chaotic. Discuss your results.

2. Using system dynamics software or a computer program, model the predator-prey example including crop consumption discussed in the Historical Note about the Chinese Cultural Revolution.

3. Develop a predator-prey model where man hunts both predator and prey equally. For example, the predator might be shark and the prey tuna.

   a. Mathematically solve for the equilibrium point.

   b. Run the model for several situations assuming no fishing. Using those same situations, gradually increase the amount of fishing (the rate of catching shark and tuna), but keep this rate less than the prey birth rate. Record all your results. What happens to the number of prey and predators? Why? Use well-written discussions with supporting work from your model.

   c. Continue increasing the amount of fishing. At what level of fishing do the predators all die? How does this level compare to the prey birth rate? Using the equilibrium point from Part a, discuss why.

   d. Alter your model to have seasonal fishing. Use a formula for the rate of fishing effort similar to that in Project 1. Run the model for several situations. Discuss your results.

Historical Note   The situation in Project 3 was observed in the Mediterranean Sea between 1914 and 1923. Limited fishing occurred during World War I. Although fishing increased after the war, so did the number of tuna.

4. Implement the model of Exercise 2. Graph the populations as time progresses. Also, graph one population against the other. Run the model for several carrying capacities. Discuss the results. Compare this system to the one without a carrying capacity.

5. Are the two sets of equilibrium points in Quick Review Question 4 stable? Discuss your answers and give supporting evidence.

6. **a.** Implement the model in Exercise 3, assuming the following:

   Krill carrying capacity is 725 million tonnes (Mt) with 1 t = 1000 kg
   Krill production is 13 Mt/yr to 3 billion t/yr.
   A female blue whale has a calf every 2 to 3 yrs.
   A blue whale consumes about 4 t/day.
   Initially there are 5000 blue whales and 650 t of krill.

   Discuss your results

   b. Find an equilibrium point. Run the model assuming an ecological disaster kills 10% of the equilibrium level of whales and 80% of the equilibrium level of krill. Describe what happens.

   c. Run the model assuming whales are almost hunted to extinction, to 1% of their equilibrium level. Describe what happens after hunting stops. Develop a table of how long it takes for the whales to return to equilibrium level starting with different initial amounts: 1%, 2%, . . . , 10% of equilibrium level.

7. Suppose a rat population is growing logistically in an area of the city. Attempts to kill off the population through poisons or trapping are not 100% successful. Moreover, killing off half might cause the population to increase rapidly. Alternatively, people could attempt to decrease the carrying capacity by cleaning up garbage and sealing trash containers. Develop models for each of the proposed solutions to the rat problem. Compare and contrast the proposals, and discuss the circumstances under which each is best. Make recommendations to the city.

8. In predator-prey systems, predators make two basic types of responses to increasing prey density. Predators may react by taking more prey or taking them faster. This response is normally quick and termed by ecologists to be a **functional response**. On the other hand, predators may also respond by increased levels of immigration (movement in) or through producing more offspring. This type of response is typically going to take more time.

   In 1959, the ecologist C. S. Holling presented a classification of three types of predator functional responses determined by the proportion of prey consumed (Holling 1959). This and the next two projects involve modeling these types.

For this project, model a **Type-I predator functional response**: The predator consumes a constant proportion of prey, regardless of prey density (density-independent). Predation rate increases in a linear fashion, until the current population of predators achieves satiation. Few good examples exist, but one is a spider feeding on insects trapped in its web.

9. Model a **Type-II predator functional response** (see Project 8): The predator consumes less as it nears **satiation**, which determines the upper limit on consumption. In this type of response, prey handling time (T_H) and search time (T_S) are separated. Predators do not handle while they are searching and do not search while they are handling. Whatever time is necessary for handling decreases the time available for searching. Predation rate increases more slowly as prey populations increase than it does in Type I. A peak occurs when predators are consuming prey as fast as they can search and eat them. A praying mantis preys on insect prey and must process his prey before he can hunt for another. This type of response is described by the **disk equation**, which follows (Holling 1959):

$$N_a/P = aNT/(1 + aT_H N)$$

where N_a = prey attacked or killed, N = prey density, P = predator density, a = attack rate constant ($= N_a/T$), T = time predators and prey exposed, T_H = prey handling time, and ($T = T_S + T_H N_a$).

10. Model a **Type-III predator functional response** (see Projects 8 and 9): The predation increases slowly at low prey density, increases rapidly at higher densities, but levels off at satiation, even if prey density continues to increase. These predators also have separate handling and search times. This type of response is typical of predators that are **generalists**. They may use alternative prey as the prey densities of their primary prey decline. For instance, a hawk might switch to squirrels if smaller rodents became scarce (Holling 1959).

11. Where would an herbivorous animal (e.g., rabbit, deer, etc.) fit in the functional response schemes described in Projects 8, 9, and 10? Develop a model for this creature.

12. If you have visited the coast of northern California, Washington State, or southeastern Alaska and spent any time looking out to sea, you have probably seen a few captivating, furry animals swimming, diving, and floating on their backs. These creatures are sea otters. Sea otters have extremely dense fur to keep them warm in cold Pacific waters, because they lack the layer of insulating blubber possessed by other marine mammals of that area. For the fur, these creatures were hunted to near extinction during the eighteenth century. The sea otters have survived many challenges during the intervening centuries.

    Sea otters are carnivorous and must also eat one-fourth to one-third of their body weight per day to maintain body temperature (adult males 65 lbs and adult females 45 lbs). Therefore, they spend much of their time (20–60 %) hunting for and eating food. Sea otters eat a variety of grazing animals, such as sea urchins, snails, crabs, abalone, mussels, and clams that live in the rich kelp forests along the coast.

When sea otters are lost from a kelp forest, the grazing prey, particularly urchins, rapidly increase in numbers and feast on the kelp forests, sending them into decline. Kelp forests, some of the most highly productive communities in the world, are extremely important, especially for sheltering and feeding fish and shellfish communities. Loss of kelp results in the loss of many species. Where otters have been reintroduced, the healthy kelp forests return. For the key role otters play in maintaining the richness and diversity of the ecosystem, they are termed **keystone predators** (Bagheera Otter Project).

Model this situation and discuss the results.

13. This project allows you to model the changes in species diversity using the intertidal community of Washington State. The intertidal zone, in this case, is a rocky area covered by seawater at high tide, but uncovered at low tide. The community structure described for this project is based on the communities as reported by R.T. Paine during the 1960s (Paine 1966, 1969). This community, like the kelp forest (see Project 12), has a keystone predator, the ochre sea star *Pisaster ochraceus*. This sea star can achieve a radius of 11 inches and is a voracious predator, preferring the delectable taste of mussels. The rest of the community is made up of various species of algae (primary producers), mussels, clams, chitons, barnacles, crustaceans, and snails. The organisms that live in this zone are specialists, adapted for the conditions that exist there. Thus, competition is intense. Below are the components you should consider in developing your model, in addition to *Pisaster.*

Molluscs (herbivores)	Molluscs (carnivores)	Crustacea (filter feeders)	Algae
Katherina tunicata (grazer)	*Nucella*	*Mitella*	*Porphyra*
Mytilus (filter feeder)		*Balanus*	*Neorhodamela*
			Corralina

The following are the feeding relationships in this community:

> *Pisaster*—feeds on *Mytilus* preferentially, but will also feeds on *K. tunicata* and *Mitella*, depending on prey densities
> *Nucella*—feeds on *Mitella* and *Balanus*
> *K. tunicata*—feeds on all species of algae listed
> *Mytilus, Mitella*, and *Balanus* filter food out of the ocean water
> All of the algae photosynthesize for energy and organic matter production.

a. Before you try to model this community, generate a diagram that relates each of these organisms by feeding relationships. The following is the succession of changes that Paine found in the community, after excluding *Pisaster* from discrete areas of the intertidal zone:

Year One: *Mitella* disappears, replaced by the other barnacle, *Balanus*.
Year Two: Both barnacles disappear. *Mytilus* (mussel) out-competes and replaces them. With no barnacles for food, the snail, *Nucella*, disappears. *Mytilus* begins crowding out the algae as well.
Years Three through Six: *Mytilus* only remains.

b. Given this scenario, what do you think the role of *Pisaster* is in this community?

c. Generate a model that describes community dynamics when *Pisaster* is present.

d. Generate a model that describes community dynamics when *Pisaster* is removed.

Answers to Quick Review Questions

1. a. Deaths
   b. The product of the number of predators and the number of prey
   c. The number of prey
   d. Advantageous
   e. The product of the number of predators and the number of prey
   f. The number of predators
2. a. Δy
   b. D. 0.01
   c. A. 2
   d. E. 1.06
   e. B. 0.02
   f. H. 15
   g. G. 100
3. 6 months
4. a. $ds/dt = 0$ and $dh/dt = 0$
   b. $s = 106$ and $h = 100$. The following discussion derives the solution. We wish to solve the following system of equations:

$$0 = 2s - 0.02hs$$
$$0 = 0.01sh - 1.06h$$

Because $s \neq 0$, we can cancel out the factor s in the first equation to obtain $0 = 2 - 0.02h$. Thus, $h = 2/0.02 = 100$ hawks (predators). Similarly, because $h \neq 0$, we can cancel out the factor h in the second equation to obtain $0 = 0.01s - 1.06$. Thus, $s = 1.06/0.01 = 106$ squirrels (prey). See the section on "Equilibrium and Stability" in Module 3.3 on "Constrained Growth" for a discussion of equilibrium.

References

Bagheera. "Keystone Species." In the Wild: Spotlight. http://www.bagheera.com/inthewild/spot_spkey.htm (accessed May 24, 2003).

Danby, J.M.A. 1997. *Computer Modeling: From Sports to Spaceflight . . . From Order to Chaos*. Richmond, Va.: Willmann-Bell: 99–119.

Greenwood, Raymond N. 1983. "Whales and Krill: A Mathematical Model." UMAP Module 610, COMAP.

Heights Productions, Inc. 2002. Commanding Heights: The Battle for the World Economy, China. http://www.pbs.org/wgbh/commandingheights/lo/countries/cn/cn_full.html

Holling, C. S. 1959. "The Components of Predation as Revealed by a Study of Small-Mammal Predation of the European Pine Sawfly." *Canadian Entomologist* 91: 293–320.

Meerschaert, Mark M. 1993. *Mathematical Modeling.* Boston, Mass.: Academic Press: 115, 181.

Otter Project. "The Otter Project." The Otter Project, Inc. http://www.otterproject.org/ (accessed May 24, 2003).

Paine, R. T. 1966. "Food Web Complexity and Species Diversity." *American Naturalist*, Vol. 100: 65–75.

———. 1969. "The Pisaster-Tegula Interaction: Prey Patches, Predator Food Preference and Intertidal Community Structure." *Ecology* Vol. 50, No. 6: 950–961.

MODULE 6.5

Modeling Malaria

Download

The text's website has available for download for various system dynamics tools a *malaria* file, which contains the model of this module.

Introduction

How important to civilization can a mosquito-borne protozoan be? Unfortunately, we cannot ask the ancient Romans, but we do have some recent evidence that implicates this tiny parasite in the fall of one of the mightiest empires of all time. Excavations in a cemetery near Lugano, Italy have uncovered at least one infant from 450 A.D. that yielded the DNA of *Plasmodium falciparum*—the deadliest of all the human malarias. Nearby, 50 other infants were buried in a relatively short period of time that also showed fingerprints of this parasite.

The death of many infants would be expected during a malaria epidemic, partially because *falciparum* induces high rates of miscarriages and infant death. The mosquitoes that transmit malaria flourish in marshy areas found in the Tiber River valley; and if malaria swept through Rome, the disease may indeed have contributed to its downfall. Even if Roman troops were not directly affected by disease, disruptions to the production and supply of food and war materials could have drastically impaired the military's ability to protect Rome.

Interestingly, around the time the infant lived, Attila's Huns were pillaging in the north of Italy en route to Rome. Though legend credits Pope Leo the Great with persuading Attila to withdraw, it is more likely that the presence of malaria in the city was even more convincing (Carroll 2001).

Malaria is a very old disease (probably prehistoric), originating in Africa, spreading as humankind migrated to other lands. The disease gets its name from an Italian word for "bad air" (McConnell 2000).

After more than 1500 years, we still have mosquitoes and malaria. In fact, malaria kills more than 2 million people each year—that is about 7000/day. More than 1 million of these are children under the age of 5. In addition to the millions who die, up to a half billion suffer the effects of malaria. Because mothers are more likely to suffer malarial relapses during pregnancy, malaria is an important cause of low-weight births and stillbirths. More than half of the miscarriages in endemic areas are caused by malaria (WHO).

Ninety per cent of malaria cases occur in Africa, south of the Sahara (NetMark).

Background Information

A **vector** is an animal that transmits a pathogen, or something that causes a disease, to another animal. Mosquitoes are the only vectors for malaria, but only 60 out of the 380 species of *Anopheles* **mosquitoes** can host malaria-causing ***Plasmodium*** (Bradley).

Three-fifths of the female *Anopheles* mosquitoes, like their sisters of other lines, are dependent on blood meals to feed their maturing eggs. While sipping blood, a *Plasmodium*-infected female mosquito injects thread-shaped, infectious agents called **sporozoites** into her human **host**. Sporozoites circulate for a time and then enter the parenchymal cells of the liver to hide out from the immune system. Here, they live for one to two weeks, multiplying asexually to produce thousands of offspring, which mature into other invasive cells, **merozoites**. Eventually, all this activity causes the parenchymal cell to break open and release merozoites into the blood. In other malaria-causing parasites, *Plasmodium vivax* and *ovale*, some of the sporozoites become dormant **hypnozoites**. Later, these mature to reinvade other liver cells, where they continue to produce more merozoites, causing recurring bouts with malaria. Interestingly, the most deadly species, *Plasmodium falciparum*, does not produce these hypnozoites (Bradley; McConnell; NIAID; Wiser).

Merozoites enter red blood cells to feed on the blood. They reproduce asexually to form more merozoites, which invade other red blood cells. This cycle continues unless stopped by the body's defenses or medicine (NIAID).

While in the red blood cells, some merozoites mature into **male** and **female gametocytes**. Upon release, these do not enter the red blood cells, but circulate, awaiting transfer to the mosquito host. The female mosquito takes her blood meal and simultaneously sucks up some of the gametocytes.

In the mosquito's stomach, the male gametocyte (sperm) and the female gametocyte (egg) fuse. The resulting **oocyst** divides to produce thousands of sporozoites. The sporozoites migrate to the salivary glands of the mosquito for their journey into a human host. Figure 6.5.1 diagrams the life cycle of *Plasmodium falciparum*.

Analysis of Problem

In the discussion that follows in this module, we consider the modeling process involving malaria (see Module 1.2 on "The Modeling Process"). We begin by analyzing the situation to identify the problem and understand its primary questions.

In this problem, we wish to investigate the progress of malaria. In particular, we consider the relationships between human and *Anopheles* mosquito populations,

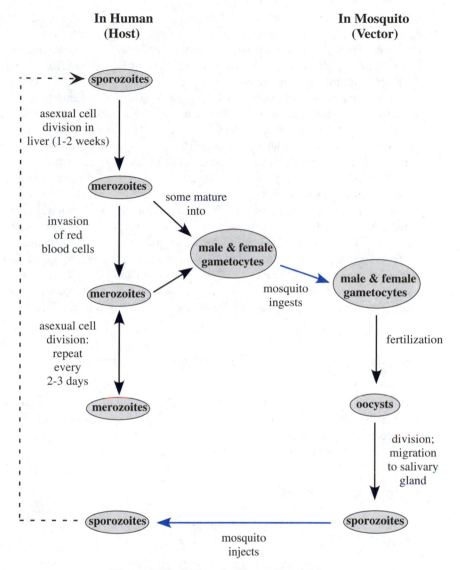

Figure 6.5.1   Life cycle *Plasmodium falciparum*

both of which are necessary for the life cycle of *Plasmodium*. Thus, with a computer simulation model, we wish to study the changing numbers of various categories of humans and mosquitoes as time progresses.

Formulating a Model: Gather Data

Many of the data on malaria are difficult to find and not dependable. Many countries with high rates of malaria are desperately poor, and the effectiveness of data collection can vary dramatically from year to year. Moreover, climate can play a significant role, with the number of cases of malaria correlating to periods of high rainfall. Also, the values associated with mosquitoes, such as numbers in each category, birth and death rates, bite probabilities, and constants of proportionality, are usually not available.

In a computational science study, such as of malaria, an interdisciplinary team approaches a problem from many directions. "Wet lab" team members conduct initial experiments and, with their "dry lab" counterparts, pose questions for the latter to consider in modeling. In formulating a model, the group may uncover the need for additional parameters, such as birth and death rates of anopheline mosquitoes. If not available from other sources, the team may decide to conduct additional experiments to collect data for empirical computations of such values.

Websites for the World Health Organization (WHO), the Southern Africa Malaria Control (Rolling Back Malaria), and other organizations do provide some enlightening and startling data concerning people. For example, all the population of Malawi lives in malarious areas. In 1999, the population of Malawi was over 10 million with a birth rate of 39.54 births/1000 population, a death rate of 23.84 deaths/1000 population, and a life expectancy of only 36.3 years (Coutsoukis1999). According to the Southern Africa Malaria Control (Country Profiles: Malawi), in 1998 there were 2,985,659 reported cases of malaria, and 19% of the deaths were attributed to malaria. 1,933,000 children under the age of five and 493,000 pregnant women were at particular risk.

Formulating a Model: Make Simplifying Assumptions

For our first model of malaria, we make several simplifying assumptions. We model the serious form of malaria that *Plasmodium falciparum* causes in which relapses do not occur. In the model, we primarily consider the number of individuals in several categories of humans and mosquitoes and ignore *Plasmodium*.

Quick Review Question 1

Considering these simplifying assumptions, give the major submodels of the malaria model.

Because the life expectancy of a human is much greater than that of a mosquito, we assume that the population of humans is closed with no births, no immigration, and no deaths except from malaria. We presume that as soon as a vector bites a human, the individual becomes a host. No immunity exists for uninfected individuals, and no incubation period occurs. Some hosts eventually become immune, others die, while still others recover and become susceptible again. We ignore the chance of relapse. Deceased individuals pass from consideration in the model.

Because of their relatively short life expectancy, we do consider mosquito births and deaths. We have the assumption that the death rates for infected and uninfected mosquitoes are identical. Similarly, we assume that all mosquitoes reproduce at the same rate. At birth, a mosquito is uninfected. As a simplification for this first version of the model, we suppose that an infected mosquito immediately becomes a host that can infect humans.

Quick Review Question 2

Based on these assumptions, list major categories of organisms for a model. In a system dynamics diagram, we represent these categories, which can accumulate individuals, as stocks (box variables).

Also, for simplification in this version of the model, except where relevant interactions between mosquitoes and humans occur, let us assume that the number of organisms in each category (uninfected humans, human hosts, immune humans, uninfected mosquitoes, mosquito vectors) expands or contracts in an unconstrained manner. In such situations, constraints, such as competition for food or predators, do not exist.

Formulating a Model: Determine Variables and Units

Based on these simplifying assumptions, we monitor three categories of humans, employing the following variables with the basic unit being one person:

> *uninfected_humans*, who are susceptible to the disease
> *human_hosts*, who have malaria and can infect mosquitoes that bite them
> *immune*, who cannot get the disease again

For the mosquito submodel, as with the human submodel, we do not count the number of dead individuals. Consequently, assuming no incubation period for *Plasmodium*, we consider the following two categories of mosquitoes:

> *uninfected_mosquitoes* that do not carry *Plasmodium*
> *vectors* that carry *Plasmodium*

We employ days as the basic unit of time, *t*.

Quick Review Question 3

This question considers the relationships among these categories of humans and mosquitoes. After completing the question, we recommend that you develop a relationship diagram with stocks (box variables) representing the categories, and with appropriate flows. Making the above simplifying assumptions, give the requested flow information by selecting from the categories *uninfected_humans*; *human_hosts*; *immune*; *uninfected_mosquitoes*; *vectors*; and undesignated "clouds," such as for deceased humans, dead mosquitoes, and unborn mosquitoes:

  a. Destination(s) of flow(s) from *uninfected_humans*
  b. Sources(s) of flow(s) to *uninfected_humans*
  c. Destination(s) of flow(s) from *human_hosts*
  d. Sources(s) of flow(s) to *immune*
  e. Sources(s) of flow(s) to undesignated "cloud(s)" for deceased humans
  f. Destination(s) of flow(s) from *uninfected_mosquitoes*
  g. Sources(s) of flow(s) to *uninfected_mosquitoes*

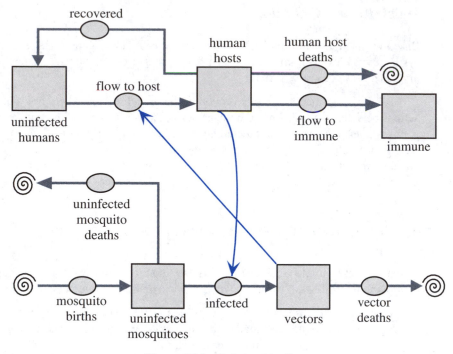

Figure 6.5.2 Relationship diagra

h. Sources(s) of flow(s) to *vectors*
i. Destination(s) of flow(s) from *vectors*
j. Sources(s) of flow(s) to undesignated "cloud(s)" for dead mosquitoes

Formulating a Model: Establish Relationships

Based on biology of the organisms and the simplifying assumptions, Figure 6.5.2 presents a relationship diagram with the two major submodels for humans and mosquitoes, stocks (box variables) representing the three human and two mosquito categories, and appropriate flows between stocks. Arrows (connectors) represent the impact of one population on the other. For example, the infected mosquito category (*vectors*) is a necessary component in an uninfected human becoming infected, a member of *human_hosts*. Similarly, for an uninfected mosquito to become to a vector, the former must bite a person in *human_hosts*.

Quick Review Question 4

One of the simplifying assumptions is that, except where interactions between mosquitoes and humans occur, the number of organisms in a category (stock or box variable in Figure 6.5.2) exhibits unconstrained expansion or contraction. Give requested differential equations that utilize this assumption. Incorporate additional proportionality constants as needed.

a. *d*(*immune*)/*dt*
b. *d*(*deceased_humans*)/*dt*
c. rate of change from *human_hosts* to *uninfected_humans*
d. *d*(deceased_vectors)/*dt*

Formulating a Model: Determine Equations and Functions

Because we assume unconstrained growth or decay except where interactions between mosquitoes and humans occur, the following flow equations with constants of proportionalities in boldface correspond to proportionalities:

> *flow_to_immune* = **immunity_rate** * *human_hosts*
> *human_host_deaths* = **malaria_induced_death_rate** * *human_hosts*
> *recovered* = **recovery_rate** * *human_hosts*
> *mosquito_births* = **mosquito_birth_rate** * *mosquitoes*, where *mosquitoes* =
>   *uninfected_mosquitoes* + *vectors*
> *uninfected_mosquito_deaths* = **mosquito_death_rate** * *uninfected_mosquitoes*
> *vector_deaths* = **mosquito_death_rate** * *vectors*

Because only uninfected mosquitoes are born and both categories of mosquitoes reproduce, *mosquito_births* is proportional to the total number of mosquitoes. Moreover, we assume that the death rate is the same for uninfected mosquitoes and vectors, so the last two equations have the same constant of proportionality, *mosquito_death_rate*.

Quick Review Question 5

Suppose for a simulation that the change in time (Δt) from one time step to another is 0.1 days.

a. Give the unit of measure for *d*(*immune*)/*dt*.
b. If at day 6 *immunity_rate* is 0.2, *human_hosts* is 500, and *immune* is 400, using Euler's Method, estimate the number of immune people at day 6.1.

For uninfected humans, we have individuals entering from and exiting to the population of human hosts. The differential equation contains a positive term for a growth component with constant *recovery_rate* while subtracting a decay term, as follows:

$$d(uninfected\_humans)/dt = (recovery\_rate)(human\_hosts)$$
$$- (infection\_rate)(uninfected\_humans) \quad (1)$$

We can break *infection_rate* into two factors, the probability that a human is bitten by a mosquito (**prob_bit**) and the probability that a mosquito is a vector (**prob_vector**). The product of these two probabilities forms the infection rate. For example, if the probability that someone is bitten by a mosquito is 60% = 0.60 and the probability that a mosquito is a vector is $20\% = 0.20$, then the probability that a human is bitten by a vector is $(0.60)(0.20) = 0.12 = 12\%$. With no presumed immunity, the infection rate is equal to this probability. Thus, the following differential equation reflects a refinement of Equation 1:

$$d(uninfected\_humans)/dt = (recovery\_rate)(human\_hosts)$$
$$- (prob\_bit)(prob\_vector)(uninfected\_humans) \quad (2)$$

The probability of a vector is the quotient of the number of vectors (*vectors*) and the total number of mosquitoes (*mosquitoes*). Thus, we have the following equation:

$$prob\_vector = vectors/mosquitoes$$

Substituting into Equation 2, the rate of change of *uninfected_humans* is as follows:

$$d(uninfected\_humans)/dt = (recovery\_rate)(human\_hosts)$$
$$- (prob\_bit)(vectors)(uninfected\_humans)/mosquitoes$$

Similar to the situation for humans, the rate of change from uninfected mosquito to vector is the product of a rate and *uninfected_mosquitoes*. We again break the rate into two factors, the probability that the mosquito bites a human (***prob_bite_human***) and the probability that a human is a host (***prob_host***). Thus, the differential equation for the rate of change from uninfected to infected mosquito (*vector_formation*) is as follows:

$$d(vector\_formation)/dt = flow\_to\_host$$
$$= (prob\_bite\_human)(prob\_host)(uninfected\_mosquitoes) \quad (3)$$

The probability that a mosquito is a vector (*prob_vector*) and the probability that a human is a host (*prob_host*) are the connections between the models for humans and mosquitoes. For ***humans*** being the total number of humans (*uninfected_humans + human_hosts + immune*), we have the following identities:

$$prob\_host = human\_hosts/humans$$
$$= human\_hosts/(uninfected\_humans + human\_hosts + immune)$$

Solving the Model

We can use a system dynamic tool to help model the spread of malaria, perform a simulation, and generate graphs and tables of the results. Figure 6.5.3 pictures a human submodel of the malaria model. A converter/variable, *humans*, stores the sum of the quantities in the three stocks (box variables) for humans. Other

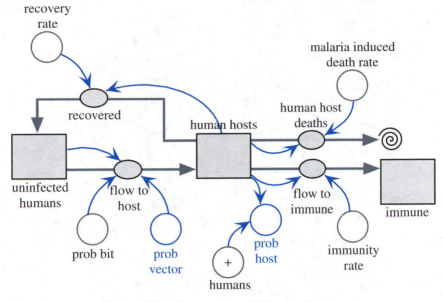

Figure 6.5.3   Human submodel for closed system

converters/variables store constants of proportionality and probabilities. For example, *immunity_rate* might store the constant 0.01 to indicate that the rate at which human hosts become immune from malaria is 1% a day. An initial value 1 for *human_hosts* would indicate that the human population has one human host at the start of the simulation.

Quick Review Question 6

Consider Figure 6.5.3.

    a. Give the number of terms in the differential equation for *d(human_hosts)/dt*.
    b. Give the number of these terms that contribute to an increase in *human_ hosts*.

    The mosquito submodel in Figure 6.5.4 also has four flows with associated converters/variables for constants of proportionality and probabilities. Similar to the human submodel, a converter/variable, *mosquitoes*, contains the sum of the populations for *uninfected_mosquitoes* and *vectors*.
    Converters/variables for the probability of a vector (*prob_vector*) and the probability of a host (*prob_host*) appear in both submodels and in color in Figures 6.5.3 and 6.5.4. We calculate *prob_vector* in the mosquito submodel and use it in the human submodel. Symmetrically, the calculation for *prob_host* is in the human submodel, while the mosquito submodel employs the result.

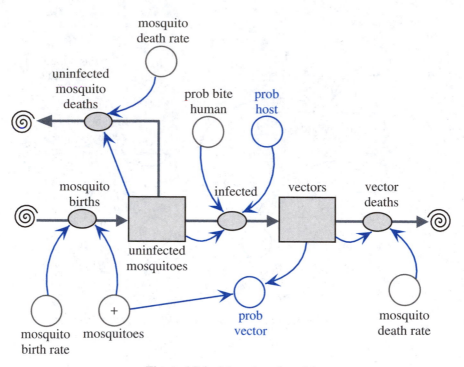

Figure 6.5.4   Mosquito submodel

Quick Review Question 7

Give the difference equation to estimate *vectors*(*t*) using Euler's Method.

We specify a simulation length of 200 days and $\Delta t = 0.0625$ days with Euler's method for the integration technique. Equation Set 6.5.1 shows parameters for one run of the simulation. We begin with equal numbers of uninfected mosquitoes and humans (300), no vectors or immune humans, and one human host. (However, we could change the units and consider, for example, the number of humans in the thousands and the number of mosquitoes in the millions.) From such a small incidence, we hope to observe the dramatic spread of the disease to become an epidemic. In this run of the simulation, for humans we make the rates of immunity, recovery to being susceptible once more, and malaria death be 1%, 30%, and 0.5% per day, respectively. We give the probability that a human is bitten by a mosquito or that a mosquito bites a human as 30%/day. For a constant number of mosquitoes, we make their birth and death rates equal, in this case 1%/day.

Equation Set 6.5.1

Parameters for one run of the simulation

> *uninfected_humans*(0) = 300
> *human_hosts*(0) = 1
> *immune*(0) = 0

$prob\_bit = 0.3$
$recovery\_rate = 0.3$
$immunity\_rate = 0.01$
$malaria\_induced\_death\_rate = 0.005$
$mosquito\_birth\_rate = 0.01$
$mosquito\_death\_rate = 0.01$
$vectors(0) = 0$
$uninfected\_mosquitoes(0) = 300$
$prob\_bite\_human = 0.3$

Verifying and Interpreting the Model's Solution

Figure 6.5.5 presents the graphs of the five stocks, and a table of values with a reporting interval of 1 day appears in Table 6.5.1. Over 80 days, we observe a dramatic drop from 300 to a minimum of about 24 uninfected mosquitoes and a corresponding rise from 0 to a maximum of about 276 vectors. Afterward, the number of uninfected mosquitoes begins to increase, while the number of vectors drops. With equal birth and death rates, the population of mosquitoes holds constant at 300, which helps to verify the solution.

Trailing the rapid increase in the number of vectors is a fast decrease in the number of uninfected humans and quick increase in the number of human hosts over about a 25-day period. The number of hosts reaches a maximum of about 119 around day 55. Then, the number of hosts gradually falls, while the number of uninfected humans continues declining but not as rapidly. Eventually, the two graphs appear almost parallel, while the graph of the number of immune humans increases in a concave-down fashion.

Extending the length of the simulation to 1500 days, we obtain the graphs of Figure 6.5.6. The numbers of uninfected humans and human hosts approach zero, and most of the surviving humans are immune. The total number of humans is reduced

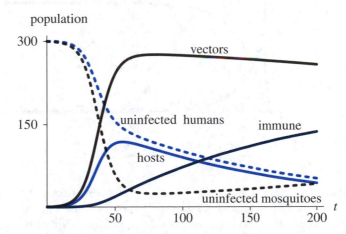

Figure 6.5.5 Graphs resulting from simulation with parameters of Equation Set 6.5.1 and time in days

Table 6.5.1
Table of Values Corresponding to Graphs of Figure 6.5.5

Time	Uninfected Humans	Human Hosts	Immune	Uninfected Mosquitoes	Vectors
0	300.00	1.00	0.00	300.00	0.00
10	299.34	1.53	0.09	297.52	2.48
20	292.08	8.20	0.48	286.58	13.42
30	261.42	35.96	2.41	239.30	60.70
40	199.40	88.69	8.61	136.33	163.67
50	156.09	116.17	19.16	59.11	240.89
60	136.89	117.66	30.97	32.17	267.83
70	125.57	111.74	42.46	25.29	274.71
80	116.63	104.46	53.28	24.15	275.85
90	108.70	97.25	63.36	24.65	275.35
100	101.44	90.44	72.75	25.65	274.35
110	94.72	84.08	81.47	26.86	273.14
120	88.48	78.16	89.58	28.19	271.81
130	82.67	72.65	97.12	29.62	270.38
140	77.28	67.53	104.13	31.13	268.87
150	72.27	62.77	110.64	32.73	267.27
160	67.61	58.35	116.69	34.41	265.59
170	63.27	54.25	122.32	36.19	263.81
180	59.24	50.43	127.56	38.06	261.94
190	55.49	46.88	132.42	40.03	259.97
200	52.00	43.58	136.94	42.10	257.90

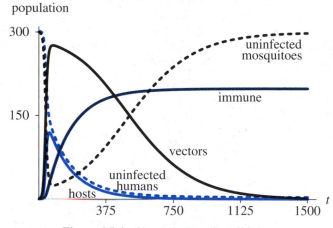

Figure 6.5.6   Simulation run for 1500 days

by about one-third to around 200. With the number of vectors tending to zero, the vast majority of the mosquitoes are uninfected. Because most humans are immune and almost no mosquito carries *Plasmodium*, malaria is virtually eradicated.

The results seem reasonable under the assumptions and simplifications. However, considerations suggest several stages of refinement.

We have extended the length of the simulation to over four years and presumed no births or deaths from causes other than malaria for humans. Also, we have not considered the incubation period for *Plasmodium*.

Another model could involve considering a different form of malaria in which an individual could have a relapse of the disease. Moreover, expectant mothers tend to have lower-birth-weight children and more miscarriages, which affects the birth and death rates for humans. Also, data show that children have a higher death rate from malaria than older humans. Consequently, other model refinements could involve expanding the number of human categories with varying death rates. Various projects consider such refinements.

Exercises

1. Discuss possible factors that could contribute to the situation in which 90% of malaria cases occur in Africa, south of the Sahara.
2. Give a differential equation for the rate of change of *human_hosts*.
3. Give a differential equation for the rate of change of *uninfected_mosquitoes*.
4. Give a differential equation for the rate of change of *vectors*.

Projects

For additional projects, see Module 7.8 on "Fueling Our Cells—Carbohydrate Metabolism."

For all model development, use an appropriate system dynamics tool.

1. Run the simulation for the following situations. Describe and explain the long-term results.
   a. Various initial values of stocks (box variables)
   b. Slightly higher birth rate than death rate for mosquitoes
   c. No human host and one vector
   d. Zero death rate for humans
   e. Probability that a human is bitten reduced by a factor of 10 to 3%
   f. Probability that a mosquito bites a human reduced by a factor of 10 to 3%
2. Refine the malaria model of this module to accommodate human births and deaths.
3. Refine the malaria model of Project 2 of this module to include a stock of tainted mosquitoes, who are infected but not yet vectors. In these mosquitoes, the *Plasmodium* protozoans are in incubation.
4. Develop an alternative implementation of the model for Project 3 that employs a conveyor for the tainted mosquitoes. Because we are considering births and deaths for mosquitoes, the inflow multiplier from the conveyor to the stock of vectors is not 1. However, the value is not $1 - mosquito\_death\_rate$, which is 0.99 for a death rate of 0.01, because the inflow multiplier applies only to the number exiting the conveyor at that time step, not over the period of incubation. Consequently, you must employ the actual, accumulated survival rate over the period of incubation as a multiplier. You can

use mathematics or your system dynamics tool to compute the actual survival rate for the number exiting the conveyor.

Because we only remove mosquitoes from the conveyor at the end of the incubation period, using the number in the conveyor to calculate the total number of mosquitoes, *mosquitoes*, results in an overestimate for *mosquitoes*. Thus, for this project, assume a constant number of mosquitoes, so *mosquitoes* is the sum of the initial values of the various mosquito stocks.

5. Develop an alternative implementation of the model for Project 3 that employs a separate stock (block variable) for each day of incubation. A portion (*mosquito_death_rate*) of the mosquitoes is siphoned off each day, and the remainder is transferred to the next day's stock or, after the final day of incubation, to the stock *vector*. With this solution, we do not need to assume a constant number of mosquitoes.

6. Model malaria caused by *Plasmodium vivax* or *ovale* in which a human host can go into remission and have relapses.

7. Refine one of the previous models to reflect a seasonal increase in the number of mosquitoes, such as in a rainy season.

8. Using one of the previous models, consider the effect on the epidemic of distribution of a prophylactic drug, such as Malazone, which travelers take to prevent malaria. Suppose everyone in the population takes the drug. Investigate varying degrees of effectiveness. Such drugs are expensive, especially relative to the economy of populations in which malaria thrives. Discuss the practicality of such treatments.

9. Using one of the previous models, consider the effect on the epidemic of using insecticides to control the mosquito population. Investigate varying degrees of insecticide effectiveness. Discuss the practicality of such an approach relative to the ecosystem.

10. Starting with the model from Project 8, consider the effect on the epidemic of a combined approach consisting of distribution of a prophylactic drug, mosquito netting, and use of insect repellant and insecticides.

11. Refine a malaria model of Projects 2–5 to consider that a person does not obtain permanent, complete immunity from malaria, but only temporary, partial immunity.

12. Refine a malaria model by expanding the number of human categories with varying contraction and death rates. In particular, data shows that *falciparum* is lethal to children under five years old. Each day, approximately 3000 children under the age of five die from malaria. In some of the worst areas, it is estimated that more than 40% of the toddlers die from the disease. More than 30% of the children in Africa get malaria by time they are three months old. However, approximately, one-eighth of the children in some countries of sub-Saharan Africa are born with sickle cell anemia, which makes it more difficult for them to contract malaria.

13. Because mothers are more likely to suffer malarial relapses during pregnancy, malaria is an important cause of low-weight births and stillbirths. More than half of the miscarriages in endemic areas are caused by malaria. Adjust the model for Project 12 to reflect this information.

14. Adjust one of the earlier models to have constrained growth with a carrying capacity for humans and a carrying capacity for mosquitoes. Examine and

discuss the effects of these changes. A logistic equation can model such constrained growth (see Module 3.3 on "Constrained Growth").

15. Write a simulation in a programming language for the model of this section.

16. Write a simulation in a programming language for one of the models in Projects 1–5.

Answers to Quick Review Questions

1. humans and mosquitoes

2. uninfected humans, human hosts, immune humans, uninfected mosquitoes, mosquito vectors

3. a. *human_hosts*
  b. *human_hosts*
  c. *uninfected_humans*, *immune*, and undesignated "cloud" for deceased humans
  d. *human_hosts*
  e. *human_hosts*
  f. *vectors* and undesignated "cloud" for dead mosquitoes
  g. undesignated "cloud" for unborn mosquitoes
  h. *uninfected_mosquitoes*
  i. undesignated "cloud" for dead mosquitoes
  j. *uninfected_mosquitoes* and *vectors*

4. a. $d(immune)/dt = immunity\_rate * human\_hosts$
  b. $d(deceased\_humans)/dt = malaria\_induced\_death\_rate * human\_hosts$
  c. rate of change from *human_hosts* to *uninfected_humans* = $recovery\_rate * human\_hosts$
  d. $d(deceased\_vectors)/dt = mosquito\_death\_rate * vectors$

5. a. people per day
  b. 410 because $immune(6.1) = immune(6) + immunity\_rate * human\_hosts(6) * \Delta t = 400 + (0.2)(500)(0.1) = 400 + 10 = 410$

6. a. 4, one for each flow entering or leaving the stock *human_hosts*.
  b. 1, because only one flow enters the stock *human_hosts*.

7. $vectors(t) = vectors(t - \Delta t) + (infected - vector\_deaths) * \Delta t$

References

Bradley, T. "Biology of Plasmodium Parasites and *Anopheles* Mosquitoes." Department of Microbiology and Immunology, University of Leicester. http://www.micro.msb.le.ac.uk/224/Bradley/Biology.html

Carroll, Rory. 2001. "Skeleton Find Links Malaria to Fall of Rome." *The Guardian*, February 21. http://news.nationalgeographic.com/news/2001/02/0221_malariarome.html

Coutsoukis, Photius. 1999. "Malawi People." http://www.photius.com/wfb1999/malawi/malawi_people.html

McConnell, Bill. 2000. "History of Malaria." *RPH Laboratory Medicine 1998–2000.* http://www.rph.wa.gov.au/labs/haem/malaria/history.html

NIAID (National Institute of Allergy and Infectious Diseases). "Life Cycle." National Institutes of Health. http://www.niaid.nih.gov/publications/malaria/life.htm

NetMark. "Key Issues—Malaria."NetMark Plus. http://www.netmarkafrica.org/keyissues/factsheet_problem.html

Southern Africa Malaria Control. "Country Profiles: Malawi." http://www.malaria.org.zw/countries/malawi.htm

———. "Southern Africa Malaria Control: Southern Africa Frontline for Taking Aim & Rolling Back Malaria in Southern Africa." http://www.malaria.org.zw

Wiser, Mark F. "Malaria." http://www.tulane.edu/~wiser/protozoology/notes/malaria.html#relapse (accessed March 12, 2003).

WHO (World Health Organization). The United Nations. http://www.who.int

7

ADDITIONAL DYNAMIC SYSTEMS PROJECTS

Overview

In the previous chapters, we have studied techniques, issues, and applications of computational science system dynamics models. Projects often extended the examples discussed and developed in the modules. We cannot overemphasize the importance of developing solutions to such projects for it is in doing modeling that we learn computational science problem-solving abilities.

This chapter provides opportunities to further this learning through additional extensive projects. Unlike earlier chapters, these modules do not include examples. Instead, each module contains sufficient background in a scientific application area for you to complete the projects. Chapter 7's modules list the prerequisite modules. Moreover, the project sections of those previous modules cross-reference the material in Chapter 7. Thus, students can work with projects in the current chapter as soon as they have covered the appropriate prerequisites or at a later time.

As with earlier projects, the projects in this chapter are well suited for teamwork. In computational science, most research and development are done using interdisciplinary teams. Thus, experiences developing models with teams, perhaps on applications out of an area of major study, are important for a student studying computational science.

Chapter 7's applications with projects involving system dynamics models are in a variety of scientific areas, including the following: radioactive chains, blood cell populations, scuba diving, the carbon cycle, global warming, the cardiovascular system, electrical circuits, carbohydrate metabolism, mercury pollution, and the economics of commercial fishing.

MODULE 7.1

Radioactive Chains—Never the Same Again

Prerequisite: Module 3.2 on "Unconstrained Growth and Decay."

Introduction

The mass $Q(t)$ of a radioactive substance decays at a rate proportional to the mass of the substance (see Section "Unconstrained Decay" in Module 3.2 on "Unconstrained Growth and Decay"). Thus, for positive **disintegration constant** or **decay constant** r, we have the following differential equation:

$$dQ/dt = -rQ(t)$$

and its difference equation counterpart:

$$\Delta Q = -r\, Q(t - \Delta t)\, \Delta t$$

In this module, we model the situation where one radioactive substance decays into another radioactive substance, forming a chain of such substances. For example, radioactive bismuth-210 decays to radioactive polonium-210, which in turn decays to lead-206. We consider the amounts of each substance as time progresses.

Modeling the Radioactive Chain

If a radioactive substance, *substanceA*, decays into substance *substanceB*, we say that *substanceA* is the **parent** of *substanceB* and that *substanceB* is the **child** of *substanceA*. If *substanceB* is also radioactive, *substanceB* is the parent of another substance, *substanceC*, and we have a **chain** of substances. Figure 7.1.1 depicts the situation where A, B, and C are the masses of radioactive substances, *substanceA*,

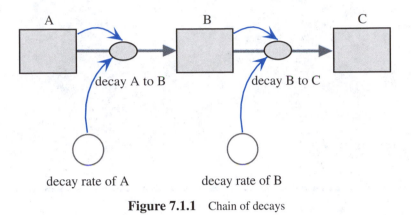

Figure 7.1.1   Chain of decays

substanceB, and *substanceC*, respectively; and different disintegration constants, *decay_rate_of_A* (*a*) and *decay_rate_of_B* (*b*), exist for each decay.

Quick Review Question 1

Suppose *A* and *B* are the masses of *substanceA* and *substanceB*, respectively, at time *t*; Δ*A* and Δ*B* are the changes in these masses; and *a* and *b* are the positive disintegration constants.

  a. Using these constants and variables along with arithmetic operators, such as minus and plus, give the difference equation for the change in the mass of *substanceA*, Δ*A*.
  b. Through disintegration of *substanceA*, *substanceB*'s mass increases, while some of *substanceB* decays to *substanceC*. Give the difference equation for the change in the mass of *substanceB*, Δ*B*.
  c. In Figure 7.1.1, where *A*, *B*, and *C* are the masses of three radioactive substances, give the formula as it appears in a systems dynamics tool's equation for the flow *decays_A_to_B*.
  d. Give the formula as it appears in a systems dynamics tool's equation for the flow *decays_B_to_C*.

The mass of *substanceA* that decays to *substanceB* is *aA*. Thus, in Figure 7.1.1, the flow *decay_A_to_B* contains the formula *decay_rate_of_A* ∗ *A*. What *substanceA* loses, *substanceB* gains. However, *substanceB* decays to *substanceC* at a rate proportional to the mass of *substanceB*, *bB*. Consequently, in Figure 7.1.1, the flow *decay_B_to_C* contains the mass that flows from one stock to another, *decay_rate_of_B* ∗ *B*. The total change in the mass of *substanceB* consists of the gain from *substanceA* minus the loss to *substanceC*:

$$\Delta B = (aA - bB)\Delta t$$

We consider the initial amounts of *substanceB* and *substanceC* to be zero.

Projects

1. a. With a system dynamics tool or a computer program, develop a model for a radioactive chain of three elements, from *substanceA* to *substanceB* to *substanceC*. Allow the user to designate constants. Generate a graph and a table for the amounts of *substanceA*, *substanceB*, and *substanceC* versus time. Answer the following questions using this model.

b. Explain the shapes of the graphs.

c. As *a* increases from 0.1 to 1, describe how the time of the maximum total radioactivity changes. The total radioactivity is the sum of the change from *substanceA* to *substanceB* and the change from *substanceB* to *substanceC*, or the total number of disintegrations. Why?

d. Observe in several cases where $a < b$ that eventually we have the following approximation:

$$\frac{B}{A} \approx \frac{a}{b - a}$$

With the ratio of the mass of *substanceB* (B) to the mass of *substanceA* (A) being almost constant, we say the system is in **transient equilibrium**. Eventually, *substanceA* and *substanceB* appear to decay at the same rate. Using the following material, verify this approximation:

Find the exact solution to the differential equation for the rate of change of A with respect to time, $dA/dt = -aA$ (see Section "Analytic Solution" in Module 3.2 on "Unconstrained Growth and Decay").

Verify that $B = \frac{aA_0}{b - a}(e^{-at} - e^{-bt})$, where A_0 is the initial mass of *substanceA*, is a solution to the differential equation for the rate of change of B with respect to time (see the difference equation for ΔB above). What number does e^{-rt} approach as t goes to infinity? For $a < b$, which is smaller, e^{-at} or e^{-bt}? Thus, for large t, B is approximately equal to what?

e. Using your model from Part a, observe in several cases where $a > b$ that the ratio of the mass of *substanceB* to the mass of *substanceA* does not approach a number. Thus, transient equilibrium (see Part d) does not occur in this case.

f. Verify the observation from Part e analytically using work similar to that in Part d.

g. If a is much smaller than b, we have $A \approx A_0$ and $B \approx \frac{aA_0}{b - a}$. With the two amounts being almost constant, we have a situation called **secular equilibrium**. Observe this phenomenon for the radioactive chain from radium-226 to radon-222 to polonium-218: $Ra^{226} \rightarrow Rn^{222} \rightarrow Po^{218}$, where the decay rate of Ra^{226}, a, is 0.00000117/day and the decay rate of Rn^{222}, b, is 0.181/day. Using your work from Part a, run the simulation for at least one year.

h. Show analytically that the approximations from Part g hold.

i. In the radioactive chain $Bi^{210} \rightarrow Po^{210} \rightarrow Pb^{206}$ (bismuth-210 to polonium-210 to lead-206), the decay rate of Bi^{210}, a, is 0.0137/day and the decay rate of Po^{210}, b, is 0.0051/day. Assuming the initial mass of Bi^{210} is 10^{-8} grams and using your model from Part a, find approximately the maximum mass of Po^{210} and when the maximum occurs.

j. In Part d, we verified that $B = \dfrac{aA_0}{b-a}(e^{-at} - e^{-bt})$. Using this result, find analytically the maximum of mass of *substanceB* and when this maximum occurs.

k. Check your approximations of Part i using your solution to Part j.

l. For the chain in Part g, find when the largest mass of Rn^{222} occurs using your solution to Part j.

m. For the chain in Part g, using your simulation of Part a, approximate the time when the largest mass of Rn^{222} occurs. How does your approximation compare with the analytical solution of Part l?

2. Develop a model for a chain of four elements. Perform simulations, observations, and analyses similar to those above. Discuss your results

Answers to Quick Review Question

1. **a.** $\Delta A = -aA\Delta t$
   b. $\Delta B = (aA - bB)\Delta t$
   c. $decay\_A\_to\_B = decay\_rate\_of\_A * A$
   d. $decay\_B\_to\_C = decay\_rate\_of\_B * B$

Reference

Horelick, Brindell, and Sinan Koont. 1979 and 1989. "Radioactive Chains: Parents and Children." *UMAP Module 234*. COMAP, Inc.

MODULE 7.2

Turnover and Turmoil—Blood Cell Populations

Prerequisite: Module 3.2 on "Unconstrained Growth and Decay."

Introduction

In a healthy individual, the count of blood cells is usually constant. However, for certain diseases, blood cell counts may oscillate, perhaps in an involved or chaotic manner. Such disorders are in a category of **dynamical diseases**, which includes HIV, forms of leukemia, and anemia. We can use modeling to study the origins, behaviors, and treatments of these diseases.

Formation and Destruction of Blood Cells

Blood is composed of fluid, called **plasma**, and **blood cells**. The following are major types of blood cells:

- **Red blood cells**, which are for oxygen transport from the lungs to tissues
- **White blood cells**, which are part of the body's defense mechanism against infections
- **Platelets**, which help the blood to clot

In a healthy individual, if a deficit for a particular type of blood cell occurs, physiological mechanisms in the body cause an increase in production of that type of cell. Similarly, if an oversupply exists, the production rate decreases. Thus, the production of new blood cells of a particular type depends on the number of blood cells of that type.

Aging, disease, or infection causes the eventual death of any cell. As with production, the number of blood cells destroyed also depends on the number of blood cells of that type.

Basic Model

Suppose x is the **number of blood cells** of a particular type, and x_i is the **number of such blood cells at time** t_i. As indicated above, the **number of such blood cells produced** (p), or the **production rate**, and the **number of such blood cells destroyed** (d), or the **destruction rate**, are functions of the number in existence, x. Thus, $p(x_i)$ and $d(x_i)$ are the production and destruction rates, respectively, for the given number of cells at time t_i.

Quick Review Question 1

Suppose the number of blood cells of a particular type at time t_i is below the normal range. For each part, give the relationship that is desirable.

a. $x_{i+1} > x_i$ $x_{i+1} < x_i$ $x_{i+1} = x_i$
b. $p(x_{i+1}) > p(x_i)$ $p(x_{i+1}) < p(x_i)$ $p(x_{i+1}) = p(x_i)$
c. $d(x_{i+1}) > d(x_i)$ $d(x_{i+1}) < d(x_i)$ $d(x_{i+1}) = d(x_i)$

For a healthy mammal, one widely accepted model for d is that the number of blood cells destroyed is directly proportional to the number of blood cells existing. Thus, for a constant of proportionality, c, called the **destruction coefficient**, we have the following model for the number of blood cells destroyed:

$$d(x_i) = cx_i \tag{1}$$

Quick Review Question 2

Indicate which of the following are true about the destruction function d and the destruction coefficient c: $c > 0$; $c < 0$; $c = 0$; $c \geq 1$; $c \leq 1$; $d(0) = 0$; the graph of d is increasing; the graph of d is decreasing; the graph of d is a line; the graph of d cannot be a line.

A model for the number of blood cells of a particular type produced is more complicated. As noted above, the production is a function of the number of blood cells of that type. Certainly, if there were no blood cells ($x = 0$), we would expect no production because the animal would be dead. As x increases, p increases rapidly to some maximum and then production decreases, tailing off to zero. One possible such graph of production versus population for a particular type of blood cell appears in Figure 7.2.1.

For a healthy person, we expect no change in the number of blood cells from one time step to another, so that $x_{i+1} = x_i, = v$, a constant called the **steady-state level**. In this case, $x_{i+1} - x_i = 0$. Thus, the production rate $p(v)$ and the destruction rate $d(v)$ are equal at v, or $p(v) = d(v)$.

Model Parameters

As noted by Gearhart and Martelli (1990), it takes about 6 days for red blood cells to reach maturity, while in a healthy person about 2.3% of the cells are destroyed per

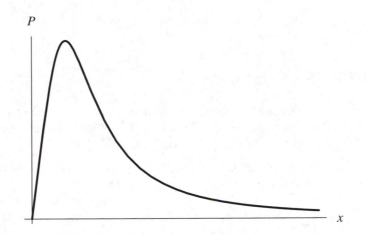

Figure 7.2.1   General graph of blood cell production (p) versus blood cell population (x)

day. Thus, for a change in time of 6 days, an approximation for the destruction coefficient is $c = (6)(0.023) = 0.14$; and the destruction function, or the number of blood cells destroyed as a function of the number of blood cells at time t_i, is as follows:

$$d(x_i) = 0.14x_i \tag{2}$$

Lasota in 1977 used the following production function, which has the shape of Figure 7.2.1 (Lasota 1977):

$$p(x) = bx^s e^{-sx/r} \tag{3}$$

where p is the number of blood cells produced as a function of the number of blood cells x and b, r, and s are positive constants. To determine these constants, use the following data (Gearhart and Martelli):

- Normal red-cell count $\approx 3.3 \times 10^{11}$ cells/kg
- Maximum red-cell production ≈ 10 times the steady-state production rate
- For a red-cell population of 75% of the steady-state level (count), red-cell production ≈ 5 times steady-state level

Because the normal red-cell count is about 3.3×10^{11} cells/kg, we can use this figure as the steady-state level or

$$v = 3.3 \times 10^{11} \text{ cells/kg}$$

At v, production equals destruction or $p(v) = d(v)$. From Equation 2

$$d(v) = 0.14v$$

so we know the following:

$$p(v) = 0.14v$$
$$p(3.3 \times 10^{11}) = 0.14 \times 3.3 \times 10^{11} = 4.62 \times 10^{10} \text{ cells/kg}$$

At $v = 3.3 \times 10^{11}$ cells/kg, production and destruction equal 4.62×10^{10} cells/kg, which is the steady-state production rate.

Quick Review Question 3

Using this information, indicate which of the following points are on the graph of the production function p: $(3.3 \times 10^{11}, 4.62 \times 10^{10})$, $(4.62 \times 10^{10}, 3.3 \times 10^{11})$, $(3.3 \times 10^{11}, 0.14)$, $(0.14, 3.3 \times 10^{11})$, $(v, 0.14v)$, $(0.14v, v)$.

According to the second bullet above, the maximum value of p is ten times the steady-state rate, or $10 \times 4.62 \times 10^{10}$ cells/kg $= 4.62 \times 10^{11}$ cells/kg. To determine where this maximum occurs, we take the first derivative of Equation 3 and set it equal to 0, solving for x (see Project 1). Keeping in mind that b, r, and s are constants, we find that the maximum occurs at $x = r$; and the maximum is as follows:

$$p(r) = br^s e^{-s} = 4.62 \times 10^{11}$$

Quick Review Question 4

Interpret the third bullet above as one or more equations:

A. $p(5v) = 0.75v$ **B.** $p(0.75v) = 5v$
C. $p(5(0.14v)) = 0.75v$ **D.** $p(0.75 \times 3.3 \times 10^{11}) = 5 \times 4.62 \times 10^{10}$
E. $p(0.75v) = 5(0.14v)$ **F.** $p(0.75 \times 3.3 \times 10^{11}) = 5 \times 3.3 \times 10^{11}$
G. $p(5 \times 4.62 \times 10^{10}) = 0.75 \times 3.3 \times 10^{11}$

Projects

For all model development, use an appropriate system dynamics tool.

1. Develop a model for a type of white blood cell (granulocyte) population using the production function by Lasota, Equation 3. Account for the (approximate) six-day maturation of cells, perhaps using a conveyor; and determine reasonable constants by referring to the "Model Parameters" section. Graph blood cells per kg, blood cells produced per kg, and blood cells destroyed per kg versus time. Discuss your results
   Find analytically where the maximum occurs and the maximum for the production function in Equation 3, and verify that your model approximately agrees with this value.
   For $r = 3$, $b = 50$, $s = 5$, and $c = 0.5$, what is the period?
2. Develop a model for a type of white blood cell (granulocyte) population using the production function by Lasota, Equation 3. Determine reasonable constants by referring to the "Model Parameters" section. For the production function p, use graphical input where $p(0) = 0$, p increases initially to a

maximum, and then decreases to 0. Graph blood cells per kg, blood cells produced per kg, and blood cells destroyed per kg versus time.

3. Develop a model for granulocytes, a type of white blood cell. Use the following production function by Mackey and Glass, where b, a, and m are positive constants (Gearhart and Martelli):

$$p(x) = \frac{ba^m x}{a^m + x^m} \tag{4}$$

The units of a are cells/kg, while b and m are unitless. Determine reasonable constants by referring to the "Model Parameters" section. Graph blood cells per kg, blood cells produced per kg, and blood cells destroyed per kg versus time.

4. Complete Project 3. Show that we can use this work to model chronic myelogenous leukemia (CML), a cancer resulting in an overproduction of the white blood cells. In CML, the white-cell count may be 150 times the normal, and counts can oscillate around the elevated level with a period of 30 to 70 days. As Gearhart and Martelli (1990) indicate, the following are parameters for a normal person:

$$c = \mu \partial \quad \text{with } \mu = 0.16/\text{day}$$

$$b = \beta \partial \quad \text{with } \beta = 1.43/\text{day}$$

$$a = 3.22 \times 10^8 \text{ cells/kg}$$

and the delay in production, ∂, is 0.68 days.

Show that increasing a can result in a gain in white blood cell count, and increasing ∂ can cause the indicated periodicity. Find appropriate values for a and ∂ that match the abnormal variations in white blood cell counts of CML.

Answers to Quick Review Questions

1. **a.** $x_{i+1} > x_i$
   b. $p(x_{i+1}) > p(x_i)$
   c. $d(x_{i+1}) < d(x_i)$
2. $c > 0$ because as the number of cells increase, the number destroyed increases. $d(0) = 0$ because if there are no cells, none can be destroyed.
   The graph of d is increasing because c and x are positive, and as the number of cells increases, the number destroyed increases.
   The graph of d is a line.
3. $(v, 0.14v)$ and $(3.3 \times 10^{11}, 4.62 \times 10^{10})$ because $p(v) = 0.14v = p(3.3 \times 10^{11})$
   $= 4.62 \times 10^{10}$ cells/kg
4. Equations B and F, $p(0.75v) = 5v$ and $p(0.75 \times 3.3 \times 10^{11}) = 5 \times 3.3 \times 10^{11}$, because $v = 3.3 \times 10^{11}$ is the steady-state level

References

Gearhart, William B., and Mario Martelli. 1990. "A Blood Cell Population Model, Dynamical Diseases, and Chaos." *UMAP Module 709*. COMAP, Inc.

Lasota, A. 1977. "Ergodic problems in biology." Astérisque (Societe Mathematique de France) 50: 239–250.

Mackey, M. C., and L. Glass. 1977. "Oscillation and chaos in physiological control systems." *Science* 197: 287–289.

MODULE 7.3

Deep Trouble—Ideal Gas Laws and Scuba Diving

Prerequisite: Module 3.2 on "Unconstrained Growth and Decay."

Pressure

Scuba divers are often under great amounts of pressure, and therefore should be very concerned with it. **Pressure**, which is the weight of matter per unit area, increases rapidly with increasing depths. For divers, the total pressure is a combination of the weight of air and water per square centimeter. **Air pressure at sea level** is about **10.1 N/cm²**, meaning that a square centimeter column of air as tall as the atmosphere (about 80 km or 50 miles) weighs about 10.1 newtons (**10.1 N/cm² = 760 millimeters of mercury (mm Hg) = 760 torr = 14.7 pounds/in²**). The atmospheric pressure at sea level is by definition equal to **one atmosphere (atm)**. Water pressure derives from the weight of water, which is considerably greater than air. As indicated in Module 4.1 on "Modeling Falling and Skydiving," the density of water at 3.98°C is 1.00000 g/cm³. Thus, at this temperature, a column of water 10 m high with a base of area 1 cm² weighs 9.81 N, as the following calculations show:

$$\text{weight} = F = ma = \left(\frac{1.00000\,\text{g}}{\text{cm}^3}\right)(10\,\text{m})(1\,\text{cm}^2)\left(\frac{9.81\,\text{m}}{\text{sec}^2}\right)$$

$$= \left(\frac{1.00000\,\text{g}}{\text{cm}^3}\right)\left(\frac{\text{kg}}{1000\,\text{g}}\right)(10\,\text{m})\left(\frac{100\,\text{cm}}{\text{m}}\right)(1\,\text{cm}^2)\left(\frac{9.81\,\text{m}}{\text{sec}^2}\right)$$

$$= (1.00000\,\text{kg})(9.81\,\text{m/sec}^2) = 9.81\,\text{N}$$

This weight of a 10 m by square centimeter column of water approximately matches the weight of the entire column of air above it. Hence, a diver at a depth of about **10 m (33 ft)** experiences **2 atmospheres of pressure**, resulting from the pressure of 1 atmosphere of air plus the equivalent of approximately an additional atmosphere of pressure from the water. Each 10 m of depth adds approximately one atmosphere of pressure to the diver.

Definitions   **Pressure** is the weight of matter per unit area. **One atmosphere (atm)** is the atmospheric pressure at sea level.

Quick Review Question 1

Determine the water pressure at 15 m in terms of each of the following units:

    a. atm
    b. N/cm^2
    c. torr

Ideal Gas

In this module, we employ several **ideal gas laws**, which describe the behaviors of an ideal gas, in models related to scuba diving. An ideal gas is one in which the volume of its atoms is insignificant in comparison to the total volume of the gas and in which atom interactions are negligible except for the energy and momentum exchanged during collisions. Remarkably, under most circumstances, the ideal gas laws model well the behaviors of real gases because of the great distances between the atoms and molecules of a gas.

Definitions   The **ideal gas laws** describe the behaviors of an ideal gas. An **ideal gas** is one in which the volume of its atoms is insignificant in comparison to the total volume of the gas and in which atom interactions are negligible except for the energy and momentum exchanged during collisions.

Dalton's Law

Scuba equipment provides air to divers at the higher pressures of deeper waters. Air normally contains 21% oxygen, 78% nitrogen, and 1% various inert gases. We must consider nitrogen and the inert gases in calculating the speeds and rest schedules for a scuba diver. Thus, scuba computations frequently group nitrogen and the inert gases together and **assume N_2 is 79% of the air**. Because the partial pressure of a gas is determined by its fractional portion in a mixture, at sea level 1 atmosphere of pressure is composed of 0.21 atmospheres of oxygen and 0.79 atmospheres of nitrogen (assuming 79% nitrogen). This relationship follows one of the ideal gas laws as proposed by **Dalton**. Although air for divers is compressed, the percentage of the gas components is the same. So, as pressure on a diver increases during a dive, the partial pressure exerted by each gas in the diver's body and tank increases. For example, a diver reaching 20 m (66 ft) experiences three atmospheres of pressure, with partial pressures of O_2 and N_2 equaling 0.63 and 2.37 atmospheres, respectively.

Dalton's Law states that the partial pressure of a gas (P_g) is the product of the fraction of the gas in the mixture (F_g) and the total pressure (P) of all gases, excluding water vapor:

$$P_g = F_g P$$

Quick Review Question 2

Determine the partial pressure in atmospheres (atm) of nitrogen in a mixture of air at 15 m.

Boyle's Law

Probably the most important gas law to divers is that of Robert Boyle, who discovered that at a particular temperature, the volume of a gas is inversely proportional to pressure. Hence, the product of pressure (P) and volume (V) yields a constant (K):

$$PV = K$$

When pressure increases at constant temperature, gas volume decreases, and vice versa. For example, assuming constant temperature, **Boyle's Law** means that if we take an air-filled balloon, which is 3 m^3 in volume at the surface of the ocean to a depth of 20 m, the balloon would shrink to a volume of 1 m^3. We can obtain this result by using the pressure at the surface $P_1 = 1$ atm with volume $V_1 = 3$ m^3. At 20 m, pressure is $P_2 = 3$ atm. Thus, assuming constant temperature, by Boyle's Law we have the following relationship:

$$P_1 V_1 = P_2 V_2$$

Substituting the values, we have $(1)(3) = 3V_2$ or $V_2 = 1$ m^3.

Boyle's Law for gas at a particular temperature is as follows:

$$PV = K$$

where P is pressure, V is volume, and K is a constant.

Quick Review Question 3

Suppose an air-filled balloon is 3 m³ in volume at the ocean's surface. Assuming constant temperature, determine its size to three decimal places at a depth of 32 m.

Human skin divers hold their breath when they dive. During descent to, say, 10 m, the air-filled lungs are reduced to one-half their surface volume. As they ascend, the lungs again expand to their normal volume. Of course, there are limits of depth for diving without the aid of scuba equipment.

Scuba divers breathe air from tanks through regulators that deliver the air at **ambient pressure**, the pressure of the surrounding water pressure. As divers at 20 m (3 atm pressure) inhale, they take in the equivalent of 3 breaths of air from the surface (1 atm pressure). Accordingly, it is important for divers to determine their **surface air consumption (SAC) rate**, so that they can calculate how long their air tanks will last at the depths they are diving.

Another important consideration that comes from breathing air at these pressures is that as divers rise from the deep, the gases in their lungs obey Boyle's Law and expand with the decreasing pressure. Rapid and extensive expansion of the equivalent of three times the normal lung volume could cause the lungs to burst. Consequently, scuba divers are always cautioned never to hold their breath, but return to the surface slowly, exhaling.

Charles' Law

Charles' Law takes into consideration pressure (P), volume (V), and temperature (T). The relationship is as follows:

$$PV = nRT$$

where T is temperature in Kelvin, n is the number of moles, and the constant $R = 0.0821$ atm/(mol K). The conversion from a Celsius temperature (T_C) to a Kelvin temperature (T_K) is the following:

$$T_K = T_C + 273.15$$

The number of moles, n, in a mass m of molecular weight M is the following:

$$n = m/M$$

The molecular mass of dry air is 29.0 g/mol, of nitrogen is 28.0 g/mol, and of oxygen is 32.0 g/mol. However, frequently in scuba diving examples, the form of the law we employ relates the pressures, volumes, and temperatures at two depths, as follows:

$$\frac{P_1 V_1}{T_1} = \frac{P_2 V_2}{T_2}$$

Charles' Law states that

$$PV = nRT$$

where P is pressure, V is volume, T is temperature in Kelvin, n is the number of moles, and the constant $R = 0.0821$ atm/(mol K). Thus, if an object has volume V_1 at pressure P_1 and Kelvin temperature T_1, then at pressure P_2 and Kelvin temperature T_2 its volume is V_2 according to the following:

$$\frac{P_1 V_1}{T_1} = \frac{P_2 V_2}{T_2}$$

The conversion from a Celsius temperature (T_C) to a **Kelvin temperature** (T_K) is the following:

$$T_K = T_C + 273.15$$

The **number of moles**, n, in a mass m of molecular weight M is the following:

$$n = m/M$$

Quick Review Question 4

Suppose a balloon is 1 m³ in 30°C degree weather on the ocean's surface. Determine the balloon's volume at 15 m in 10°C degree seawater.

Henry's Law

Divers who dive deeply for long periods of time also face another problem that follows from **Henry's Law**—the amount of any gas in a liquid at a particular temperature is a function of the partial pressure of the gas and its solubility coefficient in that liquid. Thus, divers must be concerned with the amount of nitrogen gas in their blood. For V_g being gas volume, V_L being liquid volume, s being the solubility coefficient for the gas in that liquid, and P_g being the pressure of gas, Henry's Law is as follows:

$$V_g/V_L = sP_g$$

The **solubility coefficient for nitrogen in blood** is **0.012**, and the total **volume of blood** in an adult's body is approximately **5 liters**. With the greater pressure, blood can absorb a greater volume of nitrogen, which the body cannot use.

Henry's Law for the amount of any gas in a liquid at a particular temperature is as follows:

$$V_g/V_L = sP_g$$

where V_g is gas volume, V_L is liquid volume, s is the solubility coefficient for the gas in that liquid, and P_g is the pressure of gas.

Quick Review Question 5

Determine the volume of nitrogen that can go into solution in the blood at the following pressures:

    a. 1 atm
    b. 2 atm

Rate of Absorption

The rate at which tissue (a compartment) takes up an inert gas, such as nitrogen, is proportional to the difference in the partial pressures of the gas in the lungs and the gas in the tissues:

$$dP_{tissue}/dt = k(P_{lungs} - P_{tissue})$$

where P_{lungs} is the partial pressure of the gas in the lungs, P_{tissue} is the partial pressure of the gas in the tissue, and k depends on the type of tissue. (This formula is similar to Newton's Law of Heating and Cooling in Exercise 4 of Module 3.2, "Unconstrained Growth and Decay.") We can show that

$$k = \ln 2/t_{half}$$

where t_{half} is the **half time**, or the time for the tissue to absorb or release half of the partial difference of the gas. Thus, if it takes 20 minutes for such absorption, then we have the following calculation of the constant k:

$$k = \ln2/t_{half} = \ln2/20 = 0.0346574$$

The **rate of absorption** at which tissue (a compartment) takes up an inert gas is as follows:

$$dP_{tissue}/dt = k(P_{lungs} - P_{tissue})$$

where P_{lungs} is the partial pressure of the gas in the lungs, P_{tissue} is the partial pressure of the gas in the tissue, and k is a proportionality constant, which

(continued)

depends on the type of tissue. The value of k is as follows:

$$k = \ln 2/t_{half}$$

where t_{half} is the time for the tissue to absorb or release half of the partial difference of the gas.

Quick Review Question 6

Suppose the half-time for nitrogen absorption into a certain tissue is 4 minutes, the partial pressure in the lungs is 1.58, and the partial pressure in the tissue is 1.22. Compute the following to three decimal places:

    a. The constant of proportionality, k, in the rate of absorption equation
    b. The rate at which the tissue takes up nitrogen

Decompression Sickness

Remember that the partial pressure of nitrogen in air under higher pressures is proportionately higher. Divers at 20 m for instance receive nitrogen pressures of 2.37 atmospheres, increasing its solubility in body fluids. Nitrogen gas has no role in cellular metabolism of the diver, so it accumulates in solution. The total amount of residual nitrogen is dependent on the depth and duration of the dive. Deep divers may experience **nitrogen narcosis**, a sudden feeling of euphoria that can impair judgment and lead to serious or even fatal consequences. Furthermore, such divers returning to the surface too rapidly risk **decompression sickness** ("bends"), which is not only painful, but also potentially lethal.

Decompression sickness results from the reduced solubility of nitrogen gas in the blood, causing the release of nitrogen bubbles as pressure decreases. As the diver ascends, these bubbles continue to form and to expand. Bubbles may cause joint pain and block blood vessels. Such blockages may lead to heart attack, stroke, or ruptured blood vessels in the lungs.

Professional diving organizations publish dive tables to use for calculating how much nitrogen is absorbed during dives at varying depths and durations. Generally speaking, the deeper the dive, the shorter the duration should be.

Projects

    1. Develop a scuba diving model including pressure and volume of air in the lungs. Assume that the temperature is constant, the descent rate is less than or equal to 23 m/min, and the ascent rate is no more than 12 m/min = 0.2 m/s (U.S. Navy 2004).

    2. Repeat Project 1 with comparison graphs for seawater at various locations, such as Juneau, Alaska at 10°C (50°F); Paita, Peru or Santa Monica,

Table 7.3.1
DECOM Dive Table for Dive to 39.6 m (130 ft) of Seawater

39.6 m (130 ft) Bottom	Time at			
	12.2 m (40 ft)	9.1 m (30 ft)	6.1 m (20 ft)	3.0 m (10 ft)
15			3	6
20		1	7	9
25		4	9	14
30	2	6	11	19

California at 20°C (68°F); Savannah Beach, Georgia at 30°C (86°F) (NOAA 2004).

3. Develop a model for the pressure and volume of the air in a diver's suit. Suppose initially the volume of the gas is 8 liters $= 0.008$ m^3.

4. Develop a model for the duration of scuba tank usage. Suppose the surface air consumption rate is 3.3×10^{-4} m^3/s, and the tank initially holds 12 liters (de Lara 2002). A scuba tank delivers air to a diver at ambient pressure. Consider the depth consumption rate, or consumption rate at a depth, as part of your model.

5. Develop a model for the amount of nitrogen in tissue. At about 1.5 times the partial pressure of nitrogen at surface, nitrogen is not able to go into solution in blood.

6. Repeat Project 5 and have comparative graphs with half times of 5, 10, 20, 40, and 75 minutes.

7. The DECOM dive tables give the list of decompression stops for someone who dives to 39.6 m (130 ft) of seawater for the indicated amount of time (see Table 7.3.1) (American Dive Center 2002). For example, if someone dives to 39.6 m (130 ft) of seawater and stays there 25 min, on return the diver should stop for 4 min at a depth of 9.1 m (30 ft), 9 min at 6.1 m (20 ft), and 14 min at 3.0 m (10 ft). A rate of ascent is assumed to be about 9.1 m/min (30 ft/min). Develop a model for the amount of nitrogen in the body using each of these scenarios. Run the model long enough to determine the length of time for the amount of nitrogen in the blood to return to normal.

Answers to Quick Review Questions

1. **a.** 2.5 atm
   b. 25.25 N/cm^2 = 2.5(10.1) N/cm^2
   c. 1900 torr = 2.5(760) torr
2. $1.975 = (0.79)(2.5$ atm$)$
3. 0.714 m^3 because at 32 m, $P_2 = 1$ atm $+ (32$ m$)(1$ atm$/10$ m$) = 4.2$ atm; and $P_1 V_1 = P_2 V_2$ or $(1)(3) = 4.2 V_2$ or $V_2 = 3/4.2 \approx 0.714$
4. 0.373 m^3: At 15 m, the pressure is 2.5 atm. 30°C = 303.15 K, and 10°C = 283 K. $(1)(1)/(303.15) = (2.5)(V_2)/283$ or $V_2 = 283/((2.5)(303.15)) = 0.373$

5. **a.** 0.0474 liters because $V_g = V_L s P_g = $ (5 liters)(0.012)(0.79)
   b. 0.0948 liters because $V_g = V_L s P_g = $ (5 liters)(0.012)(2 × 0.79)
6. **a.** $0.173 = k = \ln 2/4$
   b. $0.062 = dP_{tissue}/dt = k(P_{lungs} - P_{tissue}) = 0.173\ (1.58 - 1.22)$

References

American Dive Center. "Dive Tables." From "Deep Diver Independent Learning Course," American Dive Center, 1996–2002. http://www.americandivecenter.com/deep/preview/pd04.htm

Baker, Erik C. "Some Introductory 'Lessons' About Dissolved Gas Decompression Modeling." KISS Rebreather Training. http://www.rebreather.ca/Introductory Deco Lessons.pdf

Deep Ocean Diving. 2003. "Decompression Theory—Neo-Haldane Models." *Diving Science*. http://www.deepocean.net/deepocean/science03.php

de Lara, Michel. 2002. "Diving into Mathematics or Some Mathematics of Scuba Diving." 2002. École Nationale des Ponts et Chaussés. http://cermics.enpc.fr/~delara/plongee/math_diving/math_diving.html

Martin, Lawrence. 1997. "An Explanation of Pressure and the Laws of Boyle, Charles, Dalton, and Henry." *Scuba Diving Explained*, Section D. http://www.mtsinai.org/pulmonary/books/scuba/sectiond.htm

NOAA Satellite and Information Service. 2004. "Coastal Water Temperature Guide." NOAA-NESDIS-National Oceanographic Data Center. http://www.nodc.noaa.gov/dsdt/cwtg/

U.S. Navy Diving Manual. 2004. "SCUBA Diving: Water Entry and Descent." Wet Dawg: Global Headquarters for Adventure Water Sports. http://www.wetdawg.com/pages/under_tips_display.php?t=247&c=33

MODULE 7.4

What Goes Around Comes Around—The Carbon Cycle

Prerequisites: Module 3.2 on "Unconstrained Growth and Decay" for Project 1; Module 3.3 on "Constrained Growth" for Projects 2 and 3.

Introduction

Most of us are familiar with carbon in the form of the gas carbon dioxide (CO_2). However, carbon dioxide is only one form of carbon, an element with very wide distribution on the earth. Carbon combines with elements like calcium, iron, and magnesium to form rocks. Carbon-containing compounds are dissolved in the oceans and other bodies of water. All things living are made up of organic molecules, which are all carbon based. Carbon moves in varying forms among the four major **environmental subsystems** (interdependent parts of the earth's system) of the earth: **lithosphere** (ground and inside the earth), **atmosphere** (air surrounding the earth), **hydrosphere** (lakes, rivers, and oceans), and **biosphere** (all living things). Because of the importance of carbon dioxide accumulation in the atmosphere and its effect on climate, scientists are particularly interested in carbon and its movement. This movement of carbon is described as the **carbon cycle**, and as carbon is transferred from one subsystem to another it is often transformed from one form of carbon to another.

Flow Between Subsystems

Most estimates of atmospheric CO_2 fall somewhere near 2745 gigatons (1 gigaton (Gt) = 10^{15} grams). Some CO_2 is taken up by plants and converted to various organic compounds through **photosynthesis**. Most plants, animals, and some other life forms **oxidize** organic molecules to release CO_2 back into the atmosphere. Other CO_2 dissolves in seawater. Some CO_2 is also released from solution back into the atmosphere. The atmosphere, photosynthetic organisms (part of the biosphere), and the ocean (hydrosphere) all represent **reservoirs** for carbon. Soil, sediments, and

Table 7.4.1
Major Reservoirs in Carbon Cycle

Reservoir	Initial Amount of Carbon (Gt)
atmosphere	750
terrestrial biosphere	600
ocean surface	800
deep ocean	38,000
soil	1500

Table 7.4.2
Major Fluxes in Carbon Cycle

Flux	Rate (Gt C/yr)	Source	Sink
terrestrial photosynthesis	110	atmosphere	terrestrial
marine photosynthesis	40	atmosphere	ocean surface
terrestrial respiration	110	terrestrial biosphere	atmosphere
marine respiration	40	ocean surface	atmosphere
carbon dissolving	100	atmosphere	ocean surface
evaporation	100	ocean surface	atmosphere
upwelling	27	deep ocean	ocean surface
downwelling	23	ocean surface	deep ocean
marine death	4	ocean surface	deep ocean
plant death	55	terrestrial biosphere	soil
plant decay	55	soil	atmosphere

various rock formations represent other reservoirs of carbon. The transfer of carbon from one reservoir to another is usually termed a **flux** and is given as gigatons of carbon transferred per year. (1 Gt carbon is equivalent to 3.66 Gt CO_2.) Tables 7.4.1 and 7.4.2 list the major reservoirs and fluxes actively involved in the carbon cycle (Allmon et al. 2004). In Table 7.4.2, the **source** is the origin and the **sink** is the destination of the carbon flow. Carbon dioxide **gas exchange** between the atmosphere and the ocean surface, which involves gas dissolving into and evaporating from water, is in the direction of greater to lesser carbon concentration. **Upwelling** occurs when deep currents bring cool, nutrient-rich bottom ocean water to the surface. By contrast, with **downwelling** currents move ocean surface water to lower depths.

Fossil Fuels

Human activity, primarily the combustion of fossil fuels, has greatly accelerated the release of carbon dioxide from more static reservoirs into the atmosphere. Fossil fuels are essentially combinations of carbon and hydrogen (**hydrocarbons**), which are oxidized into CO_2 upon burning. Other gases are released as well—carbon monoxide, various hydrocarbons, nitrogen oxides, and sulfur dioxide. Some of these gases, like carbon dioxide, contribute further to global warming, while others contribute to severe respiratory problems, smog, and acid rain. Another anthropogenic influence that affects CO_2 concentrations is deforestation.

Projects

For all model development, use an appropriate system dynamics tool.

1. Prepare a model of the carbon cycle using the reservoirs and fluxes in Tables 7.4.1 and 7.4.2. Assume that the rates of carbon transfer from terrestrial plants to the atmosphere and vice versa are proportional to the amount of such plants. An analogous relationship exists between the atmosphere and marine life. Similarly, the rate of downwelling is proportional to the amount of carbon in the surface ocean, and the rate of upwelling is proportional to the amount of carbon in the deep ocean. However, assume that the rate of change of marine materials sinking to the deep ocean is constant. Assume that the gas exchange between the ocean surface and the atmosphere (due to carbon dissolving in the water and evaporation from the water's surface) is proportional to the difference in the quantities of carbon at the ocean surface and in the atmosphere, so that carbon can flow in either direction. Determine proportionality constants to obtain the indicated fluxes. Produce appropriate graphs, such as the quantities in the reservoirs versus time. Vary these values and discuss the results.

2. a. Modify the model you developed in Project 1 to include the effects of deforestation and fossil fuel combustion. Suppose a fossil-fuel-deposits reservoir has an initial value of 4000 Gt and fluxes for combustion and deforestation have values of 5 and 1.15 Gt C/yr, respectively. Assume that the rate of change of fossil fuel emissions has constrained growth with a carrying capacity of 15 Gt C/yr and growth rate of 0.03/yr (see Module 3.3 on "Constrained Growth"). Produce a similar model for deforestation (Houghton et al. 1999).

   b. What is the effect on carbon reservoirs of various atmospheric carbon dioxide concentrations?

   c. What is the effect of doubling the rate of deforestation or fossil fuel combustion?

   d. What is the effect of doubling both?

3. In a 1999 article from *Science*, Houghton et al. estimated that appropriate land management (e.g., reforestation, fire suppression) might offset some of the CO_2 emissions from fossil fuel consumption by 10% to 30%. Factor land management into your evolving model from Project 2 of the carbon cycle (Houghton et al. 1999; Mersereau and Zareba-Kowalska 1997).

References

Allmon, Warren, Bryan Isacks, and William White. 2004. "Modeling the Carbon Cycle." Dept. of Earth and Atmospheric Sciences, Cornell University. http://www .geo.cornell.edu/eas/education/course/descr/EAS302/302_04Lab12.pdf

Houghton, R. A., J. L. Hackler, and K. T. Lawrence. 1999. "The U.S. Carbon Budget: Contributions from Land-Use Change." *Science* 285: 574–578.

Mersereau, Martha, and Anna Zareba-Kowalska. 1997. "Changes in the Carbon Cycle Due to Deforestation and Fossil Fuel Consumption in the U.S. Using Stella Modeling." The Woodrow Wilson National Fellowship Foundation. http://www.woodrow.org/teachers/environment/institutes/1997/26/

MODULE 7.5

A Heated Debate—Global Warming

Prerequisites: Module 3.3 on "Constrained Growth;"
Module 7.4 on "What Goes Around Comes Around—The Carbon Cycle."

Greenhouse Effect

If you drive your car into an uncovered parking lot on a clear, sunny day, one of the first things you do is to look for some shade. You know from experience that if you leave your car parked in the sunlight for even a short amount of time, the temperature inside will become exceptionally high. What is happening in your automobile is what happens in a greenhouse. The visible light waves enter your car, passing through the glass. The light is absorbed by the interior of the car and is emitted as heat (infrared). This heat warms the air contained in the car to be much warmer than the air outside the car.

The heating of the earth by sunlight is often described as a result of the **greenhouse effect**. Visible light from the sun passes through the atmosphere, with more than 50% of the original solar energy reaching the earth's surface (Gow and Pidwirny 1996; Bothun 1998). Clouds and various gases and particles absorb 23% of the incident solar energy, while clouds and particles reflect another 25% (Bothun 1998). Most of the solar energy that reaches the earth's surface is absorbed, increasing the temperature of the ground or water. Energy is then radiated from the surface as heat or infrared radiation. Atmospheric gases absorb most of this radiation. In fact, 70% of atmospheric heating is realized from this energy, with the rest from the incoming light energy (Heywood 1998). The gases of the atmosphere, now warmer, begin to radiate infrared energy themselves, much of it toward the earth, which absorbs the infrared. This "natural" greenhouse effect is responsible for increasing the annual, all-latitude average temperature to 15°C. Without this effect the earth would average a chilly −20°C (Li).

Global Warming

If this so-called "greenhouse effect" makes the earth a hospitable place for human beings, why is there so much concern about it? This effect, without our help, helps to sustain life, but with our help, might become "too much of a good thing." Human activity (e.g., combustion of fossil fuel, deforestation, etc.) has gradually increased the atmospheric concentration of the greenhouse gases. From the development of James Watts' steam engine in the mid-eighteenth century, carbon dioxide (CO_2) had risen from 280 parts per million (ppm = one part in one million = mg/l) to 368 ppm in 2000 (IPCC 2001). With more absorptive gases in the atmosphere, there is greater potential for heat absorption, which can then lead to a gradual heating of the earth—**global warming**.

There is substantial evidence for this global warming. Since the late nineteenth century, the earth's surface temperatures have increased by 0.6 ± 0.2 °C, with most of that increase occurring during the last quarter of the twentieth century (0.4 ± 0.2 °C) (Waple 2004). The 1990s were the warmest decade, and 1998 was the warmest year recorded in 140 years (IPCC 2001). The vast majority of evidence leads to the conclusion that the alarming trend of global warming is largely a result of **anthropogenic** (human influence on nature) activities. Models, using various emission scenarios, predict CO_2 concentrations will reach 540 to 970 ppm by the end of the twenty-first century. Increases of these magnitudes may boost the earth's average temperature by 1.4 to 5.8°C (IPCC 2001).

Greenhouse Gases

Carbon dioxide is only one of several **greenhouse gases (GHG's)**. Greenhouse gases are atmospheric gases that absorb infrared radiation, preventing its loss to space. The most common of these gases is actually water vapor, but other naturally occurring examples include methane and nitrous oxide. Human activities add to the increase of these gases and synthetic GHG's, such as chlorofluorocarbons (CFC's) and hydrofluorocarbons (HFC's). Some of these gases have much higher absorptive power (hence, greater warming potential) than carbon dioxide. Nevertheless, carbon dioxide is the focus of most studies and concern, because CO_2 will contribute more than 50% of the increase in **radiative forcing** (increased IR absorption and warming) for the next century (IPCC 2001; Schlesinger 2001).

Consequences

Let us accept for the purpose of modeling that the accumulation of certain gases in the atmosphere results in greater absorption of heat from the earth's surface and leads to rising global temperatures. We also accept that human activities lead to an increase in the concentration of these gases. Some people might think that a 1°C increase is not very great over a century, and we would hardly live long enough to notice. Why are so many people concerned about this problem? The answer is quite complicated because temperature affects so many processes on earth, most especially climate.

Climate changes of even minor amplitude may have drastic and dramatic consequences of life on this planet. These effects will be seen globally and regionally. Global warming will result in thermal expansion of the oceans and in glacial and ice cap melting that will raise the sea level. Projections based on various emissions/warming models predict a rise of 0.09 to 0.88 meters by the end of this century (IPCC 2001). Thousands of coastal miles would be inundated by seawater. Parts of lower Manhattan, for instance, might return to the sea (Claussen 2002).

Projects

1. Assume that there is a relationship between increasing concentrations of atmospheric carbon dioxide ($[CO_2]$) and average global temperature. Add these components to your carbon cycle model of Project 2 from Module 7.4 on "What Goes Around Comes Around—The Carbon Cycle." using the following relationships for CO_2 concentration and change in temperature, respectively:

$$[CO_2] \text{ in ppm} = (\text{mass of } CO_2 \text{ in the atmosphere})/2.12 \text{ (Allmon et al.)}$$
$$\Delta T \text{ in } °C = 0.01([CO_2] - 350) \text{ (Bice)}$$

2. a. **Methane** (CH_4) is produced naturally by some anaerobic bacteria, termites, and domestic grazing animals. Human activities, such as landfills, burning, rice cultivation, coal mining, and oil/gas extraction, have drastically increased the release of this gas into the atmosphere. Some data indicate a four-fold increase in methane emissions during the twentieth century (Gow and Pidwirny 1996). Although not nearly as prevalent as CO_2 in the atmosphere, methane is an important and powerful greenhouse gas with the ability to absorb 21 times as much heat per molecule as CO_2. In 1978, the concentration was 1.52 ppm, and methane concentration increased by about 1% per year until 1990. In 2003, the concentration was about 1.77 ppm (Lawrimore 2004). Factor the effect of methane into your global warming model of Project 1. Experiment with increases and decreases in emissions.

   b. Because much of the increase in methane levels is related to the production of food, we can tie the levels of atmospheric methane to population changes. According to Kremer, world populations have increased as follows (with the years in parentheses): 720 million (1750), 1.2 billion (1850), 1.8 billion (1900), 2.5 billion(1950), and 6 billion (2004) (Kremer 1993). Corresponding methane levels were approximately 0.70 ppm (1750), 0.85 ppm (1850), 0.90 ppm (1900), 1.1 (1950), and 1.77 ppm (2004) (Etheridge et al. 1998). Factor in population changes to methane concentrations.

   c. Growth rates in various parts of the world differ significantly. For instance, Table 7.5.1 compares the population and methane concentration

Table 7.5.1
Population and Methane Concentrations

Country	Year	Population	[Methane] (ppm)
United States	1750	2,059,000	0.70
	2001	278,059,000	310
Denmark	1769	797,600	0.70
	2001	5,355,000	310
Malawi	1890	543,000	0.70
	2001	14,696,000	310

figures for Malawi, Denmark, and the United States. If world population had followed the overall growth rates of each of these countries, how might that change Part b's model and results?

3. A third greenhouse gas is **nitrous oxide (N_2O)**. Although production is much smaller, nitrous oxide absorbs 270 times as much heat as CO_2 per molecule and resides in the atmosphere for about 150 years. This gas is released with land-use conversion, combustion of fossil fuels, burning, and nitrogen fertilization of agricultural lands. The atmospheric concentration of nitrous oxide in 1998 was 314 ppb (parts per billion) (Lawrimore 2004). Factor the effect of nitrous oxide into your global warming model of Project 1 or 2. Experiment with increases and decreases in emissions.

References

Allmon, Warren, Bryan Isacks, and William White. "Modeling the Carbon Cycle." Dept. of Earth and Atmospheric Sciences, Cornell University. http://www.geo.cornell.edu/eas/education/course/descr/EAS302/302_03Lab12.pdf (accessed April 25, 2003).

Bice, Dave. "Carbon Cycle Processes." *Exploring the Dynamics of Earth Systems*, a guide to constructing and experimenting with computer models of Earth systems using *STELLA*, Carleton College. http://www.acad.carleton.edu/curricular/GEOL/DaveSTELLA/Carbon/c_cycle_processes.htm

Bothun, Greg. 1998. "Greenhouse Effect." University of Oregon. http://zebu.uoregon.edu/1998/es202/l13.html

Claussen, Eileen. 2002. "Climate Change: Myths and Realities." Swiss Re Conference *Emissions Reductions: Main Street to Wall Street*. New York: July 17. http://www.pewclimate.org/press_room/speech_transcripts/transcript_swiss_re.cfm

Etheridge, D. M., L. P. Steele, R. J. Francey, and R. J. Langenfelds. 1998. "Atmospheric Methane Between 1000 A.D. and Present: Evidence of Anthropogenic Emissions and Climatic Variability." *Journal of Geophysical Research*, 103, (D13)m: 15979–15993.

Gow, Tracy, and Michael Pidwirny. 1996. "Greenhouse Effect." *Land Use and Environmental Change in the Thompson-Okanagan*, October. http://royal.okanagan.bc.ca/mpidwirn/atmosphereandclimate/greenhouse.html

Heywood, N. C. 1998. "Energy & Heat. 1998. " University of Wisconsin, Stephens Point. http://www.uwsp.edu/geo/faculty/heywood/GEOG101/energyhe/

IPCC (Intergovernmental Panel on Climate Change). 2001. "Climate Change 2001: Synthesis Report—Summary for Policymakers." IPCC (Intergovernmental Panel on Climate Change), Working Group One, *Third Assessment Report*. Cambridge, U.K.: Cambridge University Press. http://www.ipcc.ch/pub/un/syreng/spm.pdf

Kremer, Michael. 1993. "Population Growth and Technical Change, One Million B.C. to 1990." *Quarterly Journal of Economics*, 108(3): 681–716.

Lawrimore, Jay. 2004. "Greenhouse Gases." National Oceanic and Atmospheric Administration. http://www.ncdc.noaa.gov/oa/climate/gases.html

Li, Zuotao. "The Greenhouse Effect and Enhanced Greenhouse Effect." George Mason University. http://www.science.gmu.edu/~zli/ghe.html

Schlesinger, William H. 2001. "The Carbon Cycle: Human Perturbations and Potential Management Options." *Global Climate Change: The Science, Economics and Policy Symposium*, Bush School of Government and Public Service, Texas A&M University, April 6. http://www.soc.duke.edu/~pmorgan/Schlesinger.htm

Waple, Anne. 2004. "Global Warming." National Oceanic and Atmospheric Administration. http://lwf.ncdc.noaa.gov/oa/climate/globalwarming.html

MODULE 7.6

Cardiovascular System—A Pressure-Filled Model

Prerequisite: Module 3.4 on "System Dynamics Software Tutorial 2"
or Module 3.5 on "Drug Dosage."

Circulation

As organisms assume larger dimensions, they are confronted with a number of problems resulting from this increase in size. Cells become separated from one another sometimes by great distances, and they take on specialized functions. The functional integration of these specialized cells becomes paramount for the success of the organism. One necessary adaptation is the acquisition of effective and efficient transport systems, made up of interconnected spaces and tubes that transport fluids. In multicellular animals, we refer to these systems as **circulatory systems**. Even vascular plants have such systems. As animals grew and evolved, tubular systems (**cardiovascular systems**) that included a muscular pump (**heart**) replaced primitive circulatory systems. In these animals, a fluid (**blood**) delivers oxygen, nutrients, hormones, and wastes to their proper destinations.

The human cardiovascular system is made up of a heart and two circulatory loops—the **pulmonary** (lungs) and **systemic** (rest of body) circulations. The heart consists of four chambers—right and left **atria** and right and left **ventricles**. In both circulations, blood is pumped from the ventricles through a series of tubes—arteries, arterioles, capillaries, venules, veins—and is returned to the opposite atrium of the heart. Blood leaving the left ventricle enters the arteries of the systemic circulation, which perfuses through capillaries in muscles, digestive tract, brain, and vital organs, such as the kidneys and liver. Blood is returned through venules and veins to the right atrium of the heart. This blood, low in oxygen and high in carbon dioxide, is squeezed into the right ventricle, which pumps it into the pulmonary arteries and on to the capillaries surrounding the tiny sacks in the lungs where oxygen is exchanged for carbon dioxide. Pulmonary veins return this blood to the left atrium. From there, the blood enters the left ventricle to be sent out in the systemic circulation.

Blood Pressure

Blood pressure is the hydrostatic (fluid) pressure that moves the blood through the circulation. We monitor the pressure exerted on the arteries of the systemic circuit. Blood pressure is pulsatile, not continuous, because the cardiac cycle is intermittent. We measure the highest pressure exerted as the left ventricle contracts (**systolic pressure**). Traditionally, this pressure in a healthy adult is approximately 120 millimeters of mercury (mm Hg). Likewise, the pressure in the arteries as the left ventricle relaxes (**diastolic pressure**) is approximately 80 mm Hg.

    Mean arterial pressure (*MAP*) is the average pressure during an aortic pulse cycle. For a normal resting person, *MAP* approximately obeys the following model:

$$MAP = (\text{diastolic pressure}) + \frac{(\text{systolic pressure}) - (\text{diastolic pressure})}{3}$$

Cardiac output (*CO*) and **systemic vascular resistance** (*SVR*) regulate this pressure according to the following model:

$$MAP = CO \times SVR$$

Cardiac output is the product of the **stroke volume** (*SV*) and the **heart rate** (*HR*):

$$CO = SV \times HR$$

Stroke volume, the volume of blood that the left ventricle ejects, ranges from 50 to 100 ml in a healthy adult. A number of factors influence the stroke volume, including the volume of blood returned to the ventricle and **contractility**, or the ability of heart muscle to shorten. Heart rate in a resting, healthy adult is normally between 60 and 80 beats/minute. Consequently, **cardiac output**, or the volume of blood that the left ventricle ejects over a period of time, ranges from 4 to 8 liters/minute.

Heart Rate

Cardiac output is dependent on **heart rate** (*HR*), which ranges between 60 and 80 beats per minute (bpm) in a normal heart of an adult at rest. Controlling various involuntary activities of the body, including heart rate, the **autonomic nervous system** consists of the **sympathetic** and the **parasympathetic nervous systems**. The **pacemaker** of the heart, which is in the right atrium, fires at an intrinsic rate of 100 to 115 beats per minute. The vagus nerve, which is part of the parasympathetic nervous system, can inhibit the pacemaker's normal beat to 50 bpm. The stimulatory influence of the sympathetic nerves counters the inhibitory effects of the vagus and under certain conditions can increase heartbeat to as much as 200 bpm.

Stroke Volume

As well as heart rate, the sympathetic nervous system controls stroke volume (*SV*) directly and indirectly. Directly, sympathetic stimulation causes greater contraction of the heart and, consequently, larger *SV*. The nervous system achieves this increase with an influx of calcium ions into the cardiac muscle cells. Calcium ions promote the formation of cross-bridges between the muscle fibers, increasing the strength of contraction. Likewise, epinephrine release promotes increased contractility of heart muscle. Indirectly, active sympathetic stimulation promotes **vasoconstriction** (decreasing the vessel diameter) of the veins, leading to greater **venous return** (flow of blood to the heart). Greater venous return increases end-diastolic volume and the contractile tension of the heart muscle to a more optimal length for contraction, resulting in the heart pumping out more blood.

Venous Return

Factors other than the sympathetic nervous system influence venous return. Skeletal muscle activity, respiration, and increases in blood volume also amplify venous return. In particular, salt-water balance and the vasopressin-angiotensin system (hormones that are important to fluid balance and are vasoconstrictors) influence blood volume.

Systemic Vascular Resistance

Systemic vascular resistance (*SVR*) is the resistance or impediment of the blood vessels in the systemic circulation to the flow of blood. Although there are a number of factors that regulate *SVR,* one of the most important is the diameter of the perfusing blood vessels. Increases in *SVR* are caused by numerous factors that promote **vasoconstriction**, whereas decreases are triggered by factors that encourage **vasodilation**. These factors include those that are neurohumoral (e.g., epinephrine and vasopressin promote vasoconstriction), endothelial (e.g., nitric oxide promotes vasodilation; endothelin promotes vasoconstriction), local hormones (e.g., arachidonic acid metabolites, which may promote vasoconstriction or vasodilation) and myogenic (usually promotes vasoconstriction).

Normal ranges for *SVR* are 800 to 1200 dynes sec/cm^5. From the section on "Blood Pressure" above, we see that systemic vascular resistance is the quotient of mean arterial pressure and cardiac output, or *SVR = MAP/CO*. For *MAP* in mm Hg and *CO* in l/min, we multiply the result by 79.9 to obtain the value with units of dynes sec/cm^5. We can determine this conversion factor (79.9) with the facts that 1 mm Hg = 1333.22 dyne/cm^2 and 1 ml = 1 cm^3. The definition of the force 1 **dyne** is 10^{-5} N.

Blood Flow

Blood flow through tissues is vital for the delivery of nutrients, oxygen, and chemical messages and for the removal of carbon dioxide and wastes. Regulation of blood flow is, therefore, vital to the proper functioning of those tissues. **Blood flow (Q)** through a vessel over time is equal to the **mean velocity of the blood flow (v)** times the **cross-sectional area of the vessel**. With the cross-sectional area being the area of a circle, πr^2, where r is the **radius of the vessel**, we have the following equation for blood flow:

$$Q = v\pi r^2$$

A model for the **mean velocity (v)** is the **pressure gradient (ΔP)** times the square of the radius (r^2) divided by the product of 8, the **viscosity of the blood (η)**, and the **vessel length (L)**:

$$v = \frac{\Delta P r^2}{8\eta L}$$

Viscosity indicates the degree to which the fluid resists flow. If we substitute for v in our blood flow equation, we determine blood flow in an arteriole using **Poiseuille's Equation** (Koehler 2001):

$$Q = v\pi r^2 = \frac{\Delta P r^2}{8\eta L} \cdot \pi r^2 = \frac{\pi r^4 \Delta P}{8\eta L}$$

Projects

1. **a.** Model the regulation of heart rate by the parasympathetic nervous system, holding the sympathetic system constant. Assume heart rate is linearly dependent on each system. Employ converters/variables.

   b. Extend your model to include the regulation of heart rate by **epinephrine** (adrenalin), a chemical messenger of the sympathetic nervous system. At times of stress, such as exercise, excitement, or excessive bleeding, the adrenal medulla releases this epinephrine to increase heart rate. Illustrate its action by having the adrenal medulla release the epinephrine quickly and having the epinephrine stay in the body for a few minutes before gradually diminishing at a rate proportional to the amount of epinephrine.

   c. Modify your model to include the influence of stroke volume and heart rate on cardiac output. Consider two versions of the model. One models cardiac output as the product of stroke volume and heart beat. The other has the flow out of the heart being a pulse of a volume of blood every heartbeat. Some typical parameters for a normal person are as follows: volume of blood in the body = 5000 ml; stroke volume = 70 ml; immediately after beat, volume in ventricle = 60 ml.

   d. Investigate factors controlling blood volume and modify your model for the control of cardiac output. A normal value for fluid intake and urine output is 1 ml/min. Blood volume is approximately 70 ml/kg. Consult other sources as necessary to complete the project.

2. a. Develop a simple model of mean arterial pressure, where cardiac output (*CO*) and systemic vascular resistance (*SVR*) determine mean arterial pressure (*MAP*).

    b. Modify your simple model to include the regulation of *SVR* by the ratio of vasoconstriction to vasodilation.

3. Using the components of Poiseuille's Equation (see the section on "Blood Flow"), develop a model for regulation of blood flow in the cardiovascular system. Consider both arterial and venous flow. In arterial flow, systolic and diastolic pressures determine the pressure gradient. Some possible arterial parameter values are as follows: $r = 6$ mm, $n = 0.04$ g/(cm sec), $L = 1000$ mm. Some possible venous parameter values are as follows: $r = 3.5$ mm, $n = 0.04$ g/(cm sec), $L = 10$ mm.

References

Blumenthal, Donald K. 1998. "Introduction to Cardiovascular Pharmacology." University of Utah. http://lysine.pharm.utah.edu/netpharm/netpharm_00/notes/introcv.htm

Cheatham, Michael L. 2002. "Hemodynamic Calculations." Orlando Regional Medical Center. http://www.surgicalcriticalcare.net/Lectures/hemodynamic_calculations.pdf

Gibson, Michelle. 2001. "Oxygenation, Blood and Exercise (BIO31OBE)." Physiology Online, Applied Science website, La Trobe University, Bendigo. http://www.bendigo.latrobe.edu.au/biolsc/phys/appsci/bio30obe/lecture/Lecture10.htm

Klabunde, Richard E. 2004. "Cardiovascular Physiology Concepts." Cardiovascular Physiology Concepts website. http://www.cvphysiology.com/

Koehler, Kenneth R. 2001. "Poiseuille's Equation." College Physics for Students of Biology and Chemistry, Raymond Walters College. http://www.rwc.uc.edu/koehler/biophys/3c.html

MacLeod, Rob. "Bioengineering website." University of Utah. http://www.cvrti.utah.edu/~macleod/bioen/be3202/

Rogers, James. 1999. "Cardiovascular Physiology." *World Anaesthesia Online*, Issue 10, Article 2. http://www.nda.ox.ac.uk/wfsa/html/u10/u1002_01.htm

Romstedt, Karl. 2003. "Modeling the Cardiovascular System using STELLA: A module for Computational Biology." Capital University. http://www.capital.edu/acad/as/csac/Keck/modules/Romstedt/Cardio_module.pdf

Sherwood, Lauralee. 2004. *Human Physiology: From Cells to Systems*. 5th ed. Belmont, Calif.: Brooks-Cole Publishing/Thompson Learning.

Walton, D. Brian. 1999. "Physics of Blood Flow in Small Arteries." Mathematics and Biology, Mathematics Awareness Month. University of Arizona. http://grad.math.arizona.edu/~walton/biomath/poiseuille1.htm (accessed December 12, 2003).

MODULE 7.7

Electrical Circuits—A Complete Story

Prerequisite: Module 4.2 on "Modeling Bungee Jumping."

Defibrillators

When someone has a heart attack—what doctors call a **myocardial infarction**—insufficient blood flows to a specific portion of the heart. Interruption of blood flow may occur when the coronary arteries supplying heart muscle cells with blood are obstructed by a blood clot. The muscle soon suffers from lack of oxygen and nutrients; and if the blood supply is not restored immediately, the muscle dies. The patient may experience various symptoms, including pain, as the heart is thrown into disarray. The distress of heart muscle is accompanied by electrical instability, which may lead to **ventricular fibrillation**, or chaotic electrical disturbance (Kulick 2001).

To restore orderly electrical signals during fibrillation, medical personnel may use an instrument called a **defibrillator**. This device causes a predetermined amount of current to flow across the heart. Paddles are positioned properly on a patient's chest. Flipping of a switch forms a bridge, or we say "completes an **electrical circuit**," and stored electrons can then flow from a negatively charged plate of the defibrillator's capacitor, through the patient's heart, and back to the capacitor's positive plate. This current synchronizes the depolarization of the heart muscle and helps to restore normal electrical rhythm and normal, coordinated beating.

The use of the defibrillator provides us with but one example of the utility of electrical circuits. Circuits may be simple, as in the flashlight we keep in an automobile, or very complex, such as those in sophisticated computers. In any application, completed circuits make functions possible.

Current and Potential

In an atom, **electrons** orbit a nucleus, which contains **neutrons** and **protons**. With opposite charges, electrons and protons are attracted to each other. A **coulomb (C)** is a unit of **electric charge**, Q. The charge on an electron is -1.6×10^{-19} coulombs, while a proton has a charge of $+1.6 \times 10^{-19}$ coulombs.

When a charge flows through a region, we say that a current exists. The **current** I is the rate of change of the charge with respect to time, or

$$I = \frac{dQ}{dt}$$

One **ampere** or **amp** (**A**) is the unit of current for a charge of one coulomb to pass through a region in one second, or **1 A = 1 C/s**. A charge is analogous to water, while current corresponds to the movement of the water. Similarly, a ball and a ball falling form analogies to charge and current, respectively.

> **Definitions**   A **coulomb** (**C**) is a unit of **electric charge**, Q. **Current**, I, is the rate of change of the charge with respect to time, or $I = dQ/dt$. One **ampere** or **amp** (**A**) is the unit of current for a charge of one coulomb to pass through a region in one second, or **1 A = 1 C/s**.

A metal wire is a good **conductor** of current, and an electrical **circuit** usually consists of wires and other components. By convention, we say that the **direction of the current** is opposite to the direction in which the electrons flow.

Energy must be employed to pull opposite charges apart. When together, they have potential energy. The **electronic potential** or **potential** V at a point is the potential energy per unit charge, or the work per unit charge to bring a positive charge from infinity to the point. The **potential difference** or **voltage difference** between two points A and B is the difference in potential between the points. A unit of measure of potential difference is a **volt** (**V**). Figure 7.7.1 presents the circuit symbol for the **imposed voltage** E, such as a battery. We define the **voltage** at a point A in the circuit as the voltage difference between A and a circuit reference point, the **ground** (often the negative terminal of a battery). In a circuit, current flows from a region of high voltage to one of low voltage. Electronic potential is analogous to mechanical potential energy. For example, water flows from the top of a waterfall, where it has high potential energy, to the bottom, where its potential energy is lower.

> **Definitions**   The **electronic potential** or **potential** V at a point is the potential energy per unit charge, or the work per unit charge to bring a positive charge from infinity to the point. The **potential difference** or **voltage difference** between two points A and B is the difference in potential between the points. A unit of measure of potential difference is a **volt** (**V**). The **voltage** at a point A in the circuit is the voltage difference between A and a circuit reference point, the **ground**.

Figure 7.7.1   Electrical circuit symbol for imposed voltage

Resistance

Voltage is virtually constant along a wire. However, other components in the circuit cause voltage to drop. For example, a **resistor** slows the current flow and, thus, controls the current level. A resistor is analogous to a constriction in a hose that slows the water flow. Figure 7.7.2 displays the electrical circuit symbol for a resistor. A constant **resistance** R measures the ability of a resistor to reduce the flow of charges. According to **Ohm's Law**, the voltage drop or potential change across a resistor is as follows:

$$V = IR = R\frac{dQ}{dt}$$

or

$$I = V/R$$

or

$$R = V/I$$

We measure the resistance of a resistor in **ohms** (Ω), and $1\ \Omega = 1$ V/A. A good wire has resistance much less than 1 ohm. With an incandescent light bulb, the dissipated potential from resistance appears as light and heat. For a toaster, resistance dissipates potential that results mostly in heat.

> **Definitions**   A **resistor** slows the current flow. A constant **resistance** R measures the ability of a resistor to reduce the flow of charges. **Ohm's Law** states that $V = IR = R\ dQ/dt$. A measure of resistance is 1 **ohm** (Ω) = 1 V/A.

Quick Review Question 1

Suppose a circuit has a battery with voltage 4.5 V and a resistor with resistance 100 Ω. Calculate the current through the circuit and give its units.

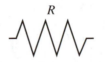

Figure 7.7.2   Electrical circuit symbol for a resistor

Figure 7.7.3   Electrical circuit symbol for a capacitor

Capacitance

A **capacitor**, whose symbol appears in Figure 7.7.3, is a component for storing charge. A simple capacitor consists of two conductors, such as metal plates, one with a positive charge and one with an equal negative charge, with an insulator between them. The potential difference can build between the two conductors. Just as a dam can hold water from a river, a capacitor can hold charge. The ability to store charge is **capacitance** (C), which we can measure in **farads** (**F**). One farad of capacitance is equivalent to having a capacitor hold a charge of 1 coulomb for a potential difference of 1 volt across its conductors, or 1 farad = 1 coulomb/volt. We have the following relationship among capacitance, charge, and voltage drop or change in potential across a capacitor:

$$C = Q/V$$

or

$$Q = CV$$

Definitions   A **capacitor** is a component for storing charge. The ability to store charge is **capacitance** (C). One **farad** (**F**) of capacitance for a capacitor is equivalent to having a capacitor hold a charge of 1 coulomb for a potential difference of 1 volt across its conductors, or 1 farad = 1 coulomb/volt.

Quick Review Question 2

Suppose a capacitor has a capacitance of 32 μF, and the voltage across the capacitor is 5000 V. Calculate the amount of charge that the capacitor stores in millicoulombs (mC).

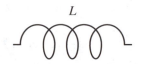

Figure 7.7.4   Electrical circuit symbol for an inductor

Inductance

A third component that reduces current is an **inductor**, a coil of wire that dampens sudden changes in current. As Figure 7.7.4 shows, the electrical symbol for an inductor suggests a coil. An inductor prevents the instantaneous increase in current and prolongs current flow. The constant **inductance L** of the coil measures the opposition to a change in current and has the following formula:

$$L = \frac{V}{dI/dt}$$

Because $I = dQ/dt$, dI/dt is the second derivative of charge with respect to time, d^2Q/dt^2, and

$$L = \frac{V}{d^2Q/dt^2}$$

A unit of measure for inductance is a **henry (H)**, which is one Vs/A. Table 7.7.1 summarizes some of the terms associated with electrical circuits along with their symbols, units, and formulas.

> **Definitions**   An **inductor** is a device that dampens sudden changes in current. A constant **inductance L** of a coil measures the opposition to a change in current and has the formula $L = V/(dI/dt)$. A unit of measure for inductance is a **henry (H)**, which is one Vs/A.

Quick Review Question 3

Suppose a large inductor has 1 H inductance, and a current of 10 A flows through the inductor. Give the voltage difference in volts if we cut off the current in 1.0 milliseconds (ms).

Circuit for Defibrillator

Figure 7.7.5 contains a circuit diagram for a defibrillator. Initially, a switch is set so that the battery can charge the capacitor. When the switch is set in the other direction, the capacitor discharges sending a surge of electricity through the heart, which

Table 7.7.1
Electrical Circuit Terms

Term	Symbol	SI Unit	Formula	Formula
Capacitance	C	Farad (F)	$C = Q/V$	
Charge	Q	Coulomb (C)	$Q = CV$	
Current	I	Amperes (A)	$I = dQ/dt$	$I = V/R$
Inductance	L	Henry (H)	$L = \dfrac{V}{dI/dt}$	$L = \dfrac{V}{d^2Q/dt^2}$
Resistance	R	Ohms (Ω)	$R = V/I$	
Voltage	V	Volt (V)	$V = IR$	$V = R\,dQ/dt$

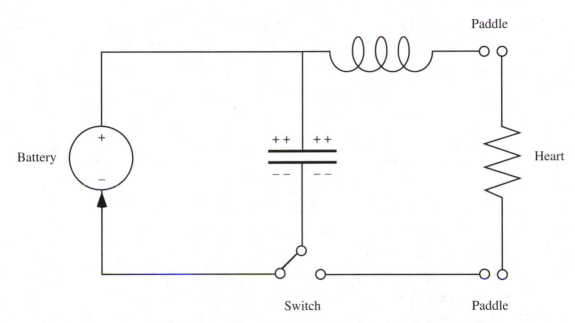

Figure 7.7.5   Circuit diagram for a defibrillator (Williams et al. 2003)

is a resistor. An inductor dampens sudden increase in current and prolongs current flow.

Kirchhoff's Voltage Law

An important connection among the components of a circuit is **Kirchhoff 's Voltage Law**, which states that in a closed loop, the sum of the changes in voltage is zero. For example, consider the **RLC circuit** (circuit with a resistor, a inductor, and a capacitor) in Figure 7.7.6 with a battery providing voltage, $E(t)$. Resistance causes a voltage drop of $IR = R\,dQ/dt$; the voltage drop due to the capacitor is Q/C; while

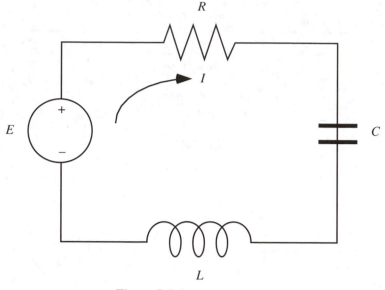

Figure 7.7.6   An *RLC* circuit

inductance causes a voltage drop of $L \, dI/dt = L \, d^2Q/dt^2$. Thus, by Kirchhoff's Voltage Law, the following equation holds:

$$E(t) - R\frac{dQ}{dt} - \frac{Q}{C} - L\frac{d^2Q}{dt^2} = 0$$

or

$$E(t) = R\frac{dQ}{dt} + \frac{Q}{C} + L\frac{d^2Q}{dt^2} \tag{1}$$

Sometimes it is convenient to express this equation using current instead of charge. Recalling that $I = dQ/dt$, we differentiate the above equation and substitute appropriately to obtain the following:

$$E'(t) - R\frac{dI}{dt} - \frac{I}{C} - L\frac{d^2I}{dt^2} = 0$$

or

$$E'(t) = R\frac{dI}{dt} + \frac{I}{C} + L\frac{d^2I}{dt^2}$$

Kirchhoff's Voltage Law   In a closed loop, the sum of the changes in voltage is zero.

Quick Review Question 4

Using Kirchhoff's Voltage Law on the defibrillator circuit diagram in Figure 7.7.5, give the equations for the following:

  a. The left loop using Q
  b. The left loop using I
  c. The right loop after the switch is thrown to complete that circuit using Q
  d. The right loop after the switch is thrown to complete that circuit using I

Kirchhoff's Current Law

Many circuits, such as the one in Figure 7.7.7, consist of several loops. **Kirchhoff's Current Law** states that the sum of the currents into a junction, such as node J_1, equals the sum of the currents out of that junction. Thus, $I_1 = I_2 + I_3$.

Quick Review Question 5

Give Kirchhoff's Current Law as it applies to the following junctions in Figure 7.7.7:

  a. Junction J_2
  b. E

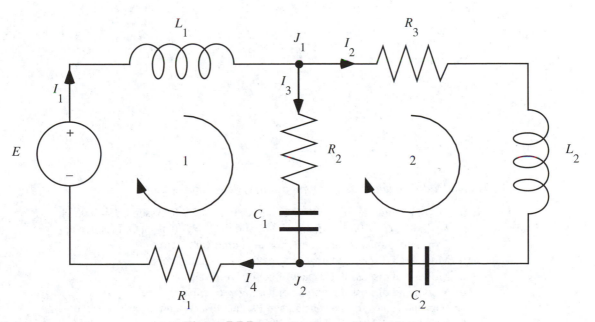

Figure 7.7.7   Circuit with more than one loop

Using Kirchhoff's Voltage and Current Laws, we obtain a system of differential equations that models the circuit. Applying Kirchhoff's Voltage Law to Loop 2 of Figure 7.7.7, we obtain the following differential equation involving current:

$$R_3 \frac{dI_2}{dt} + L_2 \frac{d^2I_2}{dt^2} + \frac{I_2}{C_2} - \frac{I_3}{C_1} - R_2 \frac{dI_3}{dt} = 0 \tag{2}$$

Because the assumed direction of the current through the components for C_1 and R_2 is opposite to that of the current through the other components in the loop, the terms involving C_1 and R_2 are negative.

Quick Review Question 6

Apply Kirchhoff's Voltage Law to Loop 1 of Figure 7.7.7 to obtain a differential equation involving current.

By applying Kirchhoff's Current Law, we can simplify the differential equations to involve fewer currents. For example, we know that I_4 and I_1 are the same and that $I_1 = I_2 + I_3$ or $I_3 = I_1 - I_2$. Taking the derivative of the latter, we have the following relationship:

$$\frac{dI_3}{dt} = \frac{dI_1}{dt} - \frac{dI_2}{dt}$$

Thus, substituting in Equation 2 and the answer to Quick Review Question 6, we have the following system of differential equations involving currents I_1 and I_2:

$$\text{Loop 1: } E'(t) = L_1 \frac{d^2I_1}{dt^2} + R_2 \frac{dI_1}{dt} - R_2 \frac{dI_2}{dt} + \frac{I_1}{C_1} - \frac{I_2}{C_1} + R_1 \frac{dI_1}{dt}$$

$$\text{Loop 2: } R_3 \frac{dI_2}{dt} + L_2 \frac{d^2I_2}{dt^2} + \frac{I_2}{C_2} - \frac{I_1}{C_1} + \frac{I_2}{C_1} - R_2 \frac{dI_1}{dt} + R_2 \frac{dI_2}{dt} = 0$$

Projects

For all model development, use an appropriate system dynamics tool.

1. **a.** Develop a model for the *RLC* circuit in Figure 7.7.6. Assume $L = 0.05$ H, $R = 20$ Ω, $C = 100$ μF, $E(t) = 100$ V, and $Q(0) = Q'(0) = 0$ C. Produce appropriate graphs, such as current and charge versus time.
   b. Model and discuss the impact of having zero inductance.
   c. Model and discuss the impact of having zero resistance.
   d. Model and discuss the impact of having zero capacitance.
   e. Vary the values for the constants in Part a. Observe and discuss the results.
   f. Referring to Module 4.2 on "Modeling Bungee Jumping," develop an analogy between an *RLC* circuit modeled by Equation 1 and a forced, damped spring-mass system (Davis 1992).

2. Repeat Project 1 assuming $L = 0.2$ H, $R = 50$ Ω, $C = 10$ μF, $E(t) = 120 \cos(120\pi t)$ $Q(0) = 10^{-6}$ C, and $Q'(0) = 0$ A.

3. **a.** Write a differential equation for the voltage applied to the heart by a de-fibrillator. This equation is piecewise, and consists of an equation for $E'(t)$ during the time when the capacitor is charging and sending no current to the heart and an equation when the capacitor is discharging and convey-ing an electrical impulse to the heart.

   b. Develop a model for a defibrillator circuit. Suppose the defibrillator has a 5000 V battery and a 32 μF capacitor. The resistance of a patient is be-tween 50 Ω and 150 Ω. Plot voltage applied to the heart versus time as well as other appropriate graphs.

4. A heart pacemaker is similar to a defibrillator. The pacemaker alternates be-tween a time, such as 4 seconds, in which the capacitor is charging and a time, such as 2 seconds, in which it is discharging and sending an electrical impulse to the heart. Suppose the pacemaker has a 12 V battery.

5. Develop a model for the circuit in Figure 7.7.7. Assume $L_1 = 0.2$ H, $L_2 = 1.0$ H, $R_1 = 10$ Ω, $R_2 = 220$ Ω, $R_3 = 330$ Ω, $C_1 = 0.1$ μF, $C_2 = 1.0$ μF, $E(t) = 117$ V, and $Q(0) = Q'(0) = 0$. Produce appropriate graphs. Discuss the re-sults.

6. Repeat Project 5 with $E(t) = 3 \cos(20\pi t)$.

7. Develop a model for a circuit of your choosing.

Answers to Quick Review Questions

1. 0.045 A because the current is $I = V/R = 4.5$ V/100 Ω = 0.045 A

2. 160 mC because $Q = CV = (32$ $\mu F)(5000$ V)$(1$ mF/$(1000$ $\mu F)) = 160$ mC

3. $V = L$ $(dI/dt) \approx L$ $(\Delta I/\Delta t) = (1$ H)$(10$ A)/$(1$ ms) $= 10/0.001 = 10{,}000$ V

4. **a.** $E(t) = Q/C$
   b. $E'(t) = I/C$
   c. $Q/C + L$ $d^2Q/dt^2 + R$ $dQ/dt = 0$
   d. $I/C + L$ $d^2I/dt^2 + R$ $dI/dt = 0$

5. **a.** $I_2 + I_3 = I_4$
   b. $I_4 = I_1$

6. $E'(t) = L_1$ $d^2I_1/dt^2 + R_2$ $dI_3/dt + I_3/C_1 + R_1$ dI_4/dt

References

Burghes, D. N., and M. S. Borrie. 1981. *Modeling with Differential Equations.* Chichester, England: Ellis Hordwood: 172.

Davis, Paul W. 1992. *Differential Equations for Mathematics, Science, and Engi-neering.* Englewood Cliffs, N.J.: Prentice-Hall: 565.

Kulick, Daniel Lee. 2001. "Heart Attack (Myocardial Infarction)." *MedicineNet.* http://www.medicinenet.com/Heart_Attack/article.htm

Ross, Clay C. 1995. *Differential Equations, An Introduction with Mathematica.* New York: Springer-Verlag: 503.

Urone, Paul Peter. 2001. *College Physics*. 2nd ed. Sacramento, Calif.: Brooks/Cole
   Publishing Co.: 893.

Williams, David J., Fiona J. McGill, and Hywel M. Jones. 2003. "Principles of
   Physical Defibrillators." *Anaesthesia and Intensive Care Medicine, Physics*.
   Abingdon, Oxon., U.K.: The Medicine Publishing Company: 23–31.

Zill, Dennis G. 2001. *A First Course in Differential Equations with Modeling Applications*. 7th ed. Pacific Grove, Calif.: Brooks/Cole Publishing Co.: 438.

MODULE 7.8

Fueling Our Cells—Carbohydrate Metabolism

Prerequisite: Module 6.3 on "Enzyme Kinetics."

Glycolysis

Carbohydrates are organic molecules composed of the elements carbon (C), hydrogen (H), and oxygen (O). Many organic molecules include these elements, but what sets carbohydrates apart is the general ratio of these elements—usually 1:2:1 (C:H:O). Carbohydrates serve as primary sources of energy for most living organisms. In fact, certain cells of many tissues, including the brain, prefer to use carbohydrates to any other energy source.

An animal may consume carbohydrates as simple molecules (sugars) or as long chains of sugars, such as starch. Once consumed, animals, using enzymes of the digestive tract, break down the larger carbohydrates to produce even more sugars. By the time these sugars (**monosaccharides**, mostly **glucose**) reach the small intestine, they are small enough to be absorbed into the blood stream and distributed to the liver and other organs of the body, where they are taken up by the cells. Depending on the cell type and the metabolic state of the animal's body, these monosaccharides may be converted into other organic constituents of the cell (e.g., fatty acids, amino acids, animal starch (**glycogen**)), or they may be broken down to produce energy.

Energy from carbohydrates and other organic food sources is obtained through gradual chemical degradation or **oxidation**. Let's define oxidation as the removal of electrons or hydrogens from a molecule. In a cell, enzymes, called **dehydrogenases**, catalyze oxidation reactions; and the molecules within a pathway that provide the electrons or hydrogens we call the **substrates**.

Most cellular monosaccharides are **glucose**, which has 6 carbons, 12 hydrogens, and 6 oxygens. The more highly **reduced** (with lots of H's) a molecule is, the better source of energy it is for the cell; and we consider glucose to be highly reduced. So in this section, we are examining the cell's sequential oxidation of glucose for energy. The complete oxidation of glucose yields carbon dioxide, water, and energy (**ATP**).

If we oxidize glucose by combustion, it yields 686 kcal/mol, all of it released as heat. Combustion in the cell is not very practical, so the cell oxidizes glucose step by step, ensuring that the cell does not burn up and that some of the energy is in a form of energy the cell can use. We term the energy made available for cellular work as **free energy**, and the cell is able to garner 275 kcal/mol of free energy from the 686 kcal available.

Glucose oxidation begins in the cytoplasm of the cell, and the initial sequence of chemical reactions is collectively referred to as **glycolysis**. The first few reactions of glycolysis essentially "prime" the molecule and require a total of 2 molecules of **ATP** (adenosine triphosphate). ATP is referred to as the "universal coupling agent", because its synthesis from $ADP + P_i$ (inorganic phosphate) "captures" some of the free energy from oxidation ($ADP + P_i + energy \rightarrow ATP$). This captured energy, released through hydrolysis of ATP ($ATP \rightarrow ADP + P_i + energy$), can then be used to power other cellular reactions.

At the end of the priming steps, the enzymes have converted glucose to another 6-carbon molecule, called *fructose 1,6-bisphosphate* (F1,6BP). F1,6BP is then processed through the oxidizing steps of glycolysis to yield 2 pyruvates, 4 ATP's, and 2 reduced coenzyme NADH's (from $NAD^+ + H$). **Pyruvates** are 3-carbon products of glycolysis, resulting from the splitting and oxidation of glucose. Note that the cell has only netted 2 ATP's from glycolysis. Although 4 were synthesized, 2 were consumed in priming glucose for oxidation.

You probably remember from basic chemistry that when something is oxidized, something else is reduced. **Coenzymes** are organic cofactors that associate with enzymes and help them catalyze. One of the most common coenzymes, often associated with dehydrogenases, is **NAD^+ (nicotinamide adenine dinucleotide)**. The enzyme removes hydrogens (H = 1 electron and 1 proton) from the glycolytic substrate and adds 2 electrons (and one proton) from oxidation to NAD^+, converting it to **NADH** (reduced coenzyme). These reduced coenzymes are particularly important for aerobic cells, because electron transport systems can reoxidize them to yield more ATP.

Recycling NAD^+'s

With Project 1, we can find that increasing the numbers of NAD^+'s available increases the ATP yield. For most cells, suddenly increasing the normal pool of cytoplasmic NAD^+'s is not really feasible, so they employ another solution—they recycle. The NADH's shed their electrons (hydrogens). The resulting NAD^+'s are reused to sustain glycolysis, allowing ATP production to continue. The recycling of coenzymes demands electron acceptors, which vary from organism to organism. Under anaerobic conditions, cells like your overexercised muscle cells will enzymatically remove the electrons (hydrogens) from the reduced coenzymes and return them to **pyruvate**, converting it to **lactate**. The enzyme that does this, called **lactate dehydrogenase**, at the same time reoxidizes NADH to NAD^+. This process is referred to as **lactate fermentation** and is common to many types of cells.

Aerobic Respiration

It might have occurred to you that 2 net ATP's is not a tremendous amount of ATP produced from glucose. At the end of this pathway, without fermentation, we have two 3-carbon molecules (pyruvates) and two molecules of NADH. In fact, the oxidation of glucose is quite unfinished, and many cells possess more elaborate pathways to complete the job. Aerobic cells, when supplied with adequate quantities of oxygen, transform pyruvate into CO_2 and H_2O. In doing so, they also generate a considerable amount of ATP. In a typical aerobic cell, pyruvate is transported into membrane-bound compartments, called **mitochondria**. These organelles contain sets of enzymes organized into a pathway often referred to as the **Krebs' Cycle**. In this pathway, the remaining electrons are removed from pyruvate and placed onto oxidized coenzymes. Very little ATP is produced directly in the Krebs' cycle. Therefore, the cell needs a way of getting the prospective energy found in all these reduced coenzymes, even those from the cytoplasm.

Within the inner membrane of the mitochondrion are sets of electron carriers, arranged into precisely structured complexes. These complexes represent **electron transport systems**, which remove and pass along the electrons and protons from all these reduced coenzymes. During the passage of the electrons, protons are actually pumped into the space outside the inner membrane, which is not very permeable to protons. This establishes an electrochemical gradient that represents a potent force. There are other protein complexes in the membrane that form channels in that membrane for protons. Attached to these channels are **ATP synthase** particles. When protons pass through the channels, they interact with these particles in such a way that they generate enough conformational changes to promote ATP synthesis. Thus, the proton gradient set up by the electron transport system has essentially powered the transfer of energy from glucose into a high-energy bond of ATP. The electrons are eventually passed on to oxygen, which serves as the terminal electron acceptor for aerobic cells.

ATP synthesis in the cytoplasm and in the Krebs' Cycle occurs by what is called **substrate-level phosphorylation**. In this type of phosphorylation, the P used to make ATP has come from an organic compound that has a higher energy level than does ATP. From one molecule of glucose, an aerobic cell produces 2 net molecules of ATP during glycolysis and 2 net molecules during the Krebs' Cycle.

Oxidative phosphorylation involves the production of ATP using the proton gradient established by the electron transport system. Pairs of electrons from NADH help to establish sufficient electrochemical gradient to power the synthesis of 3 ATP. Because the electrons from the other major coenzyme (**$FADH_2$**) enter the electron transport system at a lower level, they establish enough gradient to power the synthesis of only 2 ATP.

Projects

1. Design a simple model of glycolysis that includes the following components: glucose, NADH, ATP, pyruvate.

a. Use the following values for your model, and test it to see how many ATP's, NADH's, and pyruvates are produced.

Materials	Number of Molecules
glucose	2000
NAD^+	500
ADP	1000

b. What is the maximum number of ATP's obtained with these starting values in your model? What is the maximum number of pyruvates?

c. How many ATP's and pyruvates should glycolysis be able to produce from 2000 molecules of glucose? Why does your model yield less? Suggest ways to increase ATP production here.

2. Extend the model you developed in Project 1 to include recycling NAD^+'s.

3. Extend the model you developed in Project 1 to include **aerobic respiration**, which uses oxygen.

References

Becker Wayne M., Lewis J. Kleinsmith, and Jeff Hardin. 2003. *The World of the Cell.* 5th ed. San Francisco: Benjamin/Cummings Publishing Co.

Diwan, Joyce J. 2004. "Glycolysis and Fermentation." 2004. Instructional materials on Biochemistry of Metabolism, Rensselaer Polytechnic Institute. http://www.rpi.edu/dept/bcbp/molbiochem/MBWeb/mb1/part2/glycolysis.htm

King, Michael W. 2004. "Glycolysis." From "The Medical Biochemistry Page," Indiana University School of Medicine. http://www.indstate.edu/thcme/mwking/glycolysis.html

MODULE 7.9

Mercury Pollution—Getting on Our Nerves

Prerequisites: Module 3.2 on "Unconstrained Growth" for Project 1 a and b; Module 3.4 on "System Dynamics Software Tutorial 2" for Project 1 c; Module 6.4 on "Predator-Prey Model" for Projects 2–4.

Introduction

Many people think of rock bands when they hear the term "heavy metal." However, the "real" **heavy metals** are highly toxic elements, which generally lack any known biological function. **Mercury (Hg)** is one of these elements. Mercury's distinctive chemical and physical characteristics have been put to use in numerous commercial, industrial, and medical applications—thermometers, barometers, batteries, antiseptics, pesticides, dental restorations, fluorescent lamps. This element is also a common trace component in fossil fuels. Through all these uses, mercury has been widely dispersed in various ecosystems; and mercury pollution has become a serious problem (New Jersey 2001; Riley and Thomas).

The most significant threat to human health is through the consumption of fish contaminated with methylmercury (New Jersey 2001). **Methylmercury**, more available and more toxic than other chemical forms, is produced by the addition of a methyl group to mercury. Much of this **methylation** is accomplished by sulfate-reducing bacteria, which live at the sediment-water interface or amongst algal mats (Riley and Thomas). The bacteria are consumed by plankton, which are consumed by larger planktonic or nektonic predators. At each level of this chain of consumption, the mercury accumulates in higher and higher concentrations, increasing by up to tenfold at each level (USGS 1996). Fish may accumulate up to one million times the concentration of mercury in their aquatic environment. This accumulation is an excellent example of **biomagnification** (New Jersey 2001). For those who consume fish, the presence of this neurotoxin and possible carcinogen (Riley and Thomas) is of great importance and concern.

Mercury exists in the atmosphere primarily as **elemental Hg (Hg^0)** and as **oxidized Hg (Hg^{2+})**. The form Hg^0 is easily emitted into the atmosphere from earth.

Moreover, this elemental mercury tends to remain in the atmosphere for a year or more. Hg^{2+} is far more reactive and soluble, dissolving easily in rainwater. Some will be adsorbed to particles and aerosols. **Particulate mercury** (**dry**) and **oxidized mercury** (**wet**) are deposited on various surfaces of the earth. A portion of the mercury in the atmosphere originates from naturally-occurring emissions from the earth's surface, and the remainder is **anthropogenic**, or of human origin (Mason et al, 1994). The oxidized mercury entering the terrestrial/marine environments tends to form inorganic and organic complexes, with methylmercury being one of the most common organic forms. Because methylmercury is the most bioavailable and the most toxic, we concentrate on that form of mercury in some of our modeling. Additionally, we focus on the aquatic environment, where much of the mercury accumulates for transfer to human beings.

Projects

1. Table 7.9.1 has the pools and fluxes for the estimated "mercury budget" for preindustrial earth, while Table 7.9.2 contains the current, estimated values.
   a. Model the global mercury cycle for preindustrial and current times.
   b. In the atmosphere, elemental mercury is converted into oxidized mercury, which is either deposited dissolved in precipitation or adsorbed to particles (2%). Once in the terrestrial pool, oxidized mercury may be reduced to elemental mercury and re-emitted, be combined to form a variety of complexes (e.g., HgS) in the soil, or be methylated. Similar possibilities exist for mercury deposited in marine environments. We do not know much about the rates of these chemical conversion. Modify your "current" model to include these transformations.
   c. We know that methylation of mercury is more likely to occur under certain conditions. Methylation is favored under low pH, low oxygen, high levels of organic matter, higher temperatures, and high sulfate concentrations. Modify your model to include some of these factors.
2. **a.** Mercury movement through the food chain is often given as an example of biomagnification. Because elimination is not easy, mercury tends to accumulate. Sulfate-reducing bacteria are thought to be responsible for much of the methylation of mercury. These organisms are consumed by plankton, which are consumed by insect larvae, which may be consumed by fish fry, which are consumed by minnows, which may be consumed by small fish, which are, in turn, eaten by still bigger fish. These bigger fish are the ones usually consumed by human predators. Start with a mercury concentration of 0.6 ng/kg of mercury in the water and generate a model of biomagnification.
   b. In some parts of the world, fish is a primary part of the diet and an important source of protein. Modify your model in Part a to calculate the accumulation of mercury in the bodies of adults with varying percentages of fish in their diets. Assume that all adults weigh 65 kg and that they eat the same amount of food in kg.

Table 7.9.1
Pools and Fluxes for the Estimated "Mercury Budget" for Preindustrial Earth

Pools	($\times 10^3$ kg)	Fluxes	($\times 10^3$ kg/yr)
atmosphere	1600	terrestrial deposition	1000
mixed marine	3600	marine deposition	600
		evasion[1]	600
		natural emission(terr)[2]	1000
		riverine flow[3]	60
		particulate removal[4]	60

Table 7.9.2
Current Pools and Fluxes for the Estimated "Mercury Budget"

Pools	($\times 10^3$ kg)	Fluxes	($\times 10^3$ kg/yr)
atmosphere	5000	terrestrial deposition	3000
mixed marine	10,800	marine deposition	2000
		evasion[1]	2000
		natural emission(terr)[2]	1000
		riverine flow[3]	200
		particulate removal[4]	200
		anthropogenic (total)	4000
		atmosphere	2000
		terrestrial deposition	2000

Source: Seigel and Seigel 1997.

  1 **Evasion**—elemental mercury entering the atmosphere from the ocean

  2 **Natural emission (terr)**—mercury from natural and anthropogenic sources transferred from the terrestrial pool to the atmosphere

  3 **Riverine flow**—mercury transfer from the terrestrial pool to the ocean through runoff of streams and rivers

  4 **Particulate removal**—particles containing mercury settling to deep ocean sediments that are essentially removed from active cycling

c. Not all fish species accumulate the same amount of mercury. Table 7.9.3 contains maxima of mercury concentrations in fish species from a report by the Environmental Protection Agency (EPA) to Congress (EPA 2003). Consider that all numbers are in mg/kg dry weight. Modify your model to predict the accumulation of mercury in human adults consuming diets of different amounts of different kinds of fish.

d. Methylmercury is **lipophilic**, which means it likes fat and tends to accumulate there. Fat insulates our bodies, pads organs, and serves as energy storage depots. One might hypothesize that someone with more fat also tends to accumulate more mercury. Assuming this association, develop a model based on the **body mass index** (**BMI**). BMI, based on a mathematical relationship between height and weigh, is a commonly used method to determine the fat content of our bodies. To calculate BMI, divide weight in pounds by height in inches. Divide the results by height in inches again and then multiply by 703. The outcome may be evaluated using the common BMI categories for adults from Table 7.9.4.

Table 7.9.3
Maxima of Mercury Concentrations in Fish Species

Fish Species	Dry Weight (mg/kg)
carp	0.250
brown trout	0.418
northern pike	0.531
largemouth bass	1.369
catfish	0.890
walleye	1.383

Source: EPA 2003.

Table 7.9.4
Body Mass Index (BMI) Categories for Adults

Category	BMI
underweight	<18.5
normal weight	18.5–24.9
overweight	25–29.9
obesity	≥30

e. The U.S. EPA publishes **reference doses (RfD's)** for methylmercury. This value represents the amount of methylmercury that may be ingested on a daily basis for a lifetime with no adverse effects on health. Methylmercury is rapidly and efficiently absorbed through the gastrointestinal tract. Moreover, Methylmercury passes through the blood-brain and placental barriers. With a biological half-life in human beings of up to 80 days, we can acquire toxic amounts in small doses over a long time or through massive doses at one time. Many of the toxic effects are in the nervous system, and some of these are fatal. The RfD for methylmercury is 0.1 μg/kg body weight per day. Using a dose-conversion equation, this translates into a 1.1 μg methylmercury/kg body weight/day ingested by a 60 kg adult. Monitoring is usually done from blood or hair concentrations. Blood with 44 μg/(l of blood) or hair with 11 μg/(gram of hair) corresponds to the RfD. Incorporate this information into the model you developed in Part c. How much fish is too much?

f. Modify your latest model to include the effects of mercury toxicity in Table 7.9.5.

3. In a food chain that includes bivalve mollusks (e.g., clams, mussels, oysters), these filter feeders take in many small bits of organic matter and small organisms. Some of these organisms include bacteria and plankton. Model the accumulation of mercury in these animals. Table 7.9.6 presents the average mercury accumulation in prey (EPA 2003).

4. In a diet for shorebirds, which includes lots of shellfish, model mercury accumulation in the birds from varying mixtures of the prey in Table 7.9.6 (EPA 2003).

Table 7.9.5
Effects of Mercury Toxicity

Sympton	Hair Concentration
abnormal skin sensations (paresthesia)	250
difficulties in coordinated movement (ataxia)	400
difficulty speaking (dysarthria)	700
blindness	1000
death (LD_{50})	1600

Table 7.9.6
Average Mercury Accumulation in Prey

Prey Organism	Average Mercury Accumulation Wet Weight (mg/kg)
shrimp	0.047
clam	0.023
crab	0.117
scallop	0.042

Source: EPA 2003.

References

EPA (United States Environmental Protection Agency). 2003. *Air and Radiation.* Environmental Protection Agency, Mercury Study Report to Congress. http://www.epa.gov/oar/mercury.html (accessed January 7, 2004).

Mason, R.P., W. F. Fitzgerald, and F. M. M. Morel. 1994. "Biogeochemical cycling of elemental mercury: Anthropogenic influences." *Geochemica et Cosmochimica Acta*, 58: 3191–3198.

New Jersey Mercury Task Force. 2001. *New Jersey Mercury Task Force Report.* State of New Jersey. http://www.state.nj.us/dep/dsr/mercury_task_force.htm

Riley, D. M., and V. M. Thomas. "Mercury Pollution: Sources, Consequences, and Remedies." *PU/CEES Working Paper* No. 140. Princeton University. http://www.princeton.edu/~vmthomas/hgwp.html

Seigel, A. and H. Seigel, eds. 1997. *Metal Ions and Biological Systems: Mercury and Its Effects on Environment and Biology*, Vol. 34. New York: Marcel Dekker, Inc.

USGS (U.S. Geological Survey). 1996. South Florida Information Access. *Mercury Studies in the Florida Everglades*. U.S. Dept. of the Interior, FS-166-96. http://sofia.usgs.gov/publication/fs/166-96/printfood.html (accessed January 12, 2004).

MODULE 7.10

Managing to Eat—What's the Catch?

Prerequisite: Module 6.4 on "Predator-Prey Model."

Introduction

In 1970, fishing and aquaculture employed about 13 million worldwide, producing 65 million tons of seafood. By 1990, the numbers had increased to 28.5 million workers, hauling in almost 100 million tons (FAO 2003). Most of these fishers are in Asia. With burgeoning populations (and accompanying demand), destruction of habitat, and improved technology, many fisheries face tremendous pressures. The United Nations Food and Agriculture Organization (FAO) claims that worldwide almost 70% of the marine fish species are overfished or nearly so (FAO 2004). The National Marine Fisheries Service indicates that about one-third of the fish species in US waters are overfished (NMFS 2004). If fisheries are to remain sustainable and profitable, proper stewardship is essential. Already, according to the World Wildlife Fund, the world fishing fleet is $2\frac{1}{2}$ times bigger than necessary (Porter 1998). We need to implement management systems that insure that a gainful harvest does not exceed nature's capacity to maintain the resource.

Like many other species, Alaskan halibut began facing tremendous stresses from the fishing industry during the 1970s and 1980s. Even fishers supported the catch limits that were imposed then. By 1995, the season ran for only two days. At that time, there was open entry, and the fishing season was little more than a contest where each boat attempted to gather as large a share as possible. With shortened access times, fishing crews braved long hours and dangerous weather conditions. They cut loose tangled long lines and left them to lure and kill fish—fish that would never be harvested. Consequently, this situation resulted in lost fish, lost equipment, lost boats, and lost lives. Adding insult to injury, because all participants brought in their catches at the same time, boats had to sell the fish in a glutted market to large processors at depressed prices. For consumers, the situation meant no real market for fresh fish, so they consumed only frozen fish (ETEI; PBS).

Today, commercial halibut fishing operations are working with a closed fishery. Participant fishers must own part of the total allowable catch, called an **Individual**

Fishing Quota (IFQ). IFQ's are property that fishers actually buy and sell. This system has helped to replace the "derby fishery" that existed prior to the mid-1990s. The season is eight months long. Fishers no longer have to risk themselves and their boats during treacherous conditions. Their major income is no longer dependent on a one- or two-day venture. Although the IFQ's may have saved the halibut fishery in Alaska, fishers in other parts of the United States worry that the system favors people with more money and may lead to aggregation of quota shares into monopolies. On the other hand, most fishers have become better caretakers of this resource, if for no other reason than the IFQ's make them "owners" (ETEI; PBS).

Economics Background

Business decisions frequently involve maximizing profit. Profit in the fishing industry depends on the **cost**, or expense, of fishing and the **revenue**, or income, from fish sales. However, for an industry such as fishing, conservation of the product, the fish, for the ecosystem and future profits should be an essential part of the decision-making process.

For a **quantity** of product (q), such as metric tons of fish, the **cost function** $C(q)$ returns the total cost, or expense, of producing a quantity of q items, such as catching q metric tons of fish. Figure 7.10.1 presents a particular cost function, $C(q) = 0.01q^3 - 0.6q^2 + 13q + 35$. In this example from a very small company, the cost of producing 10 items is the corresponding value, $115, on the vertical axis. As the quantity produced increases, so does the total cost of production. Starting at about $q = 30$, this cost rises rapidly, perhaps with the company requiring new machinery to keep pace with the rising production demands or having to pay overtime to workers. Even if the company does not manufacture any product, the initial value of this cost function is a **fixed cost** of $C(0) = \$35$. Perhaps a fixed cost represents rental for warehouse space or workers' wages even when no production occurs.

> **Definitions** For a **quantity** of product (q), the **cost function** $C(q)$ returns the total cost, or expense, of producing a quantity of q items. The **fixed cost** is $C(0)$.

Quick Review Question 1

Consider the cost function $C(q) = 2000 + 50q$ for a scientific equipment company to manufacture q number of barometers.

a. Give the cost for manufacturing 100 barometers.
b. Give the fixed cost.

While cost is the total money going out, the **revenue function** $R(q)$ gives the total amount of money coming in, or income, from selling q items. Figure 7.10.2a presents an example of a revenue function, $R(q) = 12q - 0.2q^2$. Typically, with no product, no revenue exists, or the initial value is $R(0) = \$0$. In the figure's example,

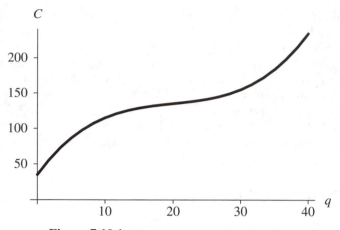

Figure 7.10.1   Example of a cost function $C(q)$

the revenue rises to a maximum of $180 for selling $q = 30$ items, or $R(30) = \$180$. Afterward, revenue decreases. Perhaps oversupply results in a drop of the **price per item**, or charge for one item. Thus, the price per item $p(q)$ is a function of quantity. In general, the revenue for producing quantity q items is the product of the price per item $p(q)$ and the number of items q as follows:

$$R(q) = p(q) \times q$$

If the price per item is constant regardless of the production quantity, the revenue function is linear, as in Figure 7.10.2b. In this example, the price of an item is $p(q) = p = \$7$; so that the revenue function is $R(q) = 7q$, a line with slope $p = 7$. Thus, the income for 10 items is $R(10) = 7 \times 10 = \$70$.

> **Definitions**   The **revenue function $R(q)$** is the total amount of income from selling q items. The **price per item $p(q)$** is the charge for one item when selling q items. Thus, the following equality holds:
>
> $$R(q) = p(q) \times q$$

Quick Review Question 2

  a. Suppose a barometer sells for $73 regardless of the number sold. Give the revenue for 100 barometers sold.
  b. Give revenue as a function of quantity q for a barometer that sells for $73 regardless of the number sold.
  c. Suppose the price of the barometer depends on quantity according to the function whose graph is in Figure 7.10.3. To the nearest dollar, determine the revenue from selling 10 barometers.
  d. To the nearest dollar, determine the revenue from selling 200 barometers.

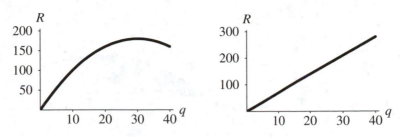

Figure 7.10.2   Examples of a revenue function $R(q)$

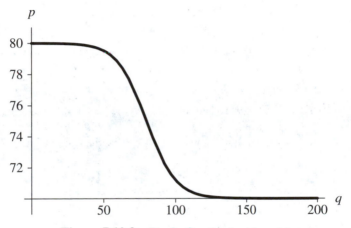

Figure 7.10.3   Graph of a price function $p(q)$

Companies or fishers are ultimately concerned with their **profit**. The **profit function** $\pi(q)$ is the profit, or the total gain, from producing and selling q items. (Economists usually employ π for the name of the profit function. This symbol is not the number $\pi \approx 3.14$.) Thus, the profit is the difference of the amount of money coming in and the amount of money going out, as indicated by the following equation:

$$\text{profit} = \text{revenue} - \text{cost}$$

or

$$\pi(q) = R(q) - C(q)$$

Thus, for revenue $R(q) = 12q - 0.2q^2$ and cost $C(q) = 0.01q^3 - 0.6q^2 + 13q + 35$, the profit function is $\pi(q) = (12q - 0.2q^2) - (0.01q^3 - 0.6q^2 + 13q + 35) = -0.01q^3 + 0.4q^2 - q - 35$. A company is working at a profit when revenue exceeds cost. Figure 7.10.4 displays in color shading the region where the company is profitable. Revenue and cost are equal, or $R(q) = C(q)$, where the graphs intersect at $q \approx 13.5$ and 34.0. For production and sales of less than 13 items or more than 34 items, cost is more than revenue, and the company is operating at a deficit.

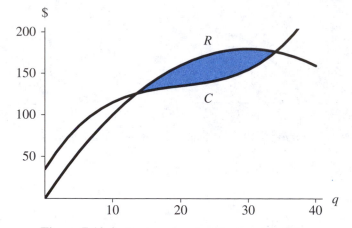

Figure 7.10.4   Region of profitability in color shading

Definition   The **profit function** $\pi(q)$ is the profit, or the total gain, from producing and selling q items. Thus, profit for selling q items is the difference in revenue and cost, so that the following equation holds:

$$\pi(q) = R(q) - C(q)$$

Quick Review Question 3

For the cost function $C(q) = 2000 + 50q$ and the revenue function $R(q) = 73q$, determine the profit function.

Certainly, a company wishes to maximize profits. From calculus, we know that to maximize (or minimize) a profit function, we set the function's derivative equal to 0; solve for the independent variable, q; and determine if a maximum does indeed occur at that q. Thus, we have the following identities when a maximum occurs:

$$\pi'(q) = R'(q) - C'(q) = 0$$

or

$$R'(q) = C'(q)$$

Economists call the derivative of the revenue function, or the instantaneous rate of change of revenue with respect to quantity, **marginal revenue**. Similarly, the derivative of the cost function, or the instantaneous rate of change of cost with respect to quantity, is **marginal cost**. Thus, a maximum profit occurs at the quantity q where marginal revenue equals marginal cost, $R'(q) = C'(q)$, and revenue exceeds cost. Because a derivative at a point is the slope of the tangent line to the curve at that point, the tangent lines to the curves are parallel at a quantity that yields a maximum profit. Figure 7.10.5 displays the revenue and cost functions with parallel tangent lines at $q \approx 25.35$. For that quantity, the profit has a maximum value of $\pi(25.35) \approx \$33.79$.

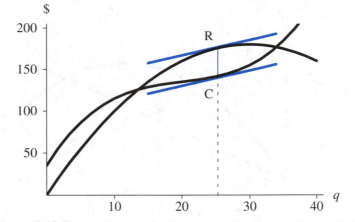

Figure 7.10.5 Maximum profit where tangent lines are parallel, $q \approx 25.35$

> **Definitions** **Marginal revenue** is the derivative of the revenue function, $R'(q)$. **Marginal cost** is the derivative of the cost function, $C'(q)$.

Quick Review Question 4

For the sale of barometers, consider cost function $C(q) = 200 + 72q$ and revenue function $R(q) = 21q^2 - q^3$, where q is the quantity in thousands of barometers and $C(q)$ and $R(q)$ are in thousands of dollars. The graphs appear in Figure 7.10.6.

    a. Give the marginal cost.
    b. Give the marginal revenue.

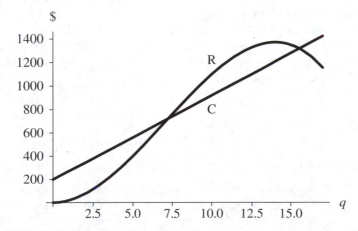

Figure 7.10.6 Cost function $C(q) = 200 + 72q$ and revenue function $R(q) = 21q^2 - q^3$

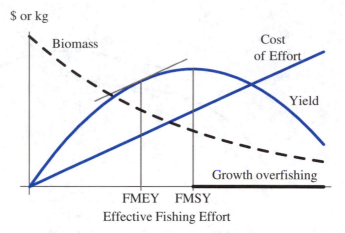

Figure 7.10.7   Gordon-Schaefer fishery production curve (Allen BioEconomics)

c. Determine the quantity of barometers sold for maximum profit.
d. Give the profit function.
e. Determine the maximum profit.

Gordon-Schaefer Fishery Production Function

In the 1950s, fishery scientist M. B. Schaeffer developed a model of biological yield, and H. Scott Gordon enhanced the model to include economics. A version of this **Gordon-Schaefer Fishery Production Curve** appears in Figure 7.10.7. The model assumes a quadratic yield function and linear cost-of-effort function. Effort involves such items as numbers of boats, traps, and days fishing. Initially, as effort increases, so does yield in dollars. As discussed above, maximum profit occurs where the derivatives of these two functions are equal, or where the tangent line to the yield function is parallel to the cost-of-effort function. We call this quantity the **Fishing Maximum Economic Yield** (**FMEY**). At the high point of the yield curve, the cost of effort is the **Fishing Maximum Sustainable Yield** (**FMSY**). After this value, increased effort results in decreased yield (Allen Bio-Economics).

Projects

For all model development, use an appropriate system dynamics tool.

1. This project concerns the economics of fishing one species.
   a. Consider the growth rate of a species of fish to be logistic when no fishing occurs (see Module 3.3 on "Constrained Growth"). Let $E(t)$ be an effort

function with respect to time t, and suppose the rate of catching fish is proportional to the product of this effort and the population of fish. Determine a differential equation for the rate of change of the population of fish with respect to time (Danby 1997).

b. Let the price per unit catch be p and the cost per unit effort be c. Write an equation for profit per unit catch.

c. Assuming that the rate of change of effort is proportional to profit, determine a differential equation for the rate of change of E with respect to time.

d. Determine the equilibrium points.

e. Develop a model involving the economics of fishing.

f. Form another version of this model with the cost per unit effort being periodic due to seasonal changes.

g. Form another version of this model with abrupt changes to the cost per unit effort. Discuss situations that could cause such changes.

h. Form another version of this model with demand and, thus, price, suddenly increasing. Discuss situations that could cause such increases.

2. Augment Project 1 to include Individual Fishing Quotas (IFQ's).

3. Table 7.10.1 contains a list of the lobster commercial catch and effort from 1942 through 1979. Adjust Project 1 to accommodate these data.

4. This project concerns the economics of fishing two species.

a. Consider the growth rates of tuna and shark to follow Lotka-Volterra's predator-prey model when no fishing occurs (see Module 6.4 on "Predator-Prey Model"). Let $E(t)$ be an effort function with respect to time t, and suppose the rate of catching each kind of animal is proportional to the product of this effort and that population. Suppose catching tuna and shark involves equal effort. Determine differential equations for the rates of change of the populations with respect to time (Danby 1997).

b. Let p_T and p_S be the price per unit catch of tuna and shark, respectively, with p_T being much greater than p_S. Let c be the cost per unit effort. Write an equation for profit per unit catch.

c. Assuming that the rate of change of effort is proportional to profit, determine a differential equation for the rate of change of E with respect to time.

d. Determine the equilibrium points.

e. Develop a model involving the economics of fishing tuna and shark.

f. Form another version of this model with the cost per unit effort being periodic due to seasonal changes.

g. Form another version of this model with abrupt changes to the cost per unit effort. Discuss situations that could cause such changes.

h. Form another version of this model with demand, and, thus, price, suddenly increasing. Discuss situations that could cause such increases.

5. Repeat Project 3 using two species of competing fish (see Module 6.1 on "Competition").

Table 7.10.1
U.S. Commercial Lobster Catch and Effort, Territorial Sea and Fishery Conservation Zone
(now called the U.S. Exclusive Economics Zone) combined (1 metric ton (mt) = 1.102311 ton)

Year	Total Catch (mt)	Total Effort (10^3 traps)	Year	Total Catch (mt)	Total Effort (10^3 traps)
1942	5577	279	1961	12,700	978
1943	7450	305	1962	13,378	1003
1944	8130	327	1963	13,731	964
1945	10,307	480	1964	14,043	1043
1946	11,012	589	1965	13,719	1163
1947	10,850	677	1966	13,399	1096
1948	9519	625	1967	12,131	1099
1949	11,183	615	1968	14,769	1168
1950	10,521	586	1969	15,327	1333
1951	11,767	517	1970	15,489	1851
1952	11,351	553	1971	15,279	1905
1953	12,749	581	1972	14,626	1858
1954	12,465	648	1973	13,152	2307
1955	13,132	701	1974	12,945	2303
1956	12,028	697	1975	13,698	2334
1957	13,679	708	1976	14,293	2305
1958	12,349	785	1977	14,434	2302
1959	13,193	898	1978	15,653	2302
1960	14,136	896	1979	16,870	2255

Source: Allen Catch.

Answers to Quick Review Questions

1. **a.** $\$7000 = C(50) = 2000 + 50(100)$
   b. $\$2000 = C(0)$
2. **a.** $\$7300 = R(100) = 73(100)$
   b. $R(q) = 73q$
   c. $\$800 = (80)(10)$
   d. $\$14,000 = (70)(200)$
3. $\pi(q) = 23q - 2000 = 73q - (2000 + 50q)$
4. **a.** $C'(q) = 72$
   b. $R'(q) = 42q - 3q^2$
   c. 12,000 because $72 = 42q - 3q^2$ or $-3q^2 + 42q - 72 = 0$, or $-3(q - 2)(q - 12) = 0$. A minimum occurs at $q = 2$, while a maximum occurs at $q = 12$.
   d. $\pi(q) = 21q^2 - q^3 - (200 + 72q) = -200 - 72q + 21q^2 - q^3$
   e. $\pi(12) = 232$, so that the profit is $\$232,000$.

References

Allen, Dick. "Catch and Effort." LobsterConservation.com. http://www.lobstercon
servation.com/catcheffort1/

————. "Fishery Bio-Economics." LobsterConservation.com. http://www.lobster conservation.com/fisherybioeconomics/

Danby, J.M.A. 1997. *Computer Modeling: From Sports to Spaceflight . . . From Order to Chaos.* Richmond, Va: Willmann-Bell: 408.

ETEI (Emissions Trading Education Initiative). "Case Study #2: Protecting Fisheries in Alaska." http://www.etei.org/case_study_2a.htm

FAO (Food and Agriculture Organization of the United Nations). 2004. Fisheries Global Information System. "FIGIS Fisheries Topic and Issues Fact Sheets." January 20. http://www.fao.org/fi/nems/news/detail_news.asp?lang=en&event_id=15151 (accessed March 15, 2004).

————. (Fisheries Global Information System). 2003. "Number of fishers doubled since 1970." Food and Agriculture Organization of the United Nations. http://www.fao.org/fi/highligh/fisher/c929.asp (accessed January 5, 2004).

Habitat Media. "Management, Overfishing, & Alaskan Halibut." Public Broadcasting Service. http://www.pbs.org/emptyoceans/eoen/halibut/

NMFS (National Marine Fisheries Service). 2004. "NOAA Fisheries—National Marine Fisheries Service." Department of Commerce. http://www.nmfs.noaa.gov/

Porter, Gareth. 1998. "Estimating Overcapacity in the Global Fishing Fleet." World Wildlife Fund. http://worldwildlife.org/

8

DATA-DRIVEN MODELS

MODULE 8.1

Computational Toolbox—Tools of the Trade: Tutorial 3

Prerequisite: Module 5.1 on "Computational Toolbox—Tools of the Trade: Tutorial 2."

Download

From the textbook's website, download Tutorial 3 in the format of your computational tool or in pdf format. We recommend that you work through the tutorial and answer all Quick Review Questions using the corresponding software.

Introduction

Various computer software tools are useful for graphing, numeric computation, and symbolic manipulation. This third computational toolbox tutorial, which is available from the textbook's website in your system of choice, prepares you to use the tool to complete projects for this and subsequent chapters. The tutorial introduces the following functions and concepts:

- List functions
- Additional graphics options
- Fitting curves to data
- Rules
- Logarithmic functions

The module gives computational examples and Quick Review Questions for you to complete and execute in the desired software system.

MODULE 8.2

Function Tutorial

Download

We recommend that you download the function tutorial in the format of your desired computational tool from the textbook's website and work through the tutorial using the software. Alternatively, you can download the corresponding tutorial in pdf format and answer the Quick Review Questions using a new file in the appropriate computational software. For the questions that do not involve using a computational tool, type the answers into the tutorial file or write the answers on a separate sheet of paper. As with other software-dependent tutorials, answers to the Quick Review Questions are not available at the end of the module. Material in the printed text, which does not depend on a particular computational tool, contains important generic information about functions.

Introduction

In this chapter, we deal with models that are driven by the data. In such a situation, we have data measurements and wish to obtain a function that roughly goes through a plot of the data points capturing the trend of the data, or **fitting the data**. Subsequently, we can use the function to find estimates at places where data does not exist or to perform further computations. Moreover, determination of an appropriate fitting function can sometimes deepen our understanding of the reasons for the pattern of the data.

In this module, we consider several important functions, some which we have already used. By being familiar with basic functions and function transformations, the modeler can sometimes more readily fit a function to the data.

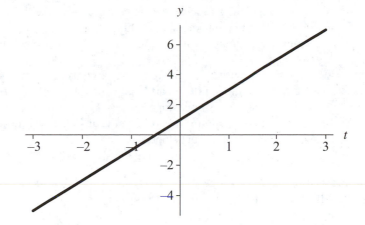

Figure 8.2.1   Graph of linear function $y = 2t + 1$

Linear Function

The concept of a linear function was essential in our discussions of the derivative and simulation techniques, such as Euler's Method. Here, we review some of the characteristics of functions whose graphs are lines.

Figure 8.2.1 presents the graph of the linear function $y = 2t + 1$. This line has y-intercept 1, because $y = 1$ when $t = 0$. Thus, the graph crosses the y-axis when $t = 0$. With data measurements where t represents time, the y-intercept indicates the initial data value. The slope of this particular line is 2, which is the coefficient of t. Consequently, when we go over 1 unit to the right, the graph rises by 2 units.

Definitions   A **linear function**, whose graph is a straight line, has the following form:

$$y = mx + b$$

The **y-intercept**, which is b, is the value of y when $x = 0$, or the place where the line crosses the y-axis. The **slope** m is the change in y over the change in x. Thus, if the line goes through points (x_1, y_1) and (x_2, y_2), the slope is as follows:

$$m = \frac{\Delta y}{\Delta x} = \frac{y_2 - y_1}{x_2 - x_1}$$

Quick Review Question 1

Use an appropriate computational tool to complete this question.

    a. Plot the above function, $f(t) = 2t + 1$, from $t = -3$ to 3.
    b. Plot f along with the equation of the line with the same slope as f but with

y-intercept 3. Distinguish between the graphs of f and the new function, such as by color, line thickness, or dashing.

   c. Copy the command from Part b, and change the second function to have a y-intercept of -3.

   d. Describe the effect that changing the y-intercept has on the graph of the line.

   e. Copy the command from Part b, and change the second function to have the same y-intercept as f but slope 3.

   f. Copy the command from Part b, and change the second function to have the same y-intercept as f but slope -3.

   g. Describe the effect that changing the slope has on the graph of the line.

Quadratic Function

In Module 2.3 on "Rate of Change" and Module 2.4 on "Fundamental Concepts of Integral Calculus," we considered a ball thrown upward off a bridge 11 m high with an initial velocity of 15 m/sec. The function of height of the ball with respect to time is the following quadratic function:

$$s(t) = -4.9t^2 + 15t + 11$$

The general form of a **quadratic function** is as follows:

$$f(x) = a_2x^2 + a_1x + a_0$$

where a_2, a_1, and a_0 are real numbers. The graph of the ball's height $s(t)$ in Figure 2.3.1 of the "Rate of Change" module is a **parabola** that is concave down. The next two Quick Review Questions develop some of the characteristics of quadratic functions.

> **Definition**   A **quadratic function** has the following form:
>
> $$f(x) = a_2x^2 + a_1x + a_0$$
>
> where a_2, a_1, and a_0 are real numbers. Its graph is a **parabola**.

Quick Review Question 2

Use an appropriate computational tool to complete this question.

   a. Plot the above function, $s(t) = -4.9t^2 + 15t + 11$, from $t = -1$ to 4.

   b. Give the command to plot $s(t)$ and another function with the same shape that crosses the y-axis at 2. Distinguish between the graphs, such as by color, line thickness, or dashing.

   c. Using calculus, determine the time t at which the ball reaches its highest point. Verify your answer by referring to the graph.

   d. What effect does changing the sign of the coefficient of t^2 have on the graph?

Quick Review Question 3

Use an appropriate computational tool to complete this question, which considers various transformations on a function. When plotting several functions together, distinguish between the curves, such as by color, line thickness, or dashing.

    a. Plot t^2, $t^2 + 3$, and $t^2 - 3$ on the same graph.
    b. Describe the effect of adding a positive number to a function.
    c. Describe the effect of subtracting a positive number from a function.
    d. Plot t^2, $(t + 3)^2$, and $(t - 3)^2$ on the same graph.
    e. Describe the effect of adding a positive number to the independent variable in a function.
    f. Describe the effect of subtracting a positive number from the independent variable in a function.
    g. Plot t^2 and $-t^2$ on the same graph.
    h. Describe the effect of multiplying a function by -1.
    i. Plot t^2, $5t^2$, and $0.2t^2$ on the same graph.
    j. Describe the effect of multiplying the function by number greater than 1.
    k. Describe the effect of multiplying the function by positive number less than 1.

Polynomial Function

Linear and quadratic functions are polynomial functions of degree 1 and 2, respectively. The general form of a **polynomial function of degree n** is as follows:

$$f(x) = a_n x^n + \cdots + a_1 x + a_0$$

where $a_n, \ldots, a_1$, and a_0 are real numbers and n is a nonnegative integer. The graph of such a function with degree greater than 1 consists of alternating hills and valleys. The quadratic function of degree 2 has one hill or valley. In general, a polynomial of degree n has at most $n - 1$ hills and valleys.

> **Definition**   A **polynomial function of degree n** has the following form:
>
> $$f(x) = a_n x^n + \cdots + a_1 x + a_0$$
>
> where $a_n, \ldots, a_1$, and a_0 are real numbers and n is a nonnegative integer.

Quick Review Question 4

Use an appropriate computational tool to complete this question.

    a. Plot the polynomial function $p(t) = t^3 - 4t^2 - t + 4$ from $t = -2$ to 5 to obtain a graph similar to Figure 8.2.2.
    b. To what value does $p(t)$ go as t goes to infinity?

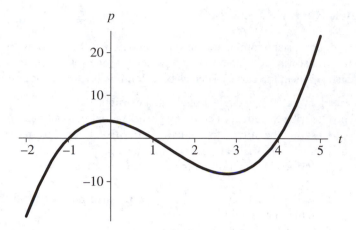

Figure 8.2.2   Graph of polynomial function $p(t) = t^3 - 4t^2 - t + 4$ from $t = -2$ to 5

c. To what value does $p(t)$ go as t goes to minus infinity?
d. Plot $p(t)$ and another function with each coefficient of t having the opposite sign as in $p(t)$. Distinguish between the curves, such as by color, line thickness, or dashing.
e. To what does the new function from Part d go as t goes to infinity?
f. To what does the new function from Part d go as t goes to minus infinity?

Square Root Function

The square root function, whose graph is in Figure 8.2.3, is increasing and concave down. Its domain and range are the set of nonnegative real numbers.

Quick Review Question 5

Use an appropriate computational tool to plot each of the following transformations of the square root function.

a. Move the graph to the right 5 units.
b. Move the graph up 3 units.

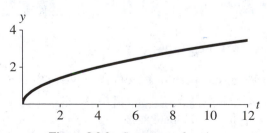

Figure 8.2.3   Square root function

    c. Rotate the graph around the *x*-axis.

    d. Double the height of each point.

Exponential Function

In Module 3.2 on "Unconstrained Growth and Decay," we considered situations where the rate of change of a quantity, such as the size of a population, is directly proportional to the size of the population, such as $dP/dt = 0.1P$ with initial population $P_0 = 100$. As we saw, the solution to this differential equation is the exponential function $P = 100e^{0.1t}$ whose graph is in Figure 3.2.2 of that module. Similarly, the solution to the differential equation $dQ/dt = -0.000120968Q$ for radioactive decay is $Q = Q_0 e^{-0.000120968t}$ with graph in that module's Figure 3.2.4. As indicated in both examples, the coefficient is the initial amount, and the coefficient of *t* is the continuous rate. For a positive rate, the function increases and is concave up; while a negative rate results in a decreasing, concave-up function.

    The base can be any positive real number, not just *e*, which is approximately 2.71828. For example, we can express $P = 100e^{0.1t}$ as an exponential function with base 2. Setting $100(2^{rt})$ equal to $100e^{0.1t}$, we cancel the 100s, take the natural logarithm of both sides, and solve for *r*, as follows:

$$100e^{0.1t} = 100(2^{rt})$$
$$0.1t = \ln(2^{rt})$$
$$0.1t = rt \ln(2)$$
$$r = 0.1/\ln(2) = 0.14427, \text{ when } t \neq 0$$
$$\text{Thus, } P = 100e^{0.1t} = 100(2^{0.14427t}).$$

> **Definition**   An **exponential function** has the following general form:
>
> $$P(t) = P_0\, a^{rt}$$
>
> where P_0, *a*, and *r* are real numbers.

Quick Review Question 6

Use an appropriate computational tool to complete this question.

    a. Define an exponential function $u(t)$ with initial value 500 and continuous rate 12%.

    b. Plot this function.

    c. On the same graph, plot exponential functions with initial value 500 and continuous rates of 12%, 13%, and 14%. Which rises the fastest?

    d. Express the function $u(t)$ as an exponential function with base 4.

Quick Review Question 7

Use an appropriate computational tool to complete this question.

    a. Define an exponential function $v(t)$ with initial value 5 and continuous rate -82%.

    b. Plot this function.

    c. Plot $v(t)$ and $v(t) + 7$ on the same graph. Distinguish between the curves, such as by color, line thickness, or dashing.

    d. What effect does adding 7 have on the graph?

    e. As t goes to infinity, what does $v(t)$ approach?

    f. As t goes to infinity, what does $v(t) + 7$ approach?

    g. Copy the answer to Part b. In the copy, plot $v(t)$ and $-v(t)$.

    h. What effect does negation (multiplying by -1) have on the graph?

    i. Copy the answer to Part g. In the copy, plot $v(t)$ and $7 - v(t)$.

    j. As t goes to infinity, what does $7 - v(t)$ approach?

    k. Give the value of $7 - v(t)$ when $t = 0$.

Quick Review Question 8

Use an appropriate computational tool to complete this question, which considers a function that has an independent variable t as a factor and as an exponent.

    a. Plot $12te^{-2t}$ from $t = 0$ to $t = 5$.

    b. Initially, with values of t close to 0, give the factor that has the most impact, t or e^{-2t}.

    c. As t gets larger, give the factor that has the most impact, t or e^{-2t}.

Logarithmic Functions

In Module 3.2 on "Unconstrained Growth and Decay," we employed the logarithmic function to obtain an analytical solution to the differential equation $dP/dt = 0.1P$ with initial population $P_0 = 100$. In that same module, the logarithmic function was useful in solving a problem to estimate the age of a mummy.

    John Napier, a Scottish baron who considered mathematics a hobby, published his invention of logarithms in 1614. Unlike most other scientific achievements, his work was not built on that of others. His highly original invention was welcomed enthusiastically, because problems of multiplication and division could be reduced to much simpler problems of addition and subtraction using logarithms.

    By definition, m is the **logarithm to the base 10**, or **common logarithm**, **of** n written as $\mathbf{log_{10}}n = m$ or $\mathbf{log\ }n = m$, provided m is the exponent of 10 such that 10^m is n or

$$\log_{10} n = m \textbf{ if and only if } n = 10^m$$

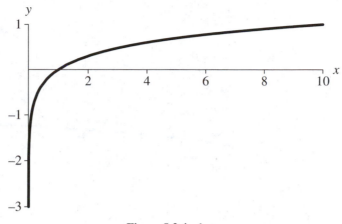

Figure 8.2.4 $\log x$

A logarithm is an exponent, in this case, an exponent of 10. Thus,

$$\log_{10} 1000 = 3 \qquad \text{because} \qquad 1000 = 10^3$$
$$\log_{10} 1{,}000{,}000 = 6 \qquad \text{because} \qquad 1{,}000{,}000 = 10^6$$
$$\log_{10} 0.01 = -2 \qquad \text{because} \qquad 0.01 = 10^{-2}$$

Because 10^m is always positive, we can only take the logarithm of positive numbers, so that the domain of a logarithmic function is the set of positive real numbers. However, the exponent m, which is the logarithm, can take on values that are positive, negative, or zero. Thus, the range of a logarithmic function is the set of all real numbers. Figure 8.2.4 shows the graph of the common logarithm. Because the logarithm is an exponent, the logarithmic function increases very slowly, and the graph is concave down.

In scientific applications, we frequently employ the **logarithm to the base e** or the **natural logarithm**. The notation for $\log_e n$ is **ln n**. Similarly to the common logarithm, we have the following equivalence:

$$\textbf{ln } \boldsymbol{n} = \boldsymbol{m} \textbf{ if and only if } \boldsymbol{n} = \boldsymbol{e^m}$$

Moreover, the graph of the natural logarithm has a similar shape to that of the common logarithm in Figure 8.2.4.

> **Definitions** The **logarithm to the base b of n**, written $\log_b n$, is m if and only if b^m is n. That is, $\log_b n = m$ is equivalent to $n = b^m$. The **common logarithm** of n, usually written **log n**, has base 10; and the **natural logarithm** of n, usually written **ln n**, has base e.

In comparing the graph of $\ln x$ to those of x and $\sqrt{x}$ in Figure 8.2.5, we see that the linear and square root functions dominate the logarithmic function, which is in color.

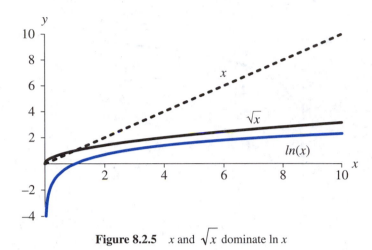

Figure 8.2.5 x and $\sqrt{x}$ dominate $\ln x$

Quick Review Question 9

Use an appropriate computational tool to complete this question.

    a. Evaluate $\log_2 8$.
    b. Write $y = \log 7$ as a corresponding equation involving an exponential function.
    c. Evaluate $\ln(e^{5.3})$.
    d. Evaluate $10^{\log(6.1)}$.

Logistic Function

In Module 3.3 on "Constrained Growth," we modeled the rate of change of a population with a carrying capacity that limited its size. The model incorporated the following differential equation with carrying capacity M, continuous growth rate r, and initial population P_0:

$$\frac{dP}{dt} = r\left(1 - \frac{P}{M}\right)P$$

The resulting analytical solution, which is a **logistic function**, is as follows:

$$P(t) = \frac{MP_0}{(M - P_0)e^{-rt} + P_0}$$

Figure 3.3.1 of the "Constrained Growth" module depicts the characteristic S-curve of this function.

Quick Review Question 10

Use an appropriate computational tool to complete this question. When plotting several functions together, distinguish between the curves, such as by color, line thickness, or dashing.

a. Plot the logistic function with initial population $P_0 = 20$, carrying capacity $M = 1000$, and instantaneous rate of change of births $r = 50\% = 0.5$ from $t = 0$ to 16 to obtain a graph as in Figure 3.3.1 of Module 3.3 on "Constrained Growth."

b. On the same graph, plot three logistic functions that each have $M = 1000$ and $r = 0.5$ but P_0 values of 20, 100, and 200.

c. What effect does P_0 have on a logistic graph?

d. On the same graph, plot three logistic functions that each have $M = 1000$ and $P_0 = 20$ but r values of 0.2, 0.5, and 0.8.

e. What effect does r have on a logistic graph?

f. On the same graph, plot three logistic functions that each have $P_0 = 20$ and $r = 0.5$ but M values of 1000, 1300, and 2000.

g. What effect does M have on a logistic graph?

Trigonometric Functions

The sine and cosine functions are employed in many models where oscillations are involved. For example, two projects in Module 6.4 on "Predator-Prey Model" considered seasonal birth rates and fishing and employed the cosine and sine functions, respectively, to achieve periodicity.

To define the trigonometric functions sine, cosine, and tangent, we consider the point (x, y) on the unit circle of Figure 8.2.6. For the angle t off the positive x-axis, with t being positive in the counterclockwise direction and negative in the clockwise direction, the definitions of these trigonometric functions are as follows:

$$\sin t = y$$
$$\cos t = x$$
$$\tan t = y/x$$

For example, if $x = 0.6$ and $y = 0.8$, then t is approximately 0.9273 radians, so that the following hold:

$$\sin(0.9273) = 0.8$$
$$\cos(0.9273) = 0.6$$
$$\tan(0.9273) = 0.8/0.6 \approx 1.33$$

For an angle of 0 radians, the opposite side, y, is zero, so that $\sin(0) = 0$. An angle of $\pi/2$ results in $(0, 1)$ being the point on the unit circle and the sine function achieving its maximum value of 1. The sine returns to 0 for the angle $\pi = 180°$.

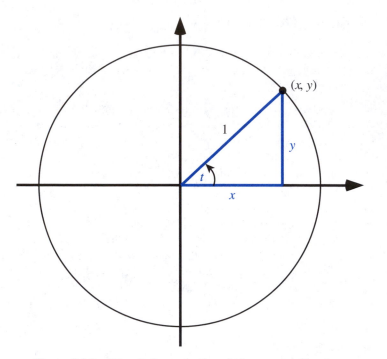

Figure 8.2.6   Triangle for evaluation of trigonometric functions

Then, sin(*t*) obtains its minimum, namely, −1, at 3π/2, where the point on the unit circle is (0, −1). At *t* = 2π = 360°, the sine function starts cycling through the same values again. Figure 8.2.7 presents one cycle of the sine function, and Figure 8.2.8 gives a cycle of the cosine function.

Quick Review Question 11

   a. Evaluate sin *t* where *x* = 0.6 and *y* = 0.8 for angle *t*.
   b. Evaluate sin(π/3) where the corresponding point on the unit circle is $\left(\dfrac{1}{2}, \dfrac{\sqrt{3}}{2}\right)$.
   c. Give the domain of the sine function.
   d. Give the range of the sine function.
   e. Give the sine's **period**, or length of time before the function starts repeating.
   f. Is sin *t* positive or negative for values of *t* in the first quadrant?

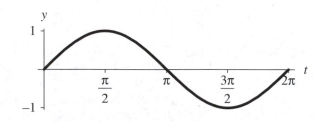

Figure 8.2.7   One cycle of the sine function

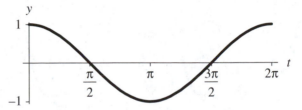

Figure 8.2.8   One cycle of the cosine function

g. Is sin t positive or negative for values of t in the second quadrant?
h. Is sin t positive or negative for values of t in the third quadrant?
i. Is sin t positive or negative for values of t in the fourth quadrant?

Quick Review Question 12

a. Evaluate $\cos(0)$.
b. Evaluate $\cos(\pi/2)$.
c. Evaluate $\cos(\pi)$.
d. Evaluate $\cos(3\pi/2)$.
e. Evaluate $\cos(\pi/3)$ where the corresponding point on the unit circle is $\left(\dfrac{1}{2}, \dfrac{\sqrt{3}}{2}\right)$.
f. Give the maximum value of $\cos t$.
g. Give the minimum value of $\cos t$.
h. Give the domain of the cosine function.
i. Give the period of the cosine function.
j. Is $\cos t$ positive or negative for values of t in the first quadrant?
k. Is $\cos t$ positive or negative for values of t in the second quadrant?
l. Is $\cos t$ positive or negative for values of t in the third quadrant?
m. Is $\cos t$ positive or negative for values of t in the fourth quadrant?

For a function of the form $f(t) = A \sin(Bt)$ or $g(t) = A \cos(Bt)$, where A and B are positive numbers, A is the **amplitude**, or maximum value of the function from the horizontal line going through the middle of the function. For example, $h(t) = 2 \sin(7t)$ has amplitude 2; the function oscillates between y values of -2 and 2. Because the period of the sine and cosine functions is 2π, the period of f and g above is $2\pi/B$. When $t = 0$, $Bt = 0$. When $t = 2\pi/B$, $Bt = B(2\pi/B) = 2\pi$. Thus, the period of $h(t) = 2 \sin(7t)$ is $2\pi/7$.

Quick Review Question 13

Use an appropriate computational tool to plot each following pair of functions, distinguishing between the curves by color, line thickness, or dashing.

a. sin t and $2 \sin(7t)$.
b. sin t and a function involving sine that has amplitude 5 and period 6π.
c. sin t and a function involving sine that has minimum value -2 and maximum value 4.

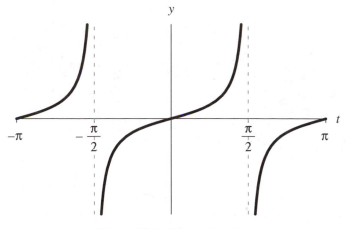

Figure 8.2.9   Tangent function

    d. sin t and a function involving sine that has amplitude 4 and crosses the t-axis at each of the following values of t: ..., $-\pi/6$, $\pi/3$, $5\pi/6$, ...

    e. cos t and a function involving cosine that has amplitude 3, period π, and maximum value 2 at $t = \pi/5$.

    f. sin($5t$) and e^{-t} sin($5t$). The latter is a function of decaying oscillations. The general form of such a function is $Ae^{-Ct}\sin(Bt)$, where A, B, and C are constants.

The tangent function is also periodic. Because tan $t = y/x$, for a corresponding point (x, y) on the unit circle (see Figure 8.2.6), tan t = sin t/cos t. The graph of this function appears in Figure 8.2.9, and the next Quick Review Question explores some of its properties.

Quick Review Question 14

    a. Evaluate tan($\pi/3$) where the corresponding point on the unit circle is $\left(\dfrac{1}{2}, \dfrac{\sqrt{3}}{2}\right)$.

    b. Evaluate tan(0).

    c. Evaluate tan(π).

    d. Evaluate tan($\pi/2$).

    e. As t approaches $\pi/2$ from values less than $\pi/2$, what does tan t approach?

    f. As t approaches $\pi/2$ from values greater than $\pi/2$, what does tan t approach?

    g. Evaluate tan($-\pi/2$).

    h. As t approaches $-\pi/2$ from values less than $-\pi/2$, what does tan t approach?

    i. As t approaches $-\pi/2$ from values greater than $-\pi/2$, what does tan t approach?

    j. Give the range of the tangent function.

    k. Give all the values between -2π and 2π for which tan t is not defined.

    l. Give an angle in the third quadrant that has the same value of tan t, where t is in the first quadrant.

    m. Give an angle in the fourth quadrant that has the same value of tan t, where t is in the second quadrant.

    n. Give the period of the tangent function.

MODULE 8.3

Empirical Models

Downloads

For several computational tools, the text's website has available for download an *8_3QRQ.pdf* file, which contains system-dependent Quick Review Questions and answers for this module.

Moreover, the text's website has available for download for various computational tools an *EmpiricalModels* file, which contains the models of this module. Table 8.3.1 lists data files that are also available on the website and where they are employed in the text. The data files are based on files from the National Institute of Standards and Technology (NIST) website on "Statistical Reference Datasets," as indicated in the "References" section. The name of each data file is as on the NIST site except that *EM* appears before the extension *.txt* or *.dat*. File names with the

Table 8.3.1
Data Files on Textbook's Website

Description File	Data File	Where Used
BoxBODEM.txt	*BoxBODEM.dat*	Project 9
DanWoodEM.txt	*DanWoodEM.dat*	"Nonlinear One-Term Model"
FilipEM.txt	*FilipEM.dat*	"Multiterm Models"
Gauss1EM.txt	*Gauss1EM.dat*	Project 5
Lanczos1EM.txt	*Lanczos1EM.dat*	Project 7
Lanczos3EM.txt	*Lanczos3EM.dat*	Project 4
MGH10EM.txt	*MGH10EM.dat*	Project 8
MGH17EM.txt	*MGH17EM.dat*	Project 6
Misra1aEM.txt	*Misra1aEM.dat*	"Solving for y in a One-Term Model"
NoInt1EM.txt	*NoInt1EM.dat*	Project 1
NorrisEM.txt	*NorrisEM.dat*	"Linear Empirical Model," Exercise 1
PontiusEM.txt	*PontiusEM.dat*	Project 2
Wampler1EM.txt	*Wampler1EM.dat*	Project 3

extension *.txt* give the file name for the data file, URL reference, original dataset name, description, reference, data in column format (y, then x), and statements in several computational tools assigning appropriate data lists to *xLst* and *yLst*. The corresponding files with the extension *.dat* store only the data in column format (x, then y), which most computational tools can read.

Introduction

Sometimes it is difficult or impossible to develop a mathematical model that explains a situation. However, if data exist, we can often use these data as the sole basis for an **empirical model**. The empirical model consists of a function that fits the data. The graph of the function goes through the data points approximately. Thus, although we cannot employ an empirical model to explain a system, we can use such a model to predict behavior where data do not exist. Data are crucial for an empirical model. We utilize data to suggest the model, to estimate its parameters, and to test the model.

> **Definition**   An **empirical model** is based only on data and is used to predict, not explain, a system. An empirical model consists of a function that captures the trend of the data.

When we derive a mathematical model through analysis of a system, we may accept a model that does not fit the data as closely as we would wish because the model explains the situation well. However, with an empirical model, the data are our only source of information about the system.

Sometimes with a derived model, which helps to explain the science, it may be difficult or impossible to differentiate or integrate a function to perform further analysis. In this case, too, we can derive an empirical model, such as a polynomial function, that is differentiable and integrable. For example, a step function might accurately model a pulsing signal, but we cannot differentiate such a function where it is discontinuous, or jumps from one step to the next. In this case, we might use trigonometric functions, which we can differentiate and integrate, in an empirical model that captures the trend of the data.

Linear Empirical Model

We begin studying empirical models by considering a National Institute of Standards and Technology (NIST) "study involving calibration of ozone monitors," where x is "NIST's measurement of ozone concentration" and y is "the customer's measurement." For the purpose of this example, we take the subset of the data shown in Table 8.3.2.

Using an appropriate computational tool, we define a set of ordered pairs, assigning the result to a variable *pts* = {(0.2, 0.1), (0.4, 0.3), (0.3, 0.3), (0.3, 0.6)}. Figure 8.3.1 shows a plot of these data.

Table 8.3.2
Subset of NIST *Norris* Dataset, Where x is "NIST's Measurement of Ozone Concentration" and y is "the Customer's Measurement"

x	y
0.2	0.1
0.4	0.3
0.3	0.3
0.3	0.6

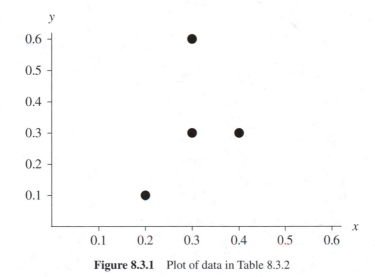

Figure 8.3.1   Plot of data in Table 8.3.2

To explain how we obtain a line that fits the data well, we need the definition of **linear combination**. The linear combination of p and q is a sum of the form $ap + bq$, where a and b are constants. For example, $3p + 7q$ is a linear combination of p and q. A linear combination of 1 and x has the form $b(1) + mx = b + mx = mx + b$, for constants b and m. For example, $-2.2x + 9.3$ is a linear combination of x and 1. We can extend the definition of linear combination to any number of terms. Thus, $4 - 3x + 19x^2$ is a linear combination of 1, x, and x^2, while $5x$ is a linear combination of just x.

Definition   For positive integer n, a **linear combination** of $x_1, x_2, \ldots, x_n$ is a sum

$$a_1 x_1 + a_2 x_2 + \cdots + a_n x_n,$$

where $a_1, a_2, \ldots,$ and a_n are constants.

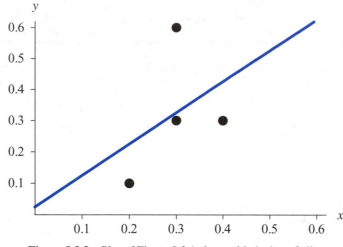

Figure 8.3.2   Plot of Figure 8.3.1 along with the best-fit line

Quick Review Question 1

List the expressions that are linear combinations of u and v.

A. $5u - 18v$ **B.** $-18v + 5u$ **C.** $7u$
D. $15uv$ **E.** $u/5 + v/3$ **F.** $5/u + 3/v$

   A computational tool usually has a function that can return an equation that is a **least-squares** fit to a list of points. In the section "Linear Regression," we discuss the algorithm, but we can use such a fit function without knowing the formulas involved. The equation $y = 0.025 + 1.0x$, which is a linear combination of 1 and x, is the least-squares linear function that best fits the data in Table 8.3.2. We can plot this line along with the original data to obtain a graph similar to Figure 8.3.2.

Quick Review Question 2

From the text's website, obtain your computational tool's *8_3QRQ.pdf* file for this system-dependent question concerning a command to obtain a least-squares line that best fits a set of points.

Predictions

   We can use the result of the least-squares linear fit, $y = 0.025 + 1.0x$, to predict y values where no data value exists as long as those values are within the range of values used to determine the formula. For example, for NIST's measurement of ozone concentration of $x = 0.34$, the predicted customer's ozone concentration measurement is $y = 0.025 + 1.0(0.34) = 0.365$. Figure 8.3.3 displays on the curve the point (0.34, 0.365), which is larger than the other points.

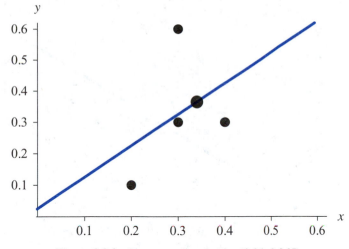

Figure 8.3.3   Larger predicted point (0.34, 0.365)

We must be careful not to employ this predictive function beyond the range of the data. With an empirical model, the data drive the model. Outside the range of the data, we cannot depend on the data behaving in a similar manner to observations within the range. For example, the ozone monitor being calibrated could be fairly accurate when measuring small concentrations but completely unreliable for large concentrations.

Linear Regression

A fit function in an appropriate computational tool returns a least-squares fit to the data. In the example above, a fit function determined that $y = 0.025 + 1.0x$ is the line that best captures the trend of the data using a technique called **linear least-squares regression** or **linear regression**. We call x the **predictor variable** and y the **response variable**. The method can find the line $y = mx + b$ that minimizes the sum of the squares of the vertical distances from the data points to the line. For example, the point that is directly above or below $(0.2, 0.1)$ on a line $y = mx + b$ is $(0.2, m(0.2) + b)$. We obtain the y value, $m(0.2) + b$, on the line by substituting the x value, 0.2, into the linear function. The difference in the y values of the point on the line and the point $(0.2, 0.1)$ is $(m(0.2) + b - 0.1)$. The lengths of the dotted lines in Figure 8.3.4 are the absolute values such differences. Linear regression finds m and b so that the sum of the squares of the vertical distances is as small as possible. Thus, for n points, $(x_1, y_1), (x_2, y_2), \ldots, (x_n, y_n)$ the method does the following, where the summation $(\sum_{i=1}^{n})$ indicates summing the squared terms for $i = 1, 2, \ldots, n$:

$$\text{minimize } \sum_{i=1}^{n}(mx_i + b - y_i)^2 = (mx_1 + b - y_1)^2 + (mx_2 + b - y_2)^2$$
$$+ \cdots + (mx_n + b - y_n)^2$$

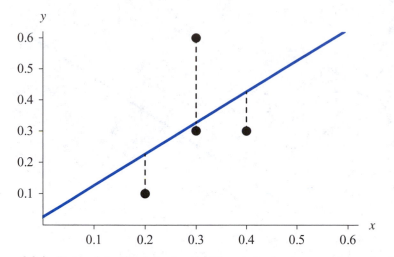

Figure 8.3.4 Data points with dashed vertical lines to the least-squares regression line

Using calculus, minimization techniques yield m and b, as follows:

$$m = \frac{n \sum x_i y_i - \sum x_i \sum y_i}{n \sum x_i^2 - (\sum x_i)^2}$$

$$b = \frac{\sum x_i^2 \sum y_i - \sum x_i y_i \sum x_i}{n \sum x_i^2 - (\sum x_i)^2}$$

Fortunately, a fit function in an appropriate computational tool performs these calculations for us.

In a fashion similar to this computation, a fit function can return the equation that is a linear combination of given functions and that yields the minimum of the sum of the squares of the vertical distances from the points to the corresponding curve. Consequently, using a fit function we can obtain nonlinear functions with multiple terms that model the data empirically.

Nonlinear One-Term Model

Table 8.3.3 presents data from NIST's *DanWood* dataset for the next example (see the section on "Downloads"). The predictor variable x is "the absolute temperature of the filament in 1000 degrees Kelvin," while the response variable y is the "energy radiated from a carbon filament lamp per cm^2 per second."

Figure 8.3.5 shows a plot of the data from Table 8.3.3. Although it is not the regression line, a faint line through the first and last points helps us to see that the con-

Table 8.3.3
Data from NIST's *DanWood* Dataset

x	y
1.309	2.138
1.471	3.421
1.490	3.597
1.565	4.340
1.611	4.882
1.680	5.660

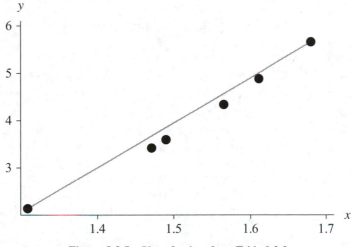

Figure 8.3.5   Plot of points from Table 8.3.3

figuration of points is slightly concave up. Such a data set whose plot is concave up or down throughout can usually be modeled effectively with a function in which only one term has the single dependent variable. To effectively determine a mathematical **one-term model** for this data using linear regression, we use a transformation of the data that appears linear.

Quick Review Question 3

From the text's website, obtain your computational tool's *8_3QRQ.pdf* file for this system-dependent question concerning commands to plot data and a line together.

We can accomplish the transformation on this data from being concave up to straight in one of two ways:

1. Extend the points to the right, stretching the distance from the y-axis to the rightmost points more than to those on the left. Thus, perform an operation

on x that results in greater values and has more effect on the larger x values than on the smaller ones.

2. Pull points down, shrinking the distance between the x-axis and the higher points more than to the lower ones. Thus, perform an operation on y that results in lesser values and has more effect on the larger y values than on the smaller ones.

For the first alternative with the above data, the operation might be to raise all x values to a power greater than 1, such as 2. To see the effect, consider the data points (1.309, 2.138) and (1.680, 5.660), where the former is to the left of the latter. Squaring both x coordinates, we find that $1.309^2 = 1.713$, while $1.680^2 = 2.822$. The difference $x^2 - x$ for the first point is $1.309^2 - 1.309 = 0.404$, but the effect on the second, rightmost point is much greater with a difference of $1.680^2 - 1.680 = 1.142$. Thus, the transformation of squaring the x coordinate where $x > 1$ stretches the rightmost points to the right even more than to the points that are further to the left.

Similarly, for the second alternative, taking the square root of the y coordinates, which are all greater than 1, gives smaller values. However, the effect is more pronounced on the larger y values. The point $(1.309, \sqrt{2.138})$ is 2.38 units lower than (1.309, 2.138), but $(1.680, \sqrt{5.660})$ is 3.28 units lower than (1.680, 5.660).

We should note that these operations perform as indicated because the coordinates are all greater than 1. Recall that for a number c between 0 and 1, c^2 is smaller than c, while $\sqrt{c}$ is larger. Moreover, when the values are negative, we cannot perform certain operations, such as taking the square root or logarithm. Also, raising a negative value to an even exponent gives a positive number. To obtain the desired results, we should reason carefully and not apply operations randomly.

Table 8.3.4 gives a sequence of transformations that have an increasingly greater impact on larger values, where the numbers are greater than 1. The transformations, such as $-1/z$, that involve a unary minus do so to maintain the same ordering of the data points. For example, (1.309, 2.138) is to the left of (1.680, 5.660). Using the transformation $1/x$, $(1/1.309, 2.138) = (0.7639, 2.138)$ is to the right of $(1/1.680, 5.660) = (0.5952, 5.660)$. However, with $-1/x$, the point $(-1/1.309, 2.138) = (-0.7639, 2.138)$ remains to the left of $(-1/1.680, 5.660) = (-0.5952, 5.660)$.

Taking the first alternative above, we pair various powers of x with the corresponding y. We plot the resulting ordered pairs in an attempt to find a graph that appears approximately linear. As Figures 8.3.6 and 8.3.7 show with the assistance of lines through the first and last points, squaring and cubing the x coordinates seem still to result in plots that are concave up. However, raising the x values to the fourth power,

Table 8.3.4
Sequence of Transformations for $z > 1$

$$\cdots -\frac{1}{z^2}, -\frac{1}{z}, -\frac{1}{\sqrt{z}}, \ln(z), \sqrt{z}, z, z^2, z^3, \cdots$$

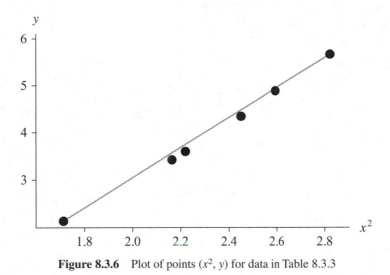

Figure 8.3.6   Plot of points (x^2, y) for data in Table 8.3.3

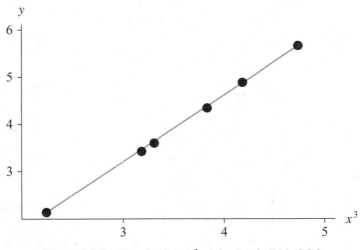

Figure 8.3.7   Plot of points (x^3, y) for data in Table 8.3.3

as in Figure 8.3.8, appears to cause a graph that is slightly concave down. If we are not satisfied with the exponent 3 or 4, we might try powers between these values. Figure 8.3.9 shows a plot of the points $(x^{3.5}, y)$.

Thus, using a fit function in an appropriate computational tool, we employ linear regression on the transformed set of points $(x^{3.5}, y)$ and obtain the following best-fit line: $y = -0.393131 + 0.988186z$. Figure 8.3.10 shows the graph of this line with the points of Figure 8.3.9.

Satisfied with this result, we still must determine the function through the original set of points and view its curve through those points. Because we fit a line

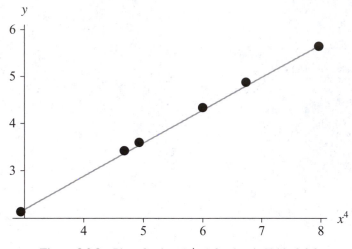

Figure 8.3.8   Plot of points (x^4, y) for data in Table 8.3.3

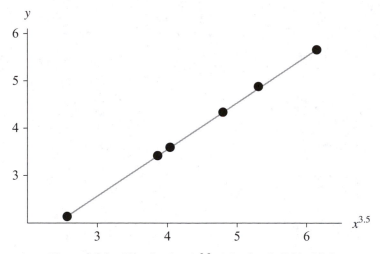

Figure 8.3.9   Plot of points $(x^{3.5}, y)$ for data in Table 8.3.3

$(y = -0.393131 + 0.988186z)$ to the transformed points of the form $(x^{3.5}, y)$, we now substitute $x^{3.5}$ for z to obtain our empirical model, which is as follows:

$$f(x) = -0.393131 + 0.988186\,x^{3.5}$$

Figures 8.3.11 and 8.3.12 present two graphs of this function along with the original data from Table 8.3.4 for different ranges of x.

The graphs indicate that our empirical model f seems to be satisfactory for x from

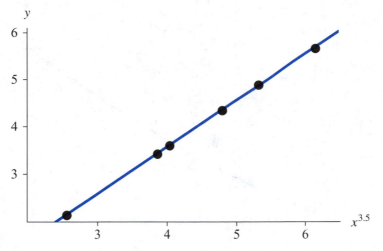

Figure 8.3.10   Graph of linear regression line with points of Figure 8.3.9

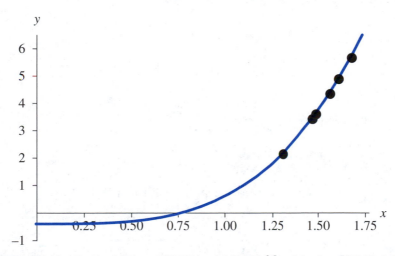

Figure 8.3.11   Graph of $f(x) = -0.393131 + 0.988186x^{3.5}$ and the data of Table 8.3.4

1.3 to 1.7. To test the model, we should collect additional data in this range and plot all the data with the graph of the model to observe how they agree. Moreover, for each newly observed x value, we should determine how closely the observed and predicted y values agree.

Clearly, other empirical models than $y = -0.393131 + 0.988186x^{3.5}$ approximate the data. Empirical modeling is an "art" as well as a science. Several metrics aid in determination of which model to use but are beyond the scope of this text.

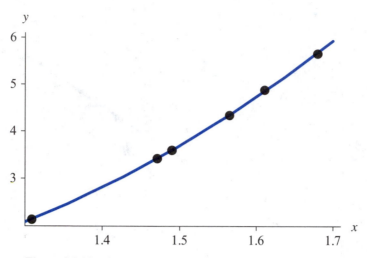

Figure 8.3.12   Graph of Figure 8.3.11 for *x* between 1.3 and 1.7

Solving for *y* in a One-Term Model

In this section, we consider empirical model development in which we make a transformation on *y* and perhaps on *x*, too, instead of just on *x* as in the previous section. The data, which Table 8.3.5 lists, are from NIST dental research in monomolecular adsorption, where *x* represents pressure and *y* volume (NIST *Misra1a* Dataset). Figure 8.3.13 displays a plot of these data with a faint line between the first and last points to emphasize concavity.

Table 8.3.5
Data from *Misra1aEM.dat*, Available on the Textbook's
Website and in the NIST *Misra1a* Dataset

x	*y*
77.6	10.07
114.9	14.73
141.1	17.94
190.8	23.93
239.9	29.61
289.0	35.18
332.8	40.02
378.4	44.82
434.8	50.76
477.3	55.05
536.8	61.01
593.1	66.40
689.1	75.47
760.0	81.78

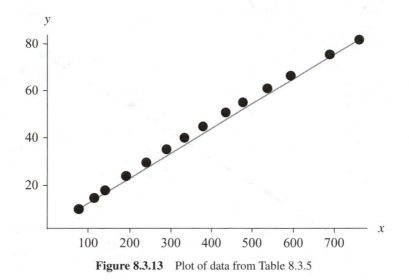

Figure 8.3.13 Plot of data from Table 8.3.5

Quick Review Question 4

Suppose *xLst* and *yLst* are lists of *x* and *y* values, respectively. Give the command in an appropriate computational tool to plot these data with large points as in Figure 8.3.13. If necessary in your tool, assign to *pts* the list of ordered pairs of corresponding *x* and *y* values but not to display the result, and then, use *pts* in a plot command.

The dots of Figure 8.3.13 are in a concave-down pattern. As with concave-up graphs having coordinates greater than 1, we can transform these data from concave up to straight in one of two ways:

1. Pull points left, shrinking the distance from the *y*-axis to the rightmost points more than to those on the left. Thus, perform an operation on *x* that results in smaller values and has more effect on the larger *x* values than on the smaller ones.
2. Extend points up, stretching the distance from the *x*-axis to the higher points more than to the lower ones. Thus, perform an operation on *y* that results in greater values and has more effect on the larger *y* values than on the smaller ones.

Using transformations from the sequence in Table 8.3.5, we can transform *x* or *y*. As Figure 8.3.14 shows, the plot of points $(x, y^{6/5})$ is close to being linear.

With a fit function, we can obtain the following equation of a line that captures the trend of these points:

$$u = -5.46747 + 0.267834z$$

Substituting $y^{6/5}$ for *u* and *x* for *z*, we obtain the following equation:

$$y^{6/5} = -5.46747 + 0.267834x$$

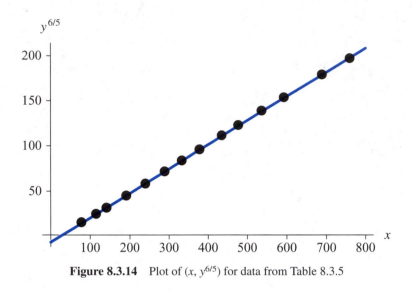

Figure 8.3.14   Plot of $(x, y^{6/5})$ for data from Table 8.3.5

After solving for y by raising each side to the 5/6 power, we can define our model, as follows:

$$f(x) = (-5.48629 + 0.267869x)^{5/6}$$

Figure 8.3.15 shows the graph of this function along with the original data. We should always plot our model with the original data to verify that the function really does capture the trend of the untransformed data.

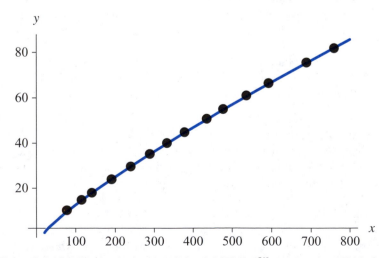

Figure 8.3.15   Graph of $y = (-5.48629 + 0.267869x)^{5/6}$ and data from Table 8.3.5

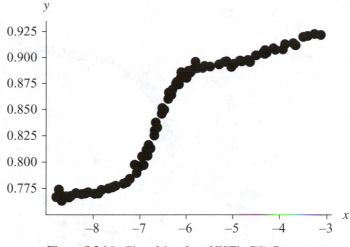

Figure 8.3.16 Plot of data from NIST's *Filip* Dataset

Multiterm Models

In a NIST study, A. Filippelli collected the data that Figure 8.3.16 displays (see the section on "Downloads"). The curve almost has the look of a logistic function. However, the right tail does not trend asymptotically to a value but keeps increasing in a wavy fashion. Thus, one possibility is to develop a polynomial empirical model. Such models are particularly useful because we can readily differentiate and integrate a polynomial for further analyses.

We can fit to the data a fourth-degree polynomial, which is a linear combination of 1, z, z^2, z^3, and z^4 with the form $y = b_0 + b_1z + b_2z^2 + b_3z^3 + b_4z^4$. Without the use of a fit function, in a process called **interpolation**, we can determine the coefficients b_i by solving five equations simultaneously. For each equation, a different data point is substituted into the general fourth-degree polynomial with z being replaced by the first coordinate and y by the second. Then, we solve the five equations simultaneously for b_i, $i = 0,1,2,3,4$. Often, as in this example, we have many more data than interpolation requires. A fit function obtains a least-squares fit of the general fourth-degree polynomial to all the data instead of interpolating through a limited number of specific points. Thus, for the remainder of this section, we continue to use a fit function. Figure 8.3.17, which is a plot of the data and the fourth-degree polynomial that a fit function returns, reveals shortcomings of the model.

Because this fourth-degree polynomial has an inadequate number of "hills" and "valleys" to represent the data, we fit higher-degree polynomials to the points. For a linear combination of expressions, 1, z, z^2, . . . , z^{10}, a fit function returns the following tenth-degree polynomial: $8.45174 + 1.36940z - 5.35707z^2 - 0.34983z^3 - 0.410472z^4 + 0.256553z^5 + 0.119554z^6 + 0.0231150z^7 + 0.00240206z^8 + 0.000131536z^9 + 2.98847 \times 10^{-6}z^{10}$. A graph of this polynomial with the data

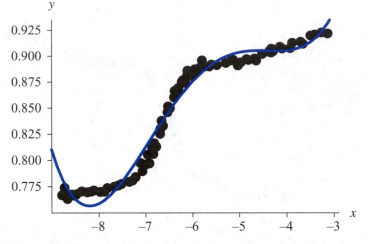

Figure 8.3.17   Plot of the fitted fourth-degree polynomial and the data

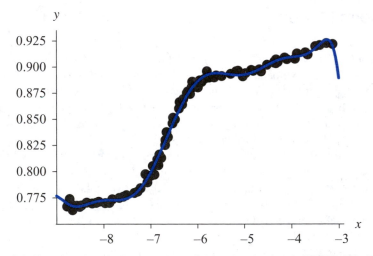

Figure 8.3.18   Plot of tenth-degree polynomial fitted to the data in NIST's *Filip* Dataset

reveals that this empirical model captures the trend of the data better than smaller-degree polynomials (see Figure 8.3.18).

As indicated previously, we must be very careful not to apply this model outside the range of the data. The danger is particularly striking when we consider polynomials. For example, Figure 8.3.19 shows the dramatic slant of this polynomial just one unit to the right and left of the graph in Figure 8.3.18.

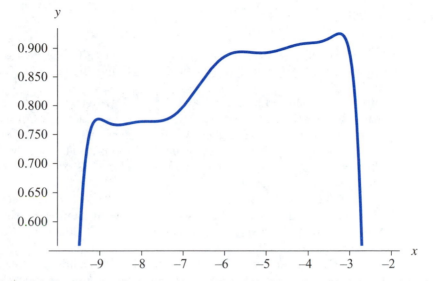

Figure 8.3.19   Plot of tenth-degree polynomial model inside and outside the range of the data

Exercises

1. Using all the data in *NorrisEM.dat*, which is available on the textbook's website, construct an empirical model using a computational tool. Compare your results to the model that was developed with a subset of the data in the section "Linear Empirical Model."

Projects

Develop a model for each of the data sets below (see the section on "Downloads").

1. *NoInt1EM*	**2.** *PontiusEM*	**3.** *Wampler1EM*
4. *Lanczos3EM*	**5.** *Gauss1EM*	**6.** *MGH17EM*
7. *Lanczos1EM*	**8.** *MGH10EM*	**9.** *BoxBODEM*

Answers to Quick Review Questions

1. The following are linear combinations of u and v:
   A. $5u - 18v = (5)u + (-18)v$
   B. $-18v + 5u = (-18)v + (5)u$
   C. $7u = (7)u + (0)v$
   E. $u/5 + v/3 = (1/5)u + (1/3)v$
2–4. The text's website has available for download for various computational tools an *8_3QRQ.pdf* file, which contains these system-dependent Quick Review Questions and answers.

References

Giordano, Frank R., Maurice D. Weir, and William P. Fox. 2003. *A First Course in Mathematical Modeling*. 3rd ed. Pacific Grove, Calif.: Brooks/Cole–Thompson Learning.

Information Technology Laboratory. "NIST Statistical Reference Datasets." National Institute of Standards and Technology. http://www.itl.nist.gov/div898/strd/

NIST Dataset Archives. "StRD Dataset BoxBOD." National Institute of Standards and Technology. http://www.itl.nist.gov/div898/strd/nls/data/boxbod.shtml. Originally from Box, G. P., W. G. Hunter, and J. S. Hunter. 1978. *Statistics for Experimenters*. New York: Wiley: 483–487.

———. "StRD Dataset DanWood." National Institute of Standards and Technology. http://www.itl.nist.gov/div898/strd/nls/data/daniel_wood.shtml. Originally from Daniel, C., and F. S. Wood. *Fitting Equations to Data*. 2000 2nd ed. New York: John Wiley and Sons: 428–431.

———. "StRD Dataset Filip." National Institute of Standards and Technology. http://www.itl.nist.gov/div898/strd/lls/data/Filip.shtml. Originally from Filippelli, A. National Institute of Standards and Technology.

———. "StRD Dataset Gauss1." National Institute of Standards and Technology. http://www.itl.nist.gov/div898/strd/nls/data/gauss1.shtml. Originally from Rust, B. 1996. National Institute of Standards and Technology.

———. "StRD Dataset Lanczos1." National Institute of Standards and Technology. http://www.itl.nist.gov/div898/strd/nls/data/lanczos1.shtml. Originally from Lancoz, C. 1956. *Applied Analysis*. Englewood Cliffs, N.J.: Prentice-Hall: 272–280.

———. "StRD Dataset Lanczos3." National Institute of Standards and Technology. http://www.itl.nist.gov/div898/strd/nls/data/lanczos3.shtml. Originally from Lancoz, C. 1956. *Applied Analysis*. Englewood Cliffs, N.J.: Prentice-Hall: 272–280.

———. "StRD Dataset MGH10." National Institute of Standards and Technology. http://www.itl.nist.gov/div898/strd/nls/data/mgh10.shtml. Originally from Meyer, R. R. 1970. "Theoretical and Computation Aspects of Nonlinear Regression." In *Nonlinear Programming*. Rosen, Mangasarian, and Ritter, eds. New York: Academic Press: 465–486.

———. "StRD Dataset MGH17." National Institute of Standards and Technology. http://www.itl.nist.gov/div898/strd/nls/data/mgh17.shtml. Originally from Osborne, M. R. 1972. "Some Aspects of Nonlinear Least Squares Calculations." In *Numerical Methods for Nonlinear Optimization*. Lootsma, ed. New York: Academic Press: 171–189.

———. "StRD Dataset Misra1a." National Institute of Standards and Technology. http://www.itl.nist.gov/div898/strd/nls/data/misra1a.shtml. Originally from Misra, D. 1978. "Dental Research Monomolecular Adsorption Study." National Institute of Standards and Technology.

———. "StRD Dataset NoInt1." National Institute of Standards and Technology. http://www.itl.nist.gov/div898/strd/lls/data/NoInt1.shtml. Originally from Eberhardt, K. National Institute of Standards and Technology.

———. "StRD Dataset Norris." National Institute of Standards and Technology. http://www.itl.nist.gov/div898/strd/lls/data/Norris.shtml. Originally from Norris, J. "Calibration of Ozone Monitors." National Institute of Standards and Technology.

————. "StRD Dataset Pontius." National Institute of Standards and Technology. http://www.itl.nist.gov/div898/strd/lls/data/Pontius.shtml. Originally from Pontius, P. "Load Cell Calibration." National Institute of Standards and Technology.

————. "StRD Dataset Wampler1." National Institute of Standards and Technology. http://www.itl.nist.gov/div898/strd/lls/data/Wampler1.shtml. Originally from Wampler, R. H. 1970. "A Report of the Accuracy of Some Widely-Used Least Squares Computer Programs." *Journal of the American Statistical Association*, 65: 549–565.

9

MONTE CARLO SIMULATIONS

MODULE 9.1

Computational Toolbox—Tools of the Trade: Tutorial 4

Prerequisite: Module 8.1 on "Computational Toolbox—Tools of the Trade: Tutorial 3."

Download

From the textbook's website, download Tutorial 4 in the format of your computational tool or in pdf format. We recommend that you work through the tutorial and answer all Quick Review Questions using the corresponding software.

Introduction

This fourth computational toolbox tutorial, which is available from the textbook's website in your system of choice, prepares you to use the system to complete projects for this and subsequent chapters. The tutorial introduces the following functions and concepts:

- Random numbers
- Modulus function
- *If* statement
- Additional list functions
- Loading a file
- Removing function definitions
- The mean and standard deviation functions
- Histograms
- Defining a package of function and constant definitions

The module gives computational examples and Quick Review Questions for you to complete and execute in the desired software system.

MODULE 9.2

Simulations

Modeling is the application of methods to analyze complex, real-world problems in order to make predictions about what might happen with various actions. When it is too difficult, time consuming, costly, or dangerous to perform experiments, the modeler might resort to **computer simulation**, or having a computer program imitate reality, in order to study situations and make decisions. Simulating a process, he or she can consider various scenarios and test the effect of each.

For example, a scientist might simulate the effects of ozone depletion on global warming. Scientists at Los Alamos National Laboratory used simulations to predict the behavior of nuclear reactions before physically testing a nuclear bomb during World War II (LANL). Lawrence Livermore National Laboratory scientists have used molecular dynamics simulations to study the total energy and other quantities associated with molecules as they interact with one another. At the same laboratory, they have studied the Greenhouse Effect, making predictions based on levels of various pollutants (LLNL). Before the Gulf War, military experts simulated a number of scenarios to test preparedness. The National Oceanographic and Atmospheric Administration performs simulations to predict the path and intensities of hurricanes (NOAA). The Boeing Company designed the Boeing 777 airplane completely using computer-aided design and tested the designs using computer simulations before construction began. Also, flight simulators allow pilots to practice emergency situations under safe conditions (Boeing).

We use simulations if one or more of the following statements is true:

- It is not feasible to do the actual experiment, as in the study of the Greenhouse Effect.
- The cost in money, time, or danger of the actual experiment is prohibitive, as with the study of nuclear reactions.
- The system does not exist yet, as in the development of an airplane.
- We want to test various alternatives, as with hurricane predictions.

Element of Chance

At the core of simulation is random number generation. The computer generates a sequence of numbers, called **random numbers** or **pseudorandom numbers**. An algorithm actually produces the numbers, so they are not really random, but they appear to be random. Because of the element of chance, we often call a simulation a **Monte Carlo simulation**, named after the gambling capital. A Monte Carlo simulation is a probabilistic model involving an element of chance. Hence, a simulation is not deterministic but is probabilistic or stochastic. Each time a simulation is run, results can vary from those of other runs.

> **Definition**   A **Monte Carlo simulation** is a probabilistic model involving an element of chance.

Disadvantages

Despite the many applications and advantages, the following are some disadvantages of computer simulations:

- The simulation may be expensive in time or money to develop.
- Because it is impossible to test every alternative, we can provide good solutions but not a best solution.
- Because a simulation is probabilistic involving an element of chance, we should be careful of our conclusions.
- The results may be difficult to verify because often we do not have real-world data.
- We cannot be sure we understand what the simulation actually does.

Genesis of Monte Carlo Simulations

Computer random number generators have been available since some of the earliest days in the development of computers. John von Neumann, who introduced the idea of storing programs as well as data in the memory of the computer, also helped to develop the first algorithm for generating pseudorandom numbers with the computer.

Born in Budapest, Hungary, in 1903, von Neumann received his Ph.D. in mathematics at the age of 22. He contributed significantly to a variety of areas: the mathematical foundation of quantum theory, logic, the theory of games, economics, nuclear weapons, meteorology, as well as theory and applications in early computer science. Many stories tell of his phenomenal memory, reasoning ability, and computational speed. He could memorize a column of the telephone book at a glance, and he had mastered calculus by age 8. Halmos wrote in *Legend of John von Neumann*,

When his electronic computer was ready for its first preliminary test, someone suggested a relatively simple problem involving powers of 2. (It was
something of this kind: what is the smallest power of 2 with the property
that its decimal digit fourth from the right is 7? This is a completely trivial
problem for a present-day computer: it takes only a fraction of a second of
machine time.) The machine and Johnny started at the same time, and
Johnny finished first. (Halmos 1973)

During World War II, physicists on the Manhattan Project developed the concept of
Monte Carlo simulation. Scientists knew the behavior of one neutron, but they did not
have a formula for how a system of neutrons would behave. Although they needed to
understand such behavior to construct dampers and shields for the atomic bomb, experimentation was too time consuming and dangerous. John von Neumann and Stanislaus Ulam developed the technique of Monte Carlo simulation to solve the problem.

Multiplicative Linear Congruential Method

Many besides von Neumann have contributed to the theory of random numbers. In
1949, D. J. Lehmer presented one of the best techniques for generating uniformly
distributed pseudorandom numbers, the **linear congruential method**.

One simple linear congruential random number generator that generates values
between 0 and 10 inclusively is as follows:

$$r_0 = 10$$
$$r_n = (7r_{n-1} + 1) \bmod 11, \text{ for } n > 0$$

The initial value in the sequence of random numbers, $r_0 = 10$, is the **seed**. The **mod**
function returns the remainder. For example, 71 *mod* 11 is 5, the remainder in the division of 11 into 71. Thus, substituting $r_0 = 10$ on the right-hand side of the second
line of the definition, the **generating function**, we calculate $r_1 = (7 \times 10 + 1) \bmod$
$11 = 5$. After we calculate one "random number," to evaluate the next, we substitute
that value into the expression on the right hand side. Consequently, the next random
number is $r_2 = (7 \times \mathbf{5} + 1) \bmod 11 = 36 \bmod 11 = 3$.

Quick Review Question 1

Using this random number generator and $r_2 = 3$, calculate the next random number,
r_3, in the sequence.

Continuing in this fashion, we obtain ten pseudorandom numbers 5, 3, 0, 1, 8, 2,
4, 7, 6, 10 before the sequence starts repeating. A maximum of 11 nonnegative integers is generated for computation with *mod* 11.

Should we desire floating point numbers between 0 and 1, we divide each number in the sequence by the **modulus**, 11, to obtain the following sequence:

$$\frac{5}{11}, \frac{3}{11}, \frac{0}{11}, \frac{1}{11}, \frac{8}{11}, \frac{2}{11}, \frac{4}{11}, \frac{7}{11}, \frac{6}{11}, \frac{10}{11}$$

or

$$0.454545, 0.272727, 0.0, 0.0909091, 0.727273, 0.181818,$$

$$0.363636, 0.636364, 0.545455, 0.909091$$

For this computation, the smallest possible pseudorandom floating point number is 0.0 and the largest is $(modulus - 1)/modulus = 10/11$. Thus, floating point numbers that we generate by dividing by the modulus are in the interval $[0.0, 1.0)$, or the interval between the two values that includes 0.0 but not 1.0.

> The general form for the **linear congruential method** to generate pseudorandom integers from 0 up to, but not including, *modulus* is as follows:
>
> $$r_0 = seed$$
> $$r_n = (multiplier \times r_{n-1} + increment) \bmod modulus, \text{ for } n > 0$$
>
> where *seed*, *modulus*, and *multiplier* are positive integers and *increment* is a nonnegative integer.

Not all choices of *multiplier* and *modulus* are good. For example, consider a similar function with a multiplier of 5, $r_0 = 10$ and $r_n = (5\, r_{n-1} + 1)\ mod\ 11$ for $n > 0$. This function only produces 5 numbers—7, 3, 5, 4, 10—before returning to 7. The random number generator should give as long a sequence as possible. Another desirable characteristic of such functions is that the sequence appears random. For example, using the function $r_0 = 10$ and $r_n = (2\, r_{n-1})\ mod\ 11$, we obtain the sequence 9, 7, 3, 6, **1, 2, 4, 8**, 5, 10. With a subsequence containing powers of 2, the sequence does not appear random.

Much research has been done to discover choices for *multiplier* and *modulus* that give the largest possible sequence that appears random. For built-in random number generators, *modulus* is often the largest integer a computer can store, such as $2^{31} - 1 = 2,147,483,647$ on some machines. For this modulus, a multiplier of 16,807 and an increment of 0 produce a sequence of $2^{31} - 2$ elements.

Different Ranges of Random Numbers

Many computational tools have generators that can produce uniformly distributed integer or real random numbers in various ranges. Other software systems have limited options, such as only a generator for nonnegative random integers or no generator at all. The previous section described the linear congruential method for generating a random integer from 0 up to the modulus, where by "up to" we do not include the modulus. We also saw how to obtain a floating point counterpart with value from 0.0 up to 1.0 by dividing by the modulus. In this section, we discuss how to obtain uniformly distributed integer or real random numbers in any range. Module 9.4 considers how to generate random numbers from other distributions.

Example 1

For this discussion, suppose that *rand* is a uniformly distributed random floating point number from 0.0 up to 1.0. Suppose, however, we need a random floating point number from 0.0 up to 5.0. Because the length of this interval is 5.0, we multiply *rand* by this value 5.0 to stretch the interval of numbers. Mathematically we have the following:

$$0.0 \leq rand < 1.0$$

Thus, multiplying by 5.0 throughout, we obtain the correct interval, as shown:

$$0.0 \leq 5.0 \; rand < 5.0$$

   If the lower bound of the range is different from 0, we add that bound. For example, if we need a random floating point number from 2.0 up to 7.0, we multiply by the length of the interval, $7.0 - 2.0 = 5.0$, to expand the range. Then, we add the lower bound, 2.0, to shift or translate the result so that the following inequalities hold:

$$2.0 \leq (7.0 - 2.0) \; rand + 2.0 < 7.0$$

or

$$2.0 \leq 5.0 \; rand + 2.0 < 7.0$$

$$2.0 \leq 5.0 \; rand + 2.0 < 7.0$$

Specifying Random Floating Point Numbers in Other Ranges

If *rand* is a random floating point number such that $0.0 \leq rand < 1.0$, then $(max - min) \; rand + min$ is a random floating point number from *min* up to *max* that satisfies the following inequality:

$$min \leq (\mathbf{max} - \mathbf{min}) \; \mathbf{rand} + \mathbf{min} < max$$

Quick Review Question 2

Suppose *rand* is a random floating point number from 0.0 up to 1.0.

   a. Give an expression to obtain a random floating point number from 14.5 up to 24.5.
   b. Give the range of random numbers for the expression $73.9 \; rand + 21.2$.

Example 2

Frequently, we need a more restricted range of random integers than from 0 up to *modulus*. For example, a simulation might require random integer temperatures between 0 and 99, inclusively. One method of restricting the range is to multiply a floating point random number between 0.0 and 1.0 by 100 (the number of integers from 0 through 99, or 99 + 1) and then return the **integer part** (the number before the decimal point). For example, suppose *rand* is 0.692871. Multiplying by 100, we obtain $100(0.692871) = 69.2871$. Truncating, we obtain an integer (69) between 0 and 99.

Sometimes we want the range of random integers to have a lower bound other than 0, for example from 100 to 500, inclusively. Because we include 100 and 500 as options, the number of integers from 100 to 500 is one more than the difference in these values, $(500 − 100 + 1) = 401$. As with the last example, we multiply this value by *rand* to expand the range. Then, we add the lower bound, 100, to the product to translate the range to start at 100 as follows:

$$100.0 \le 401 \; rand + 100 < 501.0$$

Finally, we take the integer part of the result, which we write here as applying a function *int*:

$$100 \le int(401 \; rand + 100) < 501$$

or

$$100 \le int(401 \; rand + 100) \le \mathbf{500}$$

Because the floating point numbers $(401 \; rand + 100)$ are less than 501.0, after truncation, the largest possible integer part is 500.

> **Specifying Random Integers in Other Ranges**
>
> If *rand* is a random floating point number such that $0.0 \le rand < 1.0$, then $int((max − min + 1) \; rand + min)$ is a random integer from *min* to *max*, inclusively, that satisfies the following inequality:
>
> $$min \le int(\; (max − min + 1) \; rand + min) \le max$$
>
> where *int* is a function that returns the integer part of a number.

Quick Review Question 3

Suppose *rand* is a random floating point number from 0.0 up to 1.0. Assume that *int* is a function that returns the integer part of a floating point number.

   a. Give an expression to obtain a random integer from 28 to 41, inclusively.
   b. Give the range random numbers for the expression $int(73 \; rand + 21)$.

Exercises

Evaluate Exercises 1–3.

1. $349 \bmod 7$ **2.** $4621 \bmod 100$ **3.** $11{,}382 \bmod 542$

4. Consider the following linear congruential random number generator:

$$r_0 = 8697$$
$$r_n = (229\, r_{n-1})\bmod 349, \text{ for } n > 0$$

   a. Compute the next three random numbers.
   b. From the sequence of integers in Part a, compute an appropriate sequence of floating point numbers between 0 and 1.
   c. Give the maximum number of random numbers this function can generate.

5. Repeat Exercise 4 for the following random number generator:

$$r_0 = 1021$$
$$r_n = (467 r_{n-1})\bmod 1024, \text{ for } n > 0$$

6. Repeat Exercise 4 for the following random number generator:

$$r_0 = 8367$$
$$r_n = (229 r_{n-1} + 1)\bmod 10{,}000, \text{ for } n > 0$$

7. The following guidelines for choice of modulus, multiplier, and increment have been developed through computer testing of random number generators:

   • The modulus is a positive integer.
   • The multiplier and increment are nonnegative integers less than the modulus.
   • If working on a decimal machine, choose the modulus to be a large power of 10 for easy computation of the *mod* function. Most computers, however, are binary machines so that the modulus should be a large power of 2, such as 2^{32}. Division by 2^{32} moves the binary point 32 places to the left in a binary number.
   • On a binary computer, choose *multiplier* such that *multiplier mod modulus* = 5 and 0.01*modulus* < *multiplier* < 0.99*modulus*.
   • No integer great than 1 should divide both the increment and the modulus.

   Give three choices for a multiplier that meets the suggested criteria above with a modulus of 2^{20}.

8. a. If the modulus is 2^{32} and the increment is a nonnegative integer less than 10, list the choices for the increment based on the guideline in Exercise 7 that no integer greater than 1 should divide both the modulus and the increment.

   b. List choices for a nonnegative increment less than 10 if the modulus is 10^{19}.

For the following exercises, assume that rand *is a random floating point number from 0.0 up to 1.0 and that* int *is a function that returns the integer part of a number. For each exercise, write an expression to return a random number in the given interval:*

9. Random floating point number from 0.0 up to, but not including, 20.0

10. Random floating point number from 6.0 up to, but not including, 26.0

11. Random floating point number from 35.8 up to, but not including, 73.4

12. Random floating point number from -8.0 up to, but not including, 4.0

13. Random integer between 0 and 20, inclusively; thus, in {0, 1, 2, . . . , 20}

14. Random integer between 6 and 26, inclusively

15. Random integer between 35 and 73, inclusively

16. Random integer between −8 and 4, inclusively

Projects

1. Create a file containing definitions of your own random number seeding and generating functions, as the parts below describe. Do the requested parts with an appropriate computational tool without using its built-in random number generator. Starting with Part b, you define several versions of your own random number generator, if possible, each with the name *myRandom*. In a call to *myRandom*, the parameters determine which form the computational tool uses. If your computational tool does not allow the use of the same function name for different versions of the function, use different function names. Each definition's format should mimic that of your tool's corresponding built-in function. The definitions in Parts c–h should call the function of Part b. The function *myRandom* should assign a new pseudorandom number to *myRandValue*. Employ the old value of *myRandValue* to generate the new value. Test each function thoroughly by generating and checking a number of values.

a. Give the function *mySeedRandom* two definitions that seed your random number generator by assigning a value to a variable *myRandValue*. Because the value of *myRandValue* must persist and be available to both definitions of *mySeedRandom*, depending on your computational tool, you may need to declare *myRandValue* as a global variable. First, define *mySeedRandom* that assigns an integer argument to *myRandValue*. Thus, *mySeedRandom* with an argument of 12345 assigns 12345 to *myRandValue*. Second, define *mySeedRandom,* with no argument so that *myRandValue* becomes an integer associated with the date and time of day. For testing, clear out any value of *myRandValue*, and test that *mySeedRandom* with various arguments assigns the correct value to *myRandValue*.

b. Define a function *myRandom*, as in Exercise 4a, to return a random integer between 0 and 348, inclusively. If *myRandValue* does not have a value because we did not call *mySeedRandom*, use an initial value of 1 for *myRandValue*.

For each of the following parts, define a function, if possible, called myRandom, *to return the designated type of random number. In each definition,*

directly or indirectly (by calling a function that calls this function) invoke the
function in Part b.

   c. A random floating point number from 0.0 up to 1.0, expressed as a deci-
       mal number, not a fraction
   d. A random floating point number from 0.0 up to *max*
   e. A random floating point number from *min* up to *max*
   f. 0 or 1 at random
   g. A random integer between 0 and *max*, inclusively
   h. A random integer between *min* and *max*, inclusively
2. Do Project 1 using the generator in Exercise 5.
3. Do Project 1 using the generator in Exercise 6.
4. Do Project 1 for a modulus of $2^{31} - 1$, a multiplier of 16,807, and an increment of 0.

Using a programming language, repeat the indicated project.

      5. Project 1   **6.** Project 2
      7. Project 3   **8.** Project 4

Answers to Quick Review Questions

   1. 0 because $r_2 = (7 * 3 + 1)\ mod\ 11 = 22\ mod\ 11 = 0$
   2. a. 10.0 *rand* + 14.5
      b. 21.2 up to 95.1 = 21.2 + 73.9
   3. a. $int(14\ rand + 28) = int((41 - 28 + 1)\ rand + 28)$
      b. Integers from 21 to 93, inclusively, or {21, 22, 23, . . . , 93}

References

Boeing. "The Boeing Company." http://www.boeing.com/

Halmos, P. R. 1973. "Legend of John von Neumann." *American Mathematical Monthly*, 80: 382–394.

LANL (Los Alamos National Laboratory). "Los Alamos National Laboratory Homepage." The University of California and the U.S. Department of Energy. http://www.lanl.gov/

LLNL (Lawrence Livermore National Laboratory). "Lawrence Livermore National Laboratory Homepage." The University of California and the U.S. Department of Energy. http://www.llnl.gov/

NOAA (National Oceanographic and Atmospheric Administration). "NOAA Homepage." U.S. Department of Commerce. http://www.noaa.gov/

Park, Stephen K., and Keith W. Miller. 1988. "Random Number Generators: Good Ones Are Hard to Find." *Communications of the ACM*, 31(10): 1192–1201.

MODULE 9.3

Area Through Monte Carlo Simulation

Download

For several computational tools, the text's website has available for download an *Area* file, which contains the models in this module.

Introduction

One of the fundamental problems of calculus is to find the area between a curve and the (horizontal) x-axis over a certain interval. In this example, we use computer simulation to estimate the area for functions that are above the x-axis. With integration we can compute the exact area of many functions, but with others it is impossible. For example, the graph of the function $f(x) = \sqrt{\cos^2(x)+1}$ on the interval between $x = 0$ and $x = 2$ in Figure 9.3.1 is entirely above the x-axis. Unfortunately, we cannot

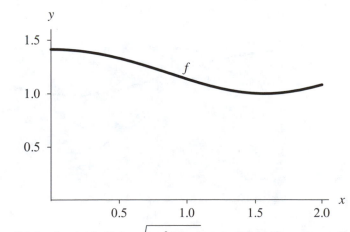

Figure 9.3.1   Graph of $f(x) = \sqrt{\cos^2(x)+1}$ on the interval between $x = 0$ and $x = 2$

use integration to find the pictured area exactly because the function does not have an antiderivative that is an elementary function. Although Monte Carlo simulation is probabilistic, the technique can model deterministic behavior, such as area under a curve. The module uses the development of a stochastic numeric integration technique to introduce some of the fundamental concepts of Monte Carlo simulation, which has so many applications in the sciences.

Throwing Darts for Area

One method to estimate the area is to enclose the region in a rectangle of known dimensions as in Figure 9.3.2. We have picked a rectangle of an arbitrary height, such as 1.5, higher than f in the interval. Then, we hypothetically throw darts at the rectangle, counting the total number of darts thrown and the number of darts that hit below the graph. To estimate the desired area, we take the proportion of dart hits below the curve times the total area of the rectangle, which we calculate as follows:

$$\text{area} \approx (\text{area of enclosing rectangle})\left(\frac{\text{\# darts below}}{\text{\# darts}}\right)$$

We can easily compute the area of the rectangle as width times height; in this case the area is $(2)(1.5) = 3.0$. If we throw 1000 darts and 778 of them hit below the graph, then $778/1000 = 0.778 = 77.8\%$ of the total lands below f. This fraction is an estimate of the portion of the rectangle that is below f; about 77.8% of the rectangle's area rests between f and the x-axis. Thus, we estimate this smaller area by taking 77.8% of the total area of the rectangle,

$$\text{area} \approx 3.0 * 778/1000 = 2.334$$

We can use a random number generator to simulate throwing the darts. As Figure 9.3.3 pictures, our "dartboard" extends from $SMALLEST\_X = 0.0$ to

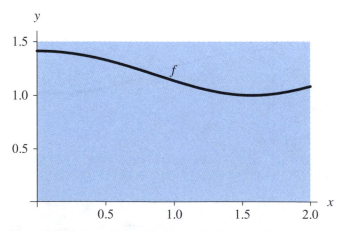

Figure 9.3.2   Area depicted in Figure 9.3.1, enclosed in a rectangle

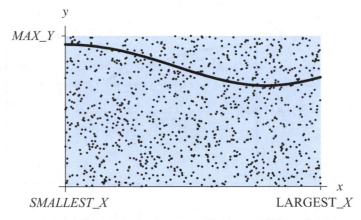

Figure 9.3.3 Bounds of "dartboard" for area in Figure 9.3.2

LARGEST_X = 2.0 horizontally and from 0.0 to *MAX_Y* = 1.5 vertically. Thus, using these boundaries, we compute the area of the rectangle as

$$\text{area of rectangle} = (2.0 - 0.0) * 1.5 = 3.0$$

The number of darts thrown, *NUMDARTS*, is given a value, such as 1000. Each time through a loop that executes *NUMDARTS* number of times, we generate a random *x* value, *randomX*, between 0.0 and 2.0. We also generate a random *y* value, *randomY*, between 0 and 1.5. Thus, (*randomX*, *randomY*) are the coordinates of where a dart hits the rectangular board. We also must determine where the dart hits in relationship to the curve. If the dart hit exactly on the curve, then its *y* coordinate would be *f*(*randomX*). For example, the random number generator might return 1.434065 for *randomX* and 0.715644 for *randomY*. Substituting the *x* value into the function *f*, we have

$$f(1.434065) = \sqrt{\cos^2(1.434065) + 1} = 1.009247$$

Since *randomY* = 0.715644 is less than *f*(1.434065) = 1.009247, as Figure 9.3.4 shows, the dart hits below the curve.

Quick Review Question 1

Consider the linear function $g(x) = 3x + 1$ from $x = 1$ to $x = 5$.

  a. Give a value of *MAX_Y* that is the height of the smallest rectangle enclosing the desired area.
  b. Suppose a random hit of a simulated dart is at location (3, 11). Does the dart hit above, below, or on the curve?
  c. Using geometry, compute the area under *g*.
  d. Compute the area of the smallest enclosing rectangle.

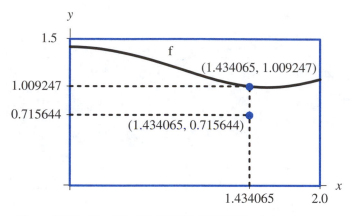

Figure 9.3.4   Dart hit at (1.434065, 0.715644) is below the curve

e. Suppose a simulation "throws" a million "darts" at the rectangle. Although any value is possible, from the following choices, indicate the most likely percentage of darts landing below the graph: 33%, 48%, 59%, 62%, 71%.

f. If 600,000 of a million darts land below the graph, estimate the area.

For each simulated hit, the computer checks the position of the strike in relationship to the function. If the hit is below the graph, the function increments a counter by 1.

Because we are using random numbers to estimate the area, we expect most runs to produce different results. The following estimates of the area under the curve from three executions with 1000 "darts" demonstrate this variation: 2.364, 2.349, and 2.334.

Measure of Quality

Theoretically, we can obtain a better estimate by throwing a larger number of darts. In this case, we define *NUMDARTS* to be larger or run the simulation several times, taking the mean (average) of the area estimates from all the runs. The latter technique has the advantage of enabling us to use the **standard deviation** (σ) of the estimates from the different executions as a measure of the quality of the overall estimate. About 68.3% of the estimates are within $\pm\sigma$ of the mean. Thus, for a mean of 6.2838 and a standard deviation of 0.442276, 68.3% of the area estimates are between $6.2838 - 0.442276 = 5.79602$ and $6.2838 + 0.442276 = 6.72608$. A small standard deviation relative to the mean indicates a certain consistency for most of the simulations and gives us more confidence in the mean as an estimate of the area.

Quick Review Question 2

a. Suppose the mean of a number of simulations to compute the area under a curve is 40.10 and the standard deviation is 0.20. Is the estimate 39.94 within one standard deviation of the mean?

b. Is it better for the standard deviation to be smaller or larger?

We should make a caution about throwing many darts: If we generate a large number of points, we have to be sure our sequence of random numbers does not start repeating one or more times, skewing the results.

Algorithm

The following algorithm gives the logic for one run of a Monte Carlo simulation to calculate the area under a curve defined by a function in one variable and whose range is nonnegative on the interval of interest.

> ***Monte Carlo Algorithm to Estimate the Area between the x-axis and f from SMALLEST_X to LARGEST_X***

$MAX\_Y \leftarrow$ number greater than or equal to all $f(x)$ in the interval
$NUMDARTS \leftarrow$ the number of darts to be thrown
$NumHitsBelow \leftarrow 0$
Do the following $NUMDARTS$ times:
    $randomX \leftarrow$ random number between $SMALLEST\_X$ and $LARGEST\_X$
    $randomY \leftarrow$ random number between 0 and MAX_Y
    if $(randomY < f(randomX))$
        increment $NumHitsBelow$ by 1

$AreaOfRectangle \leftarrow (LARGEST\_X - SMALLEST\_X) \cdot (MAX\_Y)$
$EstimatedArea \leftarrow (AreaOfRectangle) \cdot (NumHitsBelow)/NUMDARTS$

To obtain an area estimate in which we have more confidence, we run the simulation with many darts a number of times using different seeds for the random number generator and compute the mean and standard deviation of the estimates. If the standard deviation is small relative to the mean, then we accept with greater confidence the mean as the estimated area.

Implementation

An understanding of how to develop simulations in a computational tool can easily extend to implementation in other programming languages, including one of several high-level simulation languages that exist. An *Area* file, which is available for several computational tools from the text's website, and the exercises lead us through the process of estimating the area under the curve $f(x) = x^2$ between $x = 2$ and $x = 3$ using Monte Carlo simulation.

Exercises

The following exercises implement a Monte Carlo simulation for estimating the area under the curve $f(x) = x^2$ between $x = 2$ and 3. Write commands in your computational tool of choice.

1. This question defines, plots, and integrates the function.
   a. Clear old values for f and x and then define $f(x) = x^2$.
   b. Plot f from 2 to 3, showing the origin and the desired region.
   c. With calculus or an appropriate computational tool, find the value of $\int_2^3 x^2 dx$ exactly. For this function where we can integrate analytically, we can use this value to help verify our implementation.
2. For understanding and testing purposes, this question develops the code for one simulation involving only 10 darts.
   a. Generate a table/array, *dartTbl*, of ten 0s and 1s so that an entry is 1 if a random y value is less than the function at a random x value and that the element is 0 otherwise.
   b. Calculate the fraction (*fractionUnder*) of darts that hit under the curve.
   c. Calculate the area (*rectArea*) of the rectangular dartboard.
   d. Determine an estimate for the area (*area*) under the curve between 2 and 3.
3. We obtain a better estimate if we use more "darts." Revise the statements from Exercise 2 to estimate the area by throwing 1000 darts.
4. To obtain an area estimate in which we have more confidence, in this exercise we perform the simulation a number of times and calculate the mean and the standard deviation of the estimated areas.
   a. If necessary, load a statistics package.
   b. Generate a 100×100 table/array (*dartTbl*) of 0s and 1s (see Exercise 2a), where each row consists of one simulation of throwing 100 darts.
   c. Calculate the fraction (*fractionUnder*) of darts that hit under the curve for each row (simulation), placing the results in a list/vector.
   d. Find the mean and standard deviation of simulation results, and express the answers as floating point numbers.
   e. Repeat Parts b–d using columns instead of rows. In your computational tool, it might be advantageous to take the transpose of *dartTbl* and work with rows again.
   f. Calculate the fraction (*fractionUnder*) of darts that hit under the curve for the entire table/array, and find the area as a floating point number.

Projects

1. Work through an *Area* file in an appropriate computational tool to estimate the area under the curve x^2 on the interval from 2 to 3 using Monte Carlo simulation.
2. Using a copy of an *Area* file in an appropriate computational tool, repeat Project 1 for the curve e^{x^2} on the interval from 0 to 1.
3. Using the techniques of this section, develop a program to estimate π. The area of a circle is πr^2, where r is the radius. The equation of a circle of radius r with center at the origin is

$$x^2 + y^2 = r^2$$

Use a circle of radius 1. Consider the quarter of the circle in the first quadrant with $0 \leq x \leq 1$ and $0 \leq y \leq 1$, and multiply your result by 4.

4. Develop a program to estimate the volume of a sphere of radius 1 whose equation is

$$x^2 + y^2 + z^2 \leq 1$$

Consider the portion of the sphere with $x \geq 0$, $y \geq 0$, and $z \geq 0$; and multiply your result by 8. In the case of three dimensions, a point has three coordinates (x, y, z). Notice that Monte Carlo integration is useful for dimensions beyond two.

5. Using Monte Carlo techniques, estimate $\int_2^3 \sin(x^2)dx$. Note that the function is not entirely above or entirely below the x-axis, so you must adjust the algorithm in the text to estimate the integral. Recall that where a function is negative (below the x-axis), its integral is the negative of the area between the curve and the x-axis.

6. Repeat Project 5 for $\int_0^{2\pi} (1 - x^{\sin(x)})dx$.

In each of the following projects, develop a Monte Carlo simulation program in a programming language for the requested estimate.

7. Estimate the area under the curve x^2 on the interval from 2 to 3.
8. Estimate the area under the curve e^{x^2} on the interval from 0 to 1.
9. Estimate the integral in Project 5.
10. Estimate the integral in Project 6.

Answers to Quick Review Questions

1. **a.** $16 = 3(5) + 1$
   b. above because $3(3) + 1 = 10 < 11$
   c. $40 = (16 + 4)(4)/2$ is the area of the trapezoid
   d. $64 = (5 - 1)(16)$
   e. 62% because $40/64 = 62.5\%$
   f. 38.4 square units $= (64)(600,000/1,000,000) = (64)(0.6)$
2. **a.** yes, numbers between $40.10 - 0.20$ and $40.10 + 0.20$—that is, between 39.90 and 40.30—are within one standard deviation of the 40.10
   b. smaller

Reference

Einwohner, Theodore. 1983. Personal communication.

MODULE 9.4

Random Numbers from Various Distributions

Downloads

Introduction

Monte Carlo simulations are important tools in scientific work and yield solutions to problems unobtainable by other means. Moreover, where alternative solutions are possible, such simulations often provide greater precision for the same computer cost.

A Monte Carlo simulation requires the use of unbiased random numbers. The **distribution** of these numbers is a description of the portion of times each possible outcome or each possible range of outcomes occurs on the average over a great many trials. However, the distribution that a simulation requires depends on the problem. In this module, we discuss the algorithms for generating random numbers from several types of distributions.

> **Definition**   A **distribution** of numbers is a description of the portion of times each possible outcome or each possible range of outcomes occurs on the average.

Statistical Distributions

In Module 9.2 on "Simulations," we considered the linear congruential method to generate pseudorandom numbers with a uniform distribution. Suppose a specified

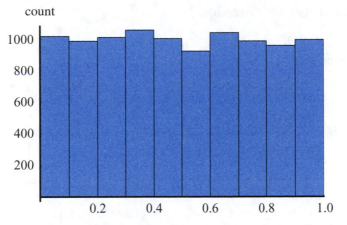

Figure 9.4.1 Histogram of 10,000 random floating point numbers, uniformly distributed from 0.0 up to 1.0

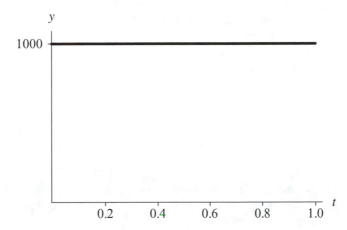

Figure 9.4.2 Horizontal line at height 1000 approximately goes across the top of the histogram in Figure 9.4.1

range is partitioned into intervals of the same length. With a **uniform distribution**, the generator is just as likely to return a value in any of the intervals. Equivalently, in a list of many such random numbers, on the average each interval contains the same number of generated values. For example, Figure 9.4.1 presents a histogram with 10 intervals of length 0.1 of a table of 10,000 random floating point numbers, uniformly distributed from 0.0 up to 1.0. As expected, approximately one-tenth of the 10,000, or 1000 numbers, appears in each subdivision. Thus, the curve across the tops of the bars is virtually a horizontal line of height 1000 (see Figure 9.4.2). As we will see, methods for generating random numbers in other distributions depend on our ability to produce random numbers with a uniform distribution.

Quick Review Question 1

Suppose we have a uniformly distributed random number generator that returns a floating point value from 0.0 to 1.0.

    a. Suppose we break the interval from 0.0 to 1.0 into 5 subintervals. If a list contains 100 random numbers, give the number of values we expect in each subinterval on the average.

    b. In a histogram of the data, as in Figure 9.4.1, give the height of each bar on the average.

    c. In general, for a list of n random numbers between 0.0 and 1.0 and for i subintervals, give the number of values we expect in each subinterval on the average.

    d. In a histogram of the data, as in Figure 9.4.1, give the height of each bar on the average.

A distribution can be **discrete** or **continuous**. To illustrate the difference between the terms discrete and continuous, a digital clock shows time in a discrete manner, from one minute to the next, while a clock with two hands indicates time in a continuous, unbroken way. Similarly, as you pass the time-and-temperature sign in front of a bank, one moment it might register 28 degrees Celsius, the next it might jump to 29 degrees. As a continuous counterpart, a thermometer outside a house might have a column of liquid, smoothly rising and falling to indicate the temperature. In a simulation of pollution, we might generate a random integer to indicate the number of dust particles in a cubic meter of air. The distribution of such values is discrete. In the same simulation, for the velocities of the particles, we might generate random floating point values that have a continuous distribution. However, as noted in Module 2.2 on "Errors," the expression of numbers in a computer is discrete. Thus, at times we employ discrete numbers to represent continuous events.

> **Definitions**   A **discrete distribution** is a distribution with discrete values. A **continuous distribution** is a distribution with continuous values.

For a discrete distribution, a **probability function** (or **density function** or **probability density function**) returns the probability of occurrence of a particular argument value. For example, $p(1382)$ might be the probability that the random number generator returns 1382, indicating 1382 dust particles. However, if a distribution is continuous, the probability of occurrence of any particular value is zero. Thus, for a continuous distribution, a probability function (or density function or probability density function) indicates the probability that a given outcome falls inside a specific range of values. The integral of the probability function from the lower to the upper bound of the range, which is the area under that portion of the curve, gives the probability that the outcome is in that range. For example, the probability that the random velocity in the x direction of a dust particle is between 3.0 and 4.0 mm/sec is the integral of the probability density function from 3.0 to 4.0. Figure 9.4.3 presents a horizontal line of height 1 that is the graph of the probability density function ($p(x) = 1$) for uniformly generated random numbers with values from 0.0 up to 1.0. The probability that a uniform random floating point number between 0.0 and 1.0 falls between 0.6

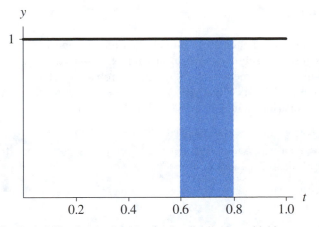

Figure 9.4.3 Probability density function for the distribution with histogram in Figure 9.4.1

and 0.8 is the integral of the function $f(x) = 1$ from 0.6 to 0.8. Thus, the probability is the area of the shaded region between 0.6 and 0.8, which is $(0.8 - 0.6)(1.0) = 0.2$. Such a random number is between 0.6 and 0.8 for $0.2 = 20\%$ of the time.

> **Definition**   For a discrete distribution, a **probability function** (or **density function** or **probability density function**) returns the probability of occurrence of a particular argument. For a continuous distribution, a **probability function** (or **density function** or **probability density function**) indicates the probability that a given outcome falls inside a specific range of values.

Quick Review Question 2

a. In the generation of random numbers to represent throws of a fair die (values 1, 2, 3, 4, 5, or 6), is the distribution discrete or continuous?

b. For the random numbers of Part a, give the value of the probability density function with an argument of 2.

c. For the probability density function with graph in Figure 9.4.3, give the probability that the random number falls between 0.2 and 0.7.

d. Can a probability density function ever have a negative value?

e. Give the value of the definite integral (i.e., area under the curve) of a probability density function over its entire range of values.

Discrete Distributions

As the following example demonstrates, if an equal likelihood of each of several discrete events exists, in a simulation we can generate a random integer to indicate the choice.

Example 1

In simulation of a pollen grain moving in a fluid, suppose at the next time step the grain is just as likely to move in any direction—north, east, south, west, up, or down—in a three-dimensional (3D) grid. A probability of 1/6 exists for the grain to move in any of the six directions. With these equal probabilities, we can generate a uniformly distributed integer between 1 and 6 to indicate the direction of movement.

To Generate Random Numbers in Discrete Distribution with Equal Probabilities for Each of *n* Events

Generate a uniform random integer from a sequence of *n* integers, where each integer corresponds to an event

Quick Review Question 3

From the text's website, download your computational tool's *9_4QRQ.pdf* file for this system-dependent question to give a command for generating an appropriate random number for Example 1.

Frequently, however, the discrete choices do not carry equal probabilities, as in the following example.

Example 2

In an initial 3D grid, suppose only 15% of the grid sites, or cells, contain pollen grains. Thus, a *probPollen* = 15% = 0.15 chance exists for a cell to contain a grain. If the location is to contain a pollen grain, we make the cell's value equal to *POLLEN* = 1; and otherwise, the cell's value becomes *EMPTY* = 0. To initialize a grid for a simulation, we must designate for each cell if the location contains pollen or not. For each cell, we need to generate a uniformly distributed random floating point number from 0.0 up to 1.0. On the average, 15% of the time this random number is less than 0.15, while 85% of the time the number is greater than or equal to 0.15 (Figure 9.4.4). Thus, to initialize the cell, if the random number is less than 0.15, we make the cell's value *POLLEN*; otherwise, we assign *EMPTY* to the cell's value. Thus, using the probabilities and cell values above, we employ the following logic to initialize each cell in the grid:

    if a random number is less than *probPollen* (i.e., pollen grain at site)
        set the cell's value to *POLLEN*
    else (i.e., no pollen grain at site)
        set the cell's value to *EMPTY*

Quick Review Question 4

From the text's website, download your computational tool's *9_4QRQ.pdf* file for this system-dependent question to give a statement to implement the pseudocode at the end of Example 2.

Figure 9.4.4   15% of floating point values between 0 and 1 are less than *probPollen* = 0.15

In many situations, more than two choices exist, as in Example 3.

Example 3

Suppose in a simulation involving animal behavior, a lab rat presses a food lever (*FOOD* = 1) 15% of the time, presses a water lever (*WATER* = 2) 20% of the time, and does neither (*NEITHER* = 3) the remainder of the time. For the simulation, we consider the range split into three parts as in Figure 9.4.5 and again generate a uniformly distributed random floating point number from 0.0 to 1.0. If the number is less than 0.15, which occurs 15% of the time, we assign *FOOD* = 1 to the mouse's action. For 20% of the time the uniformly distributed random number is greater than or equal to 0.15 and less than 0.35. With a random number in this range, we make the rat's action be *WATER* = 2. A random number is greater than or equal to 0.35 with a probability of 65%. In such a case, we assign *NEITHER* = 3 to the rat's action. Thus, with *rand* being a uniformly distributed random floating point number from 0.0 to 1.0, we employ the following logic for determination of the mouse's action:

if a random number, *rand*, is < 0.15
    the rat presses the food lever
else if *rand* < 0.35 (i.e., 0.15 ≤ *rand* < 0.35)
    the rat presses the water lever
else (i.e., 0.35 ≤ *rand*)
    the rat does neither

> ***To Generate Random Numbers in Discrete Distribution with Probabilities $p_1, p_2, \ldots, p_n$ for Events $e_1, e_2, \ldots, e_n$, Respectively, Where $p_1 + p_2 + \cdots + p_n = 1$***
>
> Generate *rand*, a uniform random floating point number in [0, 1).
> If *rand* < p_1, then return e_1
> else if *rand* < $p_1 + p_2$, then return e_2
> . . .
> else if *rand* < $p_1 + p_2 + \cdots + p_{n-1}$, then return e_{n-1}
> else return e_n

Figure 9.4.5   15% of values less than 0.15; 20% between 0.15 and 0.35; 65% between 0.35 and 1.0

Quick Review Question 5

Consider the following pseudocode that returns the direction (N, E, S, or W) a simulated animal moves:

    if a random number, *rand*, is < 0.12
        return N
    else if *rand* < 0.26
        return E
    else if *rand* < 0.69
        return S
    else
        return W

Give the probability that the animal moves in each of the following directions:

a. N **b.** E **c.** S **d.** W

Normal Distributions

A **normal** or **Gaussian distribution**, which statistics frequently employs, has a probability density function $\frac{1}{\sqrt{2\pi}\sigma}e^{-(x-\mu)^2/(2\sigma)}$, where μ is the mean and σ is the standard deviation (see Figure 9.4.6). Figure 9.4.7 displays a histogram of a set of 1000 random numbers in the Gaussian distribution with mean 0 and standard deviation 1. Without getting into a formal definition of standard deviation, 68.3% of the values in a normal distribution are within $\pm\sigma$ of the mean, μ; 95.5% are within $\pm2\sigma$ of μ; and 99.7% are within $\pm3\sigma$ of μ.

Many systems include a way to generate random numbers in a normal distribution with a given mean and standard deviation. For those that do not, we can employ the **Box-Muller-Gauss Method**. The method first generates a uniformly distributed random number a between 0 and 2π. Then, the technique computes b, the product of

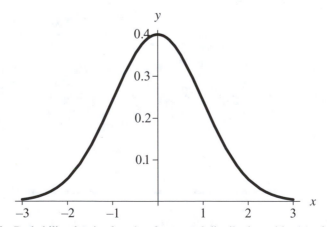

Figure 9.4.6 Probability density function for normal distribution with mean 0 and standard deviation 1

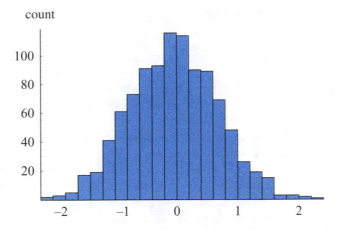

Figure 9.4.7 Histogram of normal distribution with mean 0 and standard deviation 1

the standard deviation (σ) and the square root of the negative natural logarithm of a uniformly distributed random number between 0.0 and 1.0, or $b = \sigma\sqrt{-\ln(rand)}$, where *rand* is a uniformly distributed random number between 0.0 and 1.0. The two values $b \sin(a) + \mu$ and $b \cos(a) + \mu$ are normally distributed with mean μ and standard deviation σ.

> ***Box-Muller-Gauss Method for Normal Distribution with Mean μ and Standard Deviation σ***
> compute $b \sin(a) + \mu$ and $b \cos(a) + \mu$ where
> a = a uniform random number in $[0, 2\pi)$
> *rand* = a uniform random number in $[0, 1)$
> $b = \sigma\sqrt{-\ln(rand)}$

Quick Review Question 6

Suppose for a simulation involving test scores we need random numbers in a normal distribution with mean 70 and standard deviation 8. Suppose 5.32 and 0.754 are uniformly distributed random numbers between 0 and 2π and between 0.0 and 1.0, respectively. Using these values, evaluate the following, rounding to two decimal places:

a. a
b. b
c. The normally distributed number employing sine
d. The normally distributed number employing cosine

Quick Review Question 7

From the text's website, download your computational tool's *9_4QRQ.pdf* file for system-dependent text and a question to assign to n a random number in a normal distribution with mean 70 and standard deviation 8.

Exponential Distributions

A model for unconstrained growth or decay employs an exponential function e^{rt}, where t is time and r is the growth rate or $-r$ the decay rate, respectively. Functions of the form $f(t) = |r|e^{rt}$ **with** $r < 0$ **and** $t > 0$ or $f(t) = |r|e^{rt}$ **with** $r > 0$ **and** $t < 0$ are probability density functions with areas under the curves of 1. Figure 9.4.8 contains the graph of a function in this category, $f(t) = 2e^{-2t}$. To obtain a number in such a distribution, the **Exponential Method** divides the natural logarithm of a uniformly distributed random number from 0.0 to 1.0 by the rate constant (r), that is, $\ln(rand)/r$, where $rand$ is random between 0 and 1. For example, to generate numbers in the distribution $f(t) = 2e^{-2t}$, we calculate $\ln(rand)/(-2)$. Figure 9.4.9 displays a histogram of 1000 such exponentially distributed random numbers.

We employ the same algorithm to generate random numbers from minus infinity to 0 for probability density function $f(t) = |r|e^{rt} = re^{rt}$ with $r > 0$. Figure 9.4.10 shows the graph of one such function, $f(t) = 2e^{2t}$; and Figure 9.4.11 displays a histogram of 1000 pseudorandom numbers that the algorithm $\ln(rand)/2$ generates.

> ***Exponential Method for Probability Density Function*** $|r|e^{rt}$ ***with*** $r < 0$ ***and*** $t > 0$ ***or*** $f(t) = |r|e^{rt}$ ***with*** $r > 0$ ***and*** $t < 0$
>
> compute $\textbf{\textit{ln(rand)/r}}$,
>   where $rand$ = a uniform random number in [0.0, 1.0)

Quick Review Question 8

Suppose we are performing a simulation involving a radioactive substance with initial mass 0.1 mg and decay rate 0.1.

    a. Give the probability density function.
    b. Using 0.754 as a uniformly distributed random number between 0.0 and 1.0, determine a random number to three decimal places in this exponential distribution.

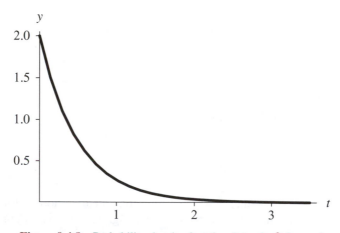

Figure 9.4.8   Probability density function $f(t) = 2e^{-2t}$ for $t > 0$

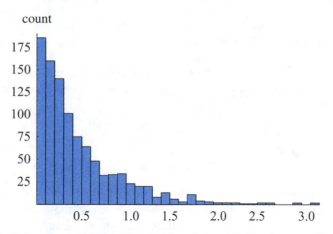

Figure 9.4.9 Histogram of 1000 random numbers ln(*rand*)/(−2), where *rand* is a uniformly generated random number in (0.0, 1.0)

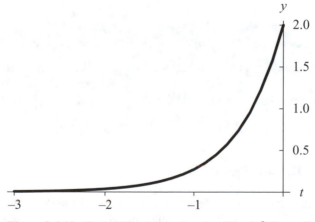

Figure 9.4.10   Probability density function $f(t) = 2e^{2t}$ for $t < 0$

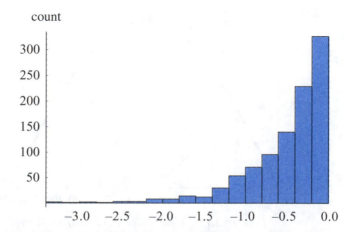

Figure 9.4.11   Histogram of 1000 numbers ln(*rand*)/2, where *rand* is a uniformly generated random number in [0.0, 1.0)

Quick Review Question 9

From the text's website, download your computational tool's *9_4QRQ.pdf* file for system-dependent text and a question about an exponential probability density function.

Rejection Method

The exercises and projects explore several methods for generating random numbers in other specific distributions. When these techniques do not apply, however, we can employ the **Rejection Method**. First, we obtain a uniformly distributed random number, *randInterval*, in the requested interval. If the probability density function at *randInterval* is greater than a uniform random number from 0.0 to an upper bound for the function, we return *randInterval*. Otherwise, we repeat the process.

> *Rejection Method for Random Numbers in Interval [a, b) for Distribution f(x)*
>
> ***if f (randInterval) > randUpperBound, then return randInterval, where***
>   *randInterval*—a uniform random number in interval [a, b)
>   *randUpperBound*—a uniform random number in [0, upper bound for f)
> else repeat the process

Quick Review Question 10

Consider the probability density function $f(x) = 2\pi \sin(4\pi x)$ from 0.0 to 0.25 (see Figure 9.4.12). For each of the following pairs of uniform random numbers in the

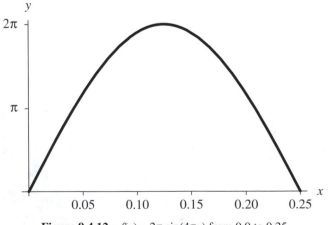

Figure 9.4.12 $f(x) = 2\pi \sin(4\pi x)$ from 0.0 to 0.25

indicated intervals, determine the random number returned in the distribution $2\pi \sin(4\pi x)$, if any:

Part	*Interval from* 0.0 *to* 0.25	*Interval from* 0.0 *to* $2\pi \sin(\pi/2) = 2\pi$
a	0.221	0.85
b	0.049	5.59
c	0.130	2.69

Exercises

Do Exercises 1–8 with an appropriate computational tool.

1. For the pseudocode at the end of Example 3, write a segment to return *FOOD, WATER,* or *NEITHER,* depending on the value of the random number.

2. This question refers to the Box-Muller-Gauss Method, which generates random numbers from a normal distribution with mean = 9 and standard deviation = 2.
   a. Assign to *a* a uniformly distributed random number between 0 and 2π.
   b. Assign to *b* the product of the standard deviation (*stdDev* = 2) and the square root of the negative natural logarithm of a uniformly distributed random number between 0.0 and 1.0.
   c. Return a list of pairs of numbers with ($b \sin(a) + mean$) as the first coordinate and ($b \cos(a) + mean$) as the second that the Box-Muller-Gauss Method produces. Be sure to use the same *a* and *b* for both members of a pair.
   d. Assign to *tblGauss* a list of 1000 random numbers in the normal distribution with mean 9 and standard deviation 2. One way of accomplishing this task is to generate a table/array of 500 ordered pairs similar to Part c and to flatten the table/array to a corresponding list of 1000 numbers.
   e. Display a histogram of these values.
   f. If available, use a built-in method to generate the table in Part d.
   g. Display a histogram of these values.

3. a. Write a statement to assign to *tblExp* a table/array of 1000 random numbers in the exponential distribution $7e^{-7t}$ using the Exponential Method. Display a histogram of *tblExp*.
   b. If available, use a built-in method to generate the table in Part a. Display a histogram of these values.

4. Using the Exponential Method, give an expression to generate numbers with the probability density function $2e^{2(t-3)}$ for $t < 3$. Display a histogram of these values.

5. The expression $\ln(rand)/(-9) + 4$, where *rand* is a random number from 0 to 1, generates pseudorandom numbers for what exponential probability density function and interval?

6. This question develops code for the Rejection Method with the probability density function $f(x) = 2\pi \sin(4\pi x)$.
   a. Define the function $f(x) = 2\pi \sin(4\pi x)$ (see Figure 9.4.12).
   b. Plot $f(x)$ from 0.0 to 0.25.

   c. Assign to variable *rand* a uniform random number from 0.0 to 0.25, the interval of interest in Figure 9.4.12.

   d. Define the function *rej* with no arguments to return a random number using the Rejection Method. If $f(rand)$ is greater than a uniform random number from 0 to $2\pi \sin(4\pi/8) = 2\pi \sin(\pi/2) = 2\pi$, which is the maximum value of $f(x)$, return *rand*. If the condition is false, we must reject *rand* and search for another candidate. To do so, we call the function *rej* again. Be sure *rand* gets a new value with each function call. (The process of a function, such as *rej*, calling itself is **recursion**.)

   e. Write a statement to generate a list of 1000 random numbers from 0.0 to 0.25 with the probability density function $f(x) = 2\pi \sin(4\pi x)$.

   f. Display a histogram of these values.

7. The **Maximum Method** is for distributions of the form nx^{n-1} with x being from 0.0 to 1.0 and n being a positive integer less than 17. The method calls for taking the maximum of n uniformly distributed random numbers.

   a. Define a function $f(x) = 3x^2$, and plot f from 0.0 to 1.0.

   b. Write a statement to return a list of three uniform random numbers in [0.0, 1.0).

   c. Write a statement to return a random number in the distribution $3x^2$ using the Maximum Method.

   d. Write a statement to assign to *tblMax* a list of 1000 numbers in the distribution $3x^2$.

   e. Display a histogram of the table in Part d.

   f. Define a function *randMax* with parameter n to return a random number in the distribution nx^{n-1} with x being a number from 0.0 to 1.0 and n being a positive integer less than 17. Use the Maximum Method.

   g. Repeat Parts d and e using *randMax* from Part f.

8. The **Root Method** is for distributions of the form nx^{n-1} with x being from 0.0 to 1.0 and n being a nonnegative number not in $\{1, 2, 3, \ldots, 16\}$. The method calls for taking the *n*th root of a uniformly distributed random number between 0.0 and 1.0.

   a. Define a function *randRoot* with parameter n to return a random number using the Root Method.

   b. Define a function $f(x) = 0.5x^{-0.5}$, and plot f from 0.0 to 1.0.

   c. Using *randRoot*, write a statement to assign to *tblRoot* a list of 1000 numbers in the distribution $0.5x^{-0.5}$, where x is between 0 and 1.

   d. Display a histogram *tblRoot* from Part c.

Do Exercises 9–14 in a programming language other than that used for the previous exercises.

  9. Exercise 1

10. Define a function to implement the Box-Muller-Gauss Method and return a random number from a normal distribution. Have the mean and standard deviation as parameters.

11. Define a function with parameter r to return a random number from an exponential distribution using the Exponential Method.

12. Define a function using the Rejection Method to return a random number from a distribution f.

13. Exercise 7f
14. Exercise 8a

Projects

1. Using a computational tool, define your own package of random number generators with continuous distributions using the following methods: Box-Muller-Gauss Method, Exponential Method, Rejection Method, Maximum Method (Exercise 7), and Root Method (Exercise 8). Do not use built-in functions other than a uniform random number generator. Test the package thoroughly.

2. Using a computational tool, define your own package of random number generators with *myRandom* definitions as in Project 4 of Module 9.2 on "Simulations" and with random number generators with continuous distributions from this module that call *myRandom* instead of a built-in uniform random number generator. Use the following methods: Box-Muller-Gauss Method, Exponential Method, Rejection Method, Maximum Method (Exercise 7), and Root Method (Exercise 8). Do not use built-in functions. Test the package thoroughly.

Using a programming language, repeat the indicated project.

3. Project 1   **4.** Project 2

Answers to Quick Review Questions

1. **a.** 20
   b. 1000
   c. *n/i*
   d. *n/i*
2. **a.** discrete
   b. $1/6 = 0.1667$
   c. $0.5 = (0.7 - 0.2)(1)$
   d. no
   e. 1
3, 4. From the text's website, download your computational tool's *9_4QRQ.pdf* file for an answer to this system-dependent question.
5. **a.** 12%
   b. $14\% = 0.26 - 0.12$
   c. $43\% = 0.69 - 0.26$
   d. $31\% = 0.1 - 0.69$
6. **a.** $a = 5.32$
   b. $b = 8\sqrt{-\ln(0.754)} = 4.25$
   c. $b \sin(a) + \mu = 4.25 \sin(5.32) + 70 = 66.51$
   d. $b \cos(a) + \mu = 4.25 \cos(5.32) + 70 = 72.43$
7. From the text's website, download your computational tool's *9_4QRQ.pdf* file for an answer to this system-dependent question.

8. **a.** $f(t) = 0.1e^{-0.1t}$
   b. $\ln(0.754)/(-0.1) = 2.824$
9. From the text's website, download your computational tool's *9_4QRQ.pdf* file for an answer to this system-dependent question.
10. **a.** 0.221 because $f(0.221) = 2\pi \sin(4\pi(0.221)) = 2.239 > 0.85$
    b. nothing because $f(0.049) = 2\pi \sin(4\pi(0.049)) = 3.629 < 5.59$
    c. 0.130 because $f(0.130) = 2\pi \sin(4\pi(0.130)) = 6.271 > 2.69$

References

Einwohner, Theodore H., and, Angela B. Shiflet. 1987. "RANDOM_NUMBER, A Syntax-Directed Package to Produce Random Numbers in User-Specified, Univariate Distributions: User's Guide." Unclassified Internal Document for Lawrence Livermore National Laboratory.

Weisstein, Eric. 2003. "Eric Weisstein's World of Mathematics." Wolfram Research. http://mathworld.wolfram.com/.

10

RANDOM WALK SIMULATIONS

MODULE 10.1

Computational Toolbox—Tools of the Trade: Tutorial 5

Prerequisite: Module 9.1 on "Computational Toolbox—Tools of the Trade: Tutorial 4."

Download

From the textbook's website, download Tutorial 5 in the format of your computational tool or in pdf format. We recommend that you work through the tutorial and answer all Quick Review Questions using the corresponding software.

Introduction

This fifth computational toolbox tutorial, which is available from the textbook's website in your system of choice, prepares you to use the system to complete projects for this and subsequent chapters. The tutorial introduces the following functions and concepts:

- Taking part of a list
- Maximum and minimum functions
- Animation

The module gives computational examples and Quick Review Questions for you to complete and execute in the desired software system.

MODULE 10.2

Random Walk

Introduction

One technique of Monte Carlo simulations that has many applications in the sciences is the random walk. **Random walk** refers to the apparently random movement of an entity. In a time-driven simulation, we depict the entity in a **cell** on a rectangular **grid**. At any time step, the entity can move, perhaps under certain constraints, at random to a neighboring cell.

> **Definition**   **Random walk** refers to the apparently random movement of an entity.

A certain type of computer simulation involving grids is a cellular automaton. **Cellular automata** are dynamic computational models that are discrete in space, state, and time. We picture space as a one-, two-, or three-dimensional **grid**, or array or lattice. A **site**, or **cell**, of the grid has a state, and the number of states is finite. **Rules**, or **transition rules**, specifying local relationships and indicating how cells are to change state, regulate the behavior of the system. An advantage of such grid-based models is that we can visualize through informative animations the progress of events. For example, we can view a simulation of the movement of ants toward a food source, the spread of fire, or the motion of gas molecules in a container. In this and the next two chapters, we consider many scientific applications involving cellular automata.

> **Definitions**   A **cellular automaton** (plural, **automata**) is a type of computer simulation that is a dynamic computational model and is discrete in space, state, and time. Space is a **grid**, or a one-, two-, or three-dimensional lattice, or array, of **sites**, or **cells**. A cell of the lattice has a state, and the number of states is finite. **Rules**, or **transition rules**, specifying local relationships and indicating how cells are to change state, regulate the behavior of the system.

A random walk cellular automaton can model **Brownian Motion**, which is the behavior of a molecule suspended in a liquid. The phenomenon bears the name of the English botanist Robert Brown. In 1827, he observed that life in pollen particles could not explain their rapid, random motion in a liquid. A generation later, the physicists Maxwell, Clausius, and Einstein explained the phenomenon as invisible liquid particles striking the visible particles, causing small movements. Because diffusion of many things, such as pollutants in the atmosphere and calcium in living bone tissue, exhibit Brownian Motion, simulations using random walks can also model these processes (Britannica 1997; Exploratorium 1995).

In genetics, random walks have been used to simulate mutation of genes. As another example, scientists use the method **Polymerase Chain Reaction (PCR)** to make many copies of particular pieces of DNA. A strand of DNA contains sequences of four bases, A, T, C, and G. Using the random walk technique in simulations, computational scientists can determine good proportions of these bases in solution to speed replication of the DNA.

Algorithm for Random Walk

Suppose that in a particular random walk simulation at each time step an entity goes at random diagonally in a NE, NW, SE, or SW direction. To move in such a direction, the entity walks east or west one unit and north or south one unit, covering a diagonal distance of $\sqrt{2}$ units.

In Algorithm 1, variables x and y store the horizontal and vertical coordinates, respectively, of the current location. The variable lst holds a list of locations in the path of the entity; and because the entity starts at the origin, we initialize lst to be a list containing the point $(0, 0)$. With n being the number of steps to be taken, a loop to produce the path executes n times. Within the loop, we generate one random integer of 0 or 1 to determine if the entity turns to the east or west by incrementing or decrementing x by 1, respectively. Then, another such "flip of the coin" dictates north with an increment of y or south with a decrement. We then append the new point (x, y) onto the developing *lst*.

After generating the list containing the path, we create and display a graphics representing the random walk. For example, we might show all the random walk locations as colored dots, the path as line segments, and the first and last points as black dots. In a section below and projects, we determine a relationship between the number of steps and the average length of the path. To aid in this endeavor, Algorithm 1 returns the distance between the final point, (x, y), and the initial one, $(x0, y0) = (0, 0)$.

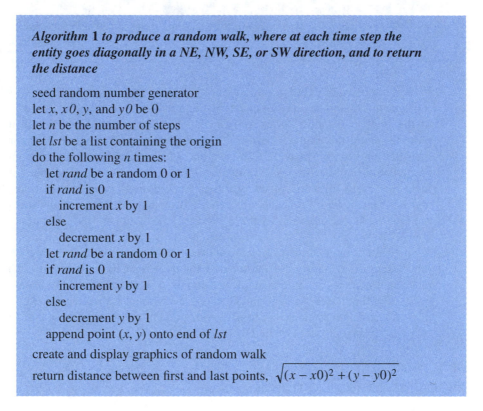

Algorithm 1 *to produce a random walk, where at each time step the entity goes diagonally in a NE, NW, SE, or SW direction, and to return the distance*

seed random number generator
let x, $x0$, y, and $y0$ be 0
let n be the number of steps
let *lst* be a list containing the origin
do the following n times:
   let *rand* be a random 0 or 1
   if *rand* is 0
      increment x by 1
   else
      decrement x by 1
   let *rand* be a random 0 or 1
   if *rand* is 0
      increment y by 1
   else
      decrement y by 1
   append point (x, y) onto end of *lst*
create and display graphics of random walk
return distance between first and last points, $\sqrt{(x - x0)^2 + (y - y0)^2}$

One execution of this code for Algorithm 1 displays a graphic similar to Figure 10.2.1. The segment also returns the distance from the starting point to the ending point. For Figure 10.2.1, the distance between these two black dots is 5.09902. Because the walk is random, each run of the code will very probably return a different value.

Quick Review Question 1

The following questions refer to the code in Algorithm 1.

   a. After execution of the loop, how many elements does *lst* have?

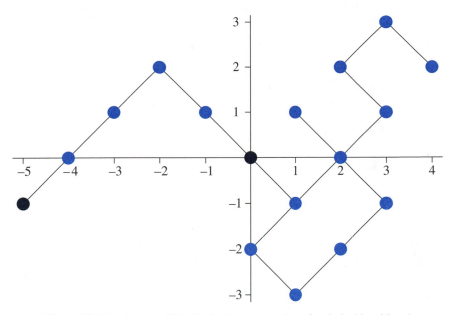

Figure 10.2.1   One possible display from execution of code in Algorithm 1

b. Is it possible for the points (3, 5) and (3, 6) to be adjacent to each other in *lst*?

Animate Path

Visualization of the path as it develops can aid in understanding the movement of the entity. Figure 10.2.2 presents several frames in such an animation.

To develop an animation, we generate a list (*lst*) of $n + 1$ points in the path. For each i going from 1 through $n + 1$, we create a graphics of the first i points of the walk, which are in a sublist of the first i points of *lst*. Thus, we generate a sequence of $n + 1$ displays that we can then animate with an appropriate computational tool.

For the animation to be consistent, we specify that each graphics have the same axes. To have axes that are the appropriate sizes, we find the minimum and maximum of all the x coordinates and do the same for the y coordinates. Thus, it is convenient to generate separate lists, *xlst* and *ylst*, of these coordinates. Afterward, we can find the appropriate minimum and maximum values and form the list of path points, *lst*. The next Quick Review Question covers some of the design, and Algorithm 2 contains the complete design to generate the graphics for the animation of one random walk of n points.

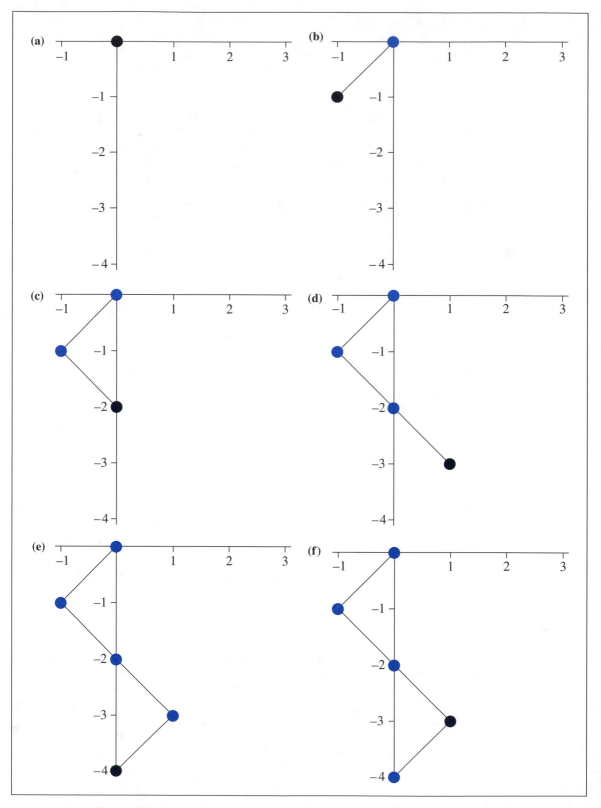

Figure 10.2.2 Several frames in an animation of the developing path from one random walk

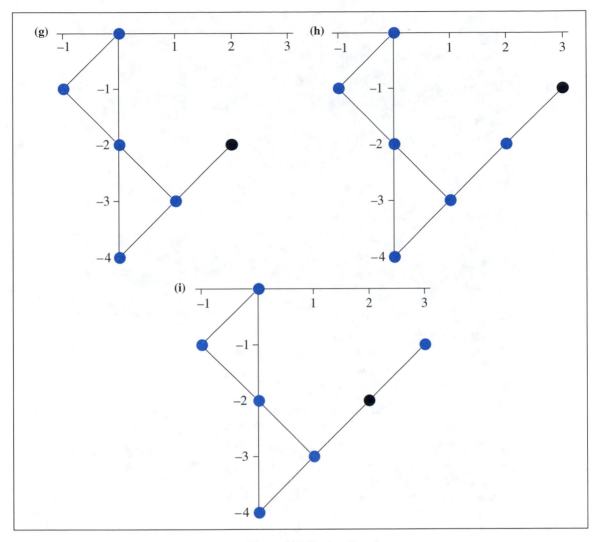

Figure 10.2.2 (*continued*)

Quick Review Question 2

This question refers to the segment from Algorithm 1 to generate a list, *lst*, of points in a random walk, where the path begins at the origin.

    a. Suppose we wish to store *x* and *y* coordinates in separate lists, *xlst* and *ylst*, respectively. Before the loop, we initialize *xlst* containing what value? (A similar statement would occur for *ylst*, and *lst* would not appear in the revised segment.)

    b. In this case, at the end of the loop, give the value to append to *xlst*.

    c. In the *x* direction of each plot of the animation, give the leftmost value.

    d. In the *y* direction of each plot of the animation, give the highest value.

Algorithm 2 *to generate the graphics for the animation of one random walk*

seed random number generator
let *x*, *x*0, *y*, and *y*0 be 0
let *n* be the number of steps
let *xlst* and *ylst* be lists containing 0
do the following *n* times:
    let *rand* be a random 0 or 1
    if *rand* is 0
       increment *x* by 1
    else
       decrement *x* by 1
    let *rand* be a random 0 or 1
    if *rand* is 0
       increment *y* by 1
    else
       decrement *y* by 1
    append *x* onto end of *xlst*
    append *y* onto end of *ylst*

let *xMin* be the minimum of *xlst*
let *xMax* be the maximum of *xlst*
let *yMin* be the minimum of *ylst*
let *yMax* be the maximum of *ylst*

let *lst* be the list of points, where for each point the first coordinate is from *xlst* and second is the corresponding member of *ylst*

do the following with *i* going from 1 through *n* + 1 and with all displays going from *xMin* to *xMax* in the *x* direction and from *yMin* to *yMax* in the *y* direction: create and display a graphics of the first *i* points of *lst*

Average Distance Covered

Because the walks are random, great variation can exist in both the paths and the final distances from the starting point. Thus, for a given number of steps (*n*), we should run the simulation many times and take the average of all the distances. To accomplish this task, we place most of the code from Algorithm 1 for one walk into the body of another loop that runs the simulation for a designated number of times (*numTests*). A variable, *sumDist*, accumulates the distances covered by the random walks. Before the loop, *sumDist* is initialized to zero; and after the loop, it is divided by *numTests* to return the average distance.

For this segment, we are not interested in producing graphics. Thus, we do not need to store all the points of the path, just the beginning point, here the origin, and the ending point, or final values of *x* and *y*. Inside the inner loop that runs one simulation, we generate two random numbers and increment or decrement x and y appropriately but do not store the point (x, y) in a list. Algorithm 3 contains a design to return the average distance covered for random walks with n steps, where the number of executions of the simulation is *numTests*. One run of this segment for $n = 25$ and *numTests* = 100 might return an average distance of 5.75278 units.

Algorithm 3 to run a random walk simulation* numTests *number of times and to return the average distance

seed random number generator
let *numTests* be the number of runs of the simulation
let $x0$ and $y0$ be 0
let n be the number of steps
let *sumDist*, the ongoing sum of distances, be 0
do the following *numTests* times:
    let x be $x0$
    let y be $y0$
    do the following n times:
        let *rand* be a random 0 or 1
        if *rand* is 0
            increment x by 1
        else
            decrement x by 1
        let *rand* be a random 0 or 1
        if *rand* is 0
            increment y by 1
        else
            decrement y by 1
    add the distance between first and last points, $\sqrt{(x - x0)^2 + (y - y0)^2}$, to *sumDist*
return the floating point average distance between first and last points, *sumDist*/*numTests*

Quick Review Question 3

This question refers to Algorithm 3.

    a. If we incorrectly move the initialization of *sumDist* inside the outer loop before the assignment of $x0$ to x, select the final value of *sumDist*:
    A. No change from current result.
    B. *sumDist* would be 0.
    C. *sumDist* would hold only the distance for the final path.
    D. *sumDist* would be undefined.

b. Select what would happen if we incorrectly move the initialization of x and y outside the outer loop before the assignment of 0 to *sumDist*:

A. The final value of *sumDist* would be 0.

B. No change from current result.

C. One path would begin where the previous path ended.

D. A run-time error would occur.

c. Suppose *numTests* is 100 and n is 25. Give the number of times one of the *if* statements in the inside loop executes.

Relationship between Number of Steps and Distance Covered

To discern a relationship between the number of steps (n) and average distance covered in a random walk, we run the code in Algorithm 3 for values of n from 1 to 50 and store each average distance in a list, *listDist*. Then, we employ the techniques of Module 8.3 on "Empirical Models" to determine the relationship. Figure 10.2.3 shows a plot of the average distances traveled versus the number of steps. Projects 3 and 4 determine a formula for this relationship.

Exercises

On the text's website, RandomWalk *files for several computational tools contain the code for Algorithms 1, 2, and 3. Complete the following exercises using your computational tool.*

1. If possible in your computational tool, revise the code of Algorithm 1 to replace the loop with a call to a function to formulate *lst*.
2. Revise the code of Algorithm 1 or Exercise 1 to have the entity go with equal probability in a north, south, east, or west direction. Hint: Choose the direction based the value of a random integer, 0, 1, 2, or 3.
3. **a.** Revise the code of Algorithm 1 to have the entity go in an easterly direction (incrementing x) with probability of 30% and in a westerly direction (decrementing x) with probability of 70%.

   b. Revise the code of Part a and have the entity go in a northerly direction (incrementing y) with probability of 45% and in a southerly direction (decrementing y) with probability of 55%.

   c. Give the probability for the entity going in each direction, NE, NW, SE, and SW.
4. Revise the code of Algorithm 1 or Exercise 1 to have the entity go in a north, south, east, or west direction with a probability of 20%, 30%, 45%, or 5%, respectively.

avg. distance

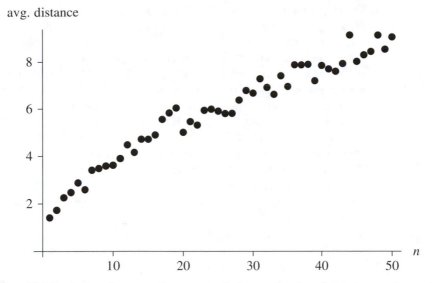

Figure 10.2.3 A plot of average distances traveled versus number of steps in a random walk

Projects

On the text's website, RandomWalk *files for several computational tools contain the code for Algorithms 1, 2, and 3. Complete the projects below using your software system.*

For additional projects, see Module 13.1 on "Polymers—Strings of Pearls" and Module 13.2 on "Solidification—Let's Make It Crystal Clear!"

1. Exercise 1
2. Exercise 2
3. Download from the text's website the data file *AverageDistances.dat* of average distances covered for step sizes from 1 to 50. Using the techniques of Module 8.3 on "Empirical Models," determine a relationship between the number of steps (n) and average distance covered in a random walk.
4. Develop code as discussed in the section on "Relationship between Number of Steps and Distance Covered" to obtain a list of average distances covered for random walks of step sizes from 1 to 50. Then, using the data the program generates, do the analysis of Project 3.
5. Develop code as discussed in the section on "Relationship between Number of Steps and Distance Covered" to obtain a list of average distances covered for random walks of step sizes from 1 to 50, where the entity travels east, west, north, or south with each step. Then, using the data the program generates, do the analysis of Project 3.
6. Develop code for Exercise 3 and run the simulation for 50 time steps. Include this code in a loop that runs the simulation 1000 or more times. Have the segment return the portion of time the entity ends on the 50th step in each of the four quadrants, NE, NW, SE, and SW. Do the figures seem to agree with your answer to Exercise 3c?

7. Develop code for Exercise 4 and run the simulation for 50 time steps. Include this code in a loop that runs the simulation 1000 or more times. Have the segment return the portion of time the entity ends on the 50th step in the north, south, east, or west direction from the starting location, the origin. On a particular run of the simulation, the 50th step could fall into one category, such as due north of the origin, or in two categories, such as north and east of the origin. Discuss the results in relationship to the probabilities of Exercise 4.

8. A hiker without a compass trying to find the way in the dark can step in any of eight directions (N, NE, E, SE, S, SW, W, NW) with each step. Studies show that people tend to veer to the right under such circumstances. Initially, the hiker is facing north. Suppose at each step probabilities of going in the indicated directions are as follows: N—19%, NE—24%, E—17%, SE—10%, S—2%, SW—3%, W—10%, NW—15%. Develop a simulation to trace a path of a hiker, and run the simulation a number of times. Describe the results. (Note that other than at the initial step, this simulation simplifies the problem by ignoring the direction in which the hiker faces.)

9. Perform a simulation of Brownian Motion of a pollen grain suspended in a liquid by generating a three-dimensional random walk. Using documentation for your computational tool, investigate how to plot three-dimensional graphics points and lines, and create a 3D graphic of the walk.

Answers to Quick Review Questions

1. **a.** $n + 1$ elements, $(0, 0)$ and the n appended points
   b. No, both coordinates are changed in the body of the loop.
2. **a.** 0
   b. x
   c. the minimum of $xlst$
   d. the maximum of $ylst$
3. **a.** C. $sumDist$ would hold only the distance for the final path.
   b. C. One path would begin where the previous path ended.
   c. 2500 because the outside loop executes 100 times and the inside loop executes 25 times.

References

Encyclopedia Britannica. 1997. "Brownian Motion." *Britannica Guide to the Nobel Prizes. Britannica Online.* http://www.britannica.com/nobel/micro/88_96.html

Exploratorium. 1995. "Brownian Motion." *Exploratorium Exhibit and Phenomena Cross-Reference.* http://www.exploratorium.edu/xref/phenomena/brownian_motion.html

11

DIFFUSION

MODULE 11.1

Computational Toolbox—Tools of the Trade: Tutorial 6

Prerequisite: Module 10.1 on "Computational Toolbox—Tools of the Trade: Tutorial 5"

Download

From the textbook's website, download Tutorial 6 in the format of your computational tool or in pdf format. We recommend that you work through the tutorial and answer all Quick Review Questions using the corresponding software.

Introduction

This sixth computational toolbox tutorial, which is available from the textbook's website in your system of choice, prepares you to use the system to complete projects for this and subsequent chapters. The tutorial introduces the following functions and concepts:

- Joining lists
- Finding the length of a list
- Visualizing a rectangular array
- Matching patterns

- Localizing variables
- Finding the position of a value in a list
- Integer parameters

The module gives computational examples and Quick Review Questions for you to complete and execute in the desired software system.

MODULE 11.2

Spreading of Fire

Downloads

For several computational tools, the text's website has available for download a *Fire* file containing the simulation this module develops and an *11_2QRQ.pdf* file containing system-dependent Quick Review Questions and answers.

Introduction

Human beings, with some justification, have considerable fear of fire. History is replete with disastrous losses of life and property from it. Nevertheless, fires in areas like the Western United States are natural, and ecologists tell us, beneficial to the plant communities there. Periodic fires help to clear the forest floor of debris and promote the growth of sturdy, fire-resistant trees. Unfortunately, expanding human populations have intruded on previously uninhabited areas, establishing their own communities in "fire-prone" zones. Furthermore, human activities, such as fire suppression, livestock grazing, and logging, have increased the possibility of hotter and more destructive fires (Wilderness Society).

During the fall of 2003, residents of Southern California faced a series of firestorms driven by powerful Santa Ana winds. After three days, the fires had destroyed over 400,000 acres and 900 homes and had killed 15 people. Hundreds of firefighters battled a chain of fires that extended from Ventura County, north of Los Angeles, east into San Bernadino County, and south to Tijuana, Mexico. A haze of toxins draped over the area like a pall (Wilson et al. 2003).

The Malibu region of above Los Angeles is dominated by the Santa Monica Mountains and canyons that run from north to south. Much of the natural vegetation is dry **chaparral**, consisting of many small, oily, woody plants that are extremely flammable. This vegetation naturally would burn every 15 to 45 years, clearing out old and dead plant materials and returning nutrients to the soil. With the prevailing dry conditions, a lit cigarette or downed power line can set off a ferocious blaze that may only stop after traveling many miles to the Pacific Ocean (Carter 1996).

Fighting fires in Southern California or anywhere else is a very risky job, where loss of life is a real possibility. Proper training is essential. In the United States the National Fire Academy, established in 1974, presents courses and programs that are intended "to enhance the ability of fire and emergency services and allied professionals to deal more effectively with fire and related emergencies." The Academy has partnered with private contractors and the U.S. Forest Service to develop a three-dimensional land fire fighting training simulator. This simulator exposes trainees to a convincing fire propagation model, where instructors can vary fuel types, environmental conditions, and topography. Responding to these variables, trainees may call for appropriate resources and construct fire lines. Instructors may continue to alter the parameters, changing fire behavior. Students can review the results of their decisions, where they can learn from their mistakes in the safety of a computer laboratory (Studebaker 2003).

This module develops a two-dimensional computer simulation for the spread of fire. The techniques can be extended to numerous other scientific examples involving contagion, such as the propagation of infectious diseases, heat diffusion, and distribution of pollution.

Initializing the System

In many simulations, we model a dynamic area under consideration with an n-by-n grid, or lattice or a two-dimensional square array, of numbers (see Figure 11.2.1). Each cell in the lattice contains a value representing a characteristic of a corresponding location. For example, in a simulation for the spread of fire, a cell can contain a value of 0, 1, or 2 indicating an empty cell, a cell with a non-burning tree, or a cell with a burning tree, respectively. Table 11.2.1 lists these values and meanings along with associated constants **EMPTY**, **TREE**, and **BURNING** that have values of **0**, **1**, and **2**,

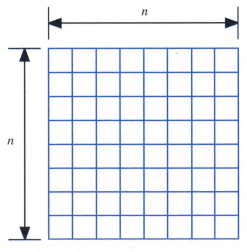

Figure 11.2.1   Cells to model area

TABLE 11.2.1
Cell Values with Associated Constants and their Meanings

Value	Constant	Meaning
0	*EMPTY*	The cell is empty ground containing no tree.
1	*TREE*	The cell contains a tree that is not burning.
2	*BURNING*	The cell contains a tree that is burning.

respectively. We initialize these constants at beginning and employ the descriptive names throughout the program. Thus, the code is easier to understand and to change.

To initialize this discrete stochastic system, we employ the following two probabilities:

> *probTree*—The probability that a tree (burning or not burning) initially occupies a site. Thus, *probTree* is the initial tree density measured as a percentage.
> *probBurning*—If a site has a tree, the probability that the tree is initially burning, or that the grid site is *BURNING*. Thus, *probBurning* is the fraction of the trees that are burning when the simulation begins.

Using the probabilities and cell values above, we employ the following logic to initialize each cell in the grid for the forest.

Cell Initialization Algorithm

if a random number is less than *probTree*	// tree at site
if another random number is less than *probBuring*	// tree is burning
assign *BURNING* to the cell	
else	// tree is not burning
assign *TREE* to the cell	
else	// no tree at site
assign *EMPTY* to the cell	

Quick Review Question 1

From the text's website, download your computational tool's *11_2QRQ.pdf* file for this system-dependent question that completes the code to initialize *forest* to be an $n \times n$ array.

Updating Rules

At each simulation iteration, we apply a function ***spread*** to each cell site to determine its value—*EMPTY*, *TREE*, or *BURNING*—at the next time step. The cell's value at the next instant depends on the cell's current value (***site***) and the values of its neighbors to the north (***N***), east (***E***), south (***S***), and west (***W***). Thus, we use five parameters—*site*, *N*, *E*, *S*, and *W*—for *spread*. In a call to this function, each argument is *EMPTY*, indicating an empty cell with no tree, *TREE* for a non-burning tree, or *BURNING* for a burning tree in that location. Figure 11.2.2 pictures the cells that

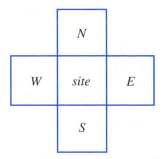

Figure 11.2.2   Cells that determine a site's next value

determine a site's next value. For this simulation, the state of a diagonal cell to the northeast, southeast, southwest, or northwest does not have an impact on a site's value at the next iteration. Thus, the term **neighbor** refers to the cells directly to the north, east, south, and west of the site's cell. These four neighbors along with the site itself comprise what is called the **von Neumann neighborhood** of a site.

> **Definitions**   In a two-dimensional grid, the **von Neumann neighbor-hood** of a site is the set of cells directly to the north, east, south, and west of the site and the site itself. The former four cells are the site's **neighbors**.

Updating rules apply to different situations: If a site is empty (cell value *EMPTY*), it remains empty at the next time step. If a tree grows at a site (cell value *TREE*), at the next instant the tree may or may not catch fire (value *BURNING* or *TREE*, respectively) due to fire at a neighboring site or to a lightning strike. A burning tree (cell value *BURNING*) always burns down, leaving an empty site (value *EMPTY*) for the next time step. We consider each situation separately.

Quick Review Question 2

From the text's website, download your computational tool's *11_2QRQ.pdf* file for this system-dependent question that develops *spread*'s rule for the situation where a site does not contain a tree at this or any time step.

When a tree is burning, the first argument, which is the site's value, is *BURNING*. Regardless of its neighbors' situations, the tree burns down, so that at the next iteration of the simulation the site's value becomes *EMPTY*. Thus, the relevant rule for the *spread* function has a first argument of *BURNING*; each of the other four arguments are immaterial; and the function returns value of *EMPTY*.

Quick Review Question 3

From the text's website, download your computational tool's *11_2QRQ.pdf* file for this system-dependent question that develops *spread*'s rule for the situation where a site contains a burning tree.

To develop this dynamic, discrete stochastic system, we employ the following additional probabilities:

> *probImmune*—The probability of immunity from catching fire. Thus, if a site contains a tree (site value of *TREE*) and fire threatens the tree, *probImmune* is the probability that the tree will not catch fire at the next time step.
> *probLightning*—The probability of lightning hitting a site

When a tree is at a location (site value of *TREE*), at the next iteration the tree might be burning due to one of two causes, a burning tree at a neighboring site or a lightning strike at the site itself. Even if one of these situations occurs, the tree at the site might not catch fire. Separate rules apply to the two causes for fire.

For the first situation involving a neighboring burning tree, we employ the following logic:

> if *site* is *TREE* and (*N*, *E*, *S*, or *W* is *BURNING*)
>> if a random number between 0.0 and 1.0 is less than *probImmune*
>>> return *TREE*
>> else
>>> return *BURNING*

Thus, even if a tree has the potential to burn because of a neighboring burning tree, it may not. Because of conditions, such as dry weather, such a tree has a probability of *probImmune* of not burning.

Quick Review Question 4

From the text's website, download your computational tool's *11_2QRQ.pdf* file for this system-dependent question that develops *spread*'s rule for the situation where a site contains a non-burning tree that may catch fire because a neighboring site contains a burning tree.

A tree might also catch fire because of a lightning strike. The probability that the tree is struck by lightning is *probLightning*. However, with a probability of *probImmune* the tree will not burn even if struck by lightning. In contrast, the probability that the tree is not immune to fire is $(1 - probImmune)$. For example, if the probability of immunity (*probImmune*) is $0.4 = 40\%$, then a $(1 - 0.4) = 0.6 = 60\%$ chance exists for the tree not to be immune from burning. For the tree to catch fire due to lightning, it must be hit and not be immune. Thus, lightning causes a tree to catch fire with the probability that is the product *probLightning* $*$ $(1 - probImmune)$. For example, if a $0.2 = 20\%$ chance exists for a lightning strike at the site of a tree, the tree burns with a probability of $(0.2)(0.6) = 0.12 = 12\%$. Two things must happen: Lightning must strike, and the tree must not be immune from burning.

Quick Review Question 5

From the text's website, download your computational tool's *11_2QRQ.pdf* file for this system-dependent question that completes *spread*'s rule for the situation where a site contains a non-burning tree that may be hit by lightning and burn.

Periodic Boundary Conditions

We must be able to apply the function *spread* to every grid point, such as in Figure 11.2.1, including those on the boundaries of the first and last rows and the first and last columns. However, the *spread* function has parameters for the grid point (*site*) and its neighbors (*N, E, S, W*). Thus, to apply *spread* we extend the boundaries by one cell. Several choices exist for values in those cells:

- Give every extended boundary cell the value *EMPTY*, as indicated in color in Figure 11.2.3. Thus, the boundary insulates. We call this situation **absorbing boundary conditions**. In the case of the spread of fire, the boundary is similar to a firebreak or an area with no trees; proximity to such a boundary cell cannot cause an internal tree to catch fire.
- Give every extended boundary cell the value of its immediate neighbor. Thus, the values on the original first row occur again on the new first row, which serves as a boundary. Similar situations occur on the last row and the first and last columns. (See Figure 11.2.4.) In the case of the spread of fire, the boundary tends to propagate the current local situation.
- Wrap around the north-south values and the east-west values in a fashion similar to a donut, or a torus. Extend the north boundary row with a copy of the original south boundary row, and extend the south boundary with a copy of the original north boundary row. Similarly, expand the column boundaries on the east and west sides. Thus, for a cell on the north boundary, its neighbor to the north is the corresponding cell to the south. (See Figure 11.2.5.) Such conditions are called **periodic boundary conditions**. In the case of the fire simulation, the area is a

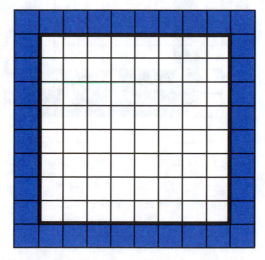

Figure 11.2.3   Grid with extended boundaries with each cell on an extended boundary having a value of *EMPTY*

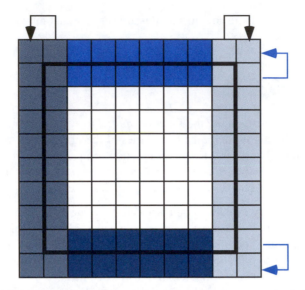

Figure 11.2.4 Grid with extended boundaries with each cell on an extended boundary having the value of its immediate neighbor in the original grid

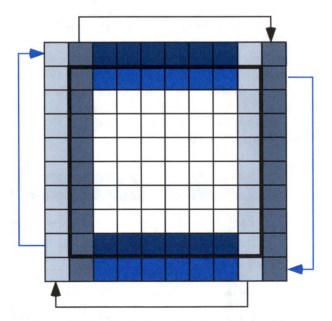

Figure 11.2.5 Grid with periodic boundary conditions

closed environment with the situation at one boundary effecting its opposite boundary cells.

In the application of spreading fire, we choose to employ periodic boundary conditions. First, we attach new first and last rows, as in Figure 11.2.6, by concatenating the original grid's last row, the original grid, and the first row to create a new lattice, *latNS*.

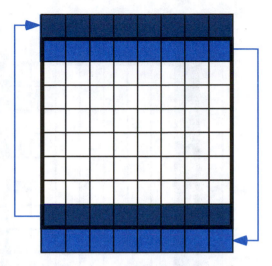

Figure 11.2.6   Grid from Figure 11.2.1 extended by having a new first row that is a copy of the last row on the original grid and having a new last row that is a copy of the first row on the original grid

Quick Review Question 6

From the text's website, download your computational tool's *11_2QRQ.pdf* file for this system-dependent question that extends a grid as in Figure 11.2.6 by attaching the last row to the beginning and the first row to the end of the original grid to form a new grid, *latNS*.

To extend the grid with periodic boundary conditions in the east and west directions, we concatenate the last column of *latNS* from Quick Review Question 6, *latNS*, and the first column of *latNS* (Figure 11.2.7). For some computational tools, it is easier to first transpose the lattice *latNS*, perform the same manipulation with the rows as in Quick Review Question 6, and then transpose the resulting lattice.

Quick Review Question 7

From the text's website, download your computational tool's *11_2QRQ.pdf* file for this system-dependent question that extends a lattice as in Figure 11.2.7.

To consolidate these tasks, we define a function ***extendLat1*** using periodic boundary conditions to extend by one cell in each direction the square lattice. Pseudocode for the function follows:

***extendLat1*(*lat*)**
    Function to accept a grid and to return a grid extended one cell in each direction with periodic boundary conditions
Pre*: lat* is a grid.

(continued)

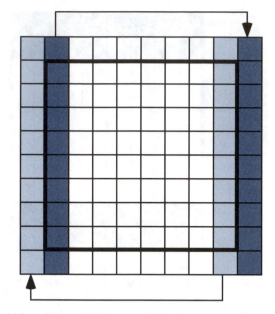

Figure 11.2.7   Grid from Figure 11.2.6 expanded by having a new first column that is a copy of the last column and a new last column that is a copy of the first column

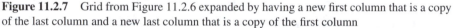

(continued)

Post: A grid extended one cell in each direction with periodic boundary conditions was returned.

Algorithm

latNS ← concatenation of last row of *lat*, *lat*, and first row of *lat*

return concatenation of last column of *latNS*, *latNS*, and first column of *latNS*

Applying a Function to Each Grid Point

After extending the grid by one cell in each direction using periodic boundary conditions, we apply the function *spread* to each internal cell and then discard the boundary cells. We define a function ***applyExtended*** that takes an extended square lattice (*latExt*) and returns the internal lattice with *spread* applied to each site. Figure 11.2.8 depicts an extended grid with the internal grid, which is a copy of the original lattice, in color. The length of *latExt* is its number of rows, which equals its number of columns. As Figure 11.2.8 depicts with row and column numbering starting at 1, the number of rows (n) or columns of the returned lattice is two less than the number of rows or columns of *latExt*. We apply the function *spread*, which has parameters *site*, *N*, *E*, *S*, and *W*, to each internal cell in lattice *latExt*. These internal cells are in rows 2 through $n + 1$ and columns 2 through $n + 1$. We added the boundary rows and columns to eliminate the different cases for cells without one or more neighbors. Thus, for i going from 2 through $n + 1$ and for j going from 2 through

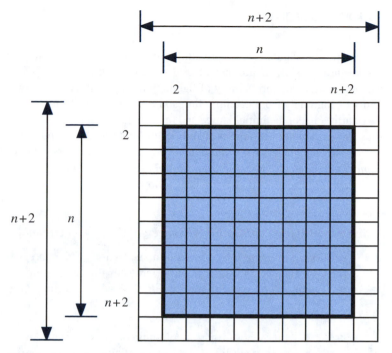

Figure 11.2.8   Internal grid in color that is a copy of the original grid (see Figure 11.2.1) embedded in an extended grid

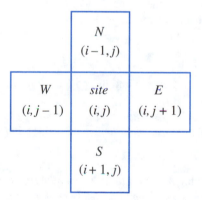

Figure 11.2.9   Indices for a lattice site and its neighbors

$n + 1$, *applyExtended* obtains a cell value for a new $n \times n$ lattice as the application of *spread* to a site with coordinates i and j and its neighbors with corresponding coordinates as in Figure 11.2.9.

Quick Review Question 8

From the text's website, download your computational tool's *11_2QRQ.pdf* file for this system-dependent question that develops the function *applyExtended*.

Simulation Program

To perform the simulation of spreading fire, we define a function *fire* with parameters *n*, the grid size, or number of grid rows or columns; *probTree*; *probBurning*; *chanceLightning*, the probability of lightning hitting a site; *chanceImmune*, the probability of immunity from catching fire; and *t*, the number of time steps. The function *fire* returns a list of the initial lattice and the next *t* lattices in the simulation. The functions *spread* and *fire* need the probabilities of lightning and immunity. To avoid having so many parameters for *spread*, in *fire* we assign parameters *chanceLightning* and *chanceImmune* to global variables *probLightning* and *probImmune*, respectively, for use by *spread*. Pseudocode for *fire* is as follows:

fire(n, probTree, probBurning, chanceLightning, chanceImmune, t)

Function to return a list of grids in a simulation of the spread of fire in a forest, where a cell value of *EMPTY* indicates the cell is empty; *TREE*, the cell contains a non-burning tree; and *BURNING*, a burning tree

Pre

n is the size (number of rows or columns) of the square grid and is positive.
probTree is the probability that a site is initially occupied by tree.
probBurning is the probability that a tree is burning initially.
chanceLightning is the probability of lightning hitting a site.
chanceImmune is the probability of a tree being immune from catching fire.
t is the number of time steps
spread is the function for the updating rules at each grid point.

Post

A list of the initial grid and the grid at each time step of the simulation was returned.
Global variable *probLightning* has the value of *chanceLightning*.
Global variable *probImmune* has the value of *chanceImmune*.

Algorithm

global variable *probLightning* ← *chanceLightning*
global variable *probImmune* ← *chanceImmune*
initialize *forest* to be an *n*-by-*n* grid of values, *EMPTY* (no tree), *TREE* (non-burning tree), or *BURNING* (burning tree), where *probTree* is the probability of a tree and *probBurning* is the probability that the tree is burning
grids ← list containing *forest*
do the following *t* times:
    forestExtended ← *extendLat1*(*forest*)
    forest ← call *applyExtended* to return $n \times n$ grid with *spread* applied to each internal cell of *forestExtended*
    grids ← the list with *forest* appended onto the end of *grids*
return *grids*

Quick Review Question 9

From the text's website, download your computational tool's *11_2QRQ.pdf* file for this system-dependent question that implements the loop in the *fire* function.

Display Simulation

Visualization helps us understand the meaning of the grids. For each lattice in the list returned by *fire*, we generate a graphic for a rectangular grid with yellow representing an empty site; green, a tree; and burnt orange, a burning tree. The function **showGraphs** with parameter *graphList* containing the list of lattices from the simulation produces these figures. We animate the sequence of graphics to view the changing forest scene.

Quick Review Question 10

From the text's website, download your computational tool's *11_2QRQ.pdf* file for this system-dependent question that develops the function *showGraphs* that produces a graphic corresponding to each simulation lattice in a list (*graphList*).

Figure 11.2.10 displays several frames of a fire sequence in which empty cells are white, burning cells are in color, and cells with nonburning trees are gray. Clearly, different initial seeds result in different sequences. This simulation employs the parameters $n = 50$, *probTree* $= 0.8$, *probBurning* $= 0.0005$, *chanceLightning* $= 0.00001$, *chanceImmune* $= 0.25$, and $t = 50$. The initial graphic displays one fire toward the bottom of the grid. At time step $t = 2$, a lightning strike starts a fire at an isolated location toward the top of the grid. Subsequent frames show both fires spreading to neighboring cells. Grids for times starting at $t = 14$ reveal the influence of periodic boundary conditions as the fire at the bottom spreads to the top of the grid, and vice versa.

Exercises

On the text's website, Fire *files for several computational tools contain the code for the simulation of the module. Complete the exercises below using your computational tool.*

For Exercises 1–3, write update rules for spread, *where "neighbor" refers to a location in the von Neumann neighborhood other than the site itself. Revise grid values as necessary.*

1. A tree takes two time steps to burn completely.
2. A tree catches on fire from neighboring trees with a probability proportional to the number of neighbors on fire.
3. A tree grows instantaneously in a previously empty cell with a probability of *probGrow*.

4. Describe changes to the code to include diagonal elements as neighbors as well.
5. Write a function to extend a grid using absorbing boundary conditions.

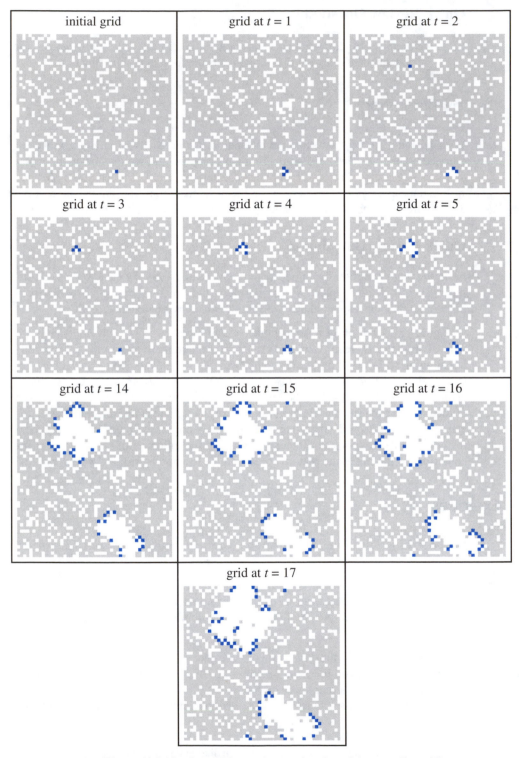

Figure 11.2.10   Several frames in an animation of the spreading of fire

6. Write a function to extend a grid so that every extended boundary cell has the value of its immediate neighbor in the original grid.

7. Write the code to assign to variables *NE*, *SE*, *SW*, and *NW* the values to the northeast, southeast, southwest, and northwest, respectively, of a site in the lattice *latExt*.

8. Suppose a lattice *g* has values for a forest grid, where a cell can be empty (value *EMPTY* = 0), a tree with the value (1 through 4) indicating the level of maturity from young to old, or a burning tree with the value indicating the intensity of the fire (5 for less intense or 6 for intense). Write code to show a graphic representing *g* with yellow for an empty cell, a different level of green from pale to full green representing the age of a tree, light red for a less intense fire and full red for an intense fire. Use constants, such as *EMPTY*, for the cell values.

Projects

On the text's website, Fire *files for several computational tools contain the code for the simulation of the module. Complete the projects below using your computational tool.*

For additional projects, see Module 13.3 on "Foraging—Finding a Way to Eats," Module 13.4 on "Pit Vipers—Hot Bodies, Dead Meat," Module 13.5 on "Mushroom Fairy Rings," Module 13.6 on "Spread of Disease— 'Gesundheit!,'" Module 13.7 on "HIV—The Enemy Within," Module 13.8 on "Predator-Prey—'Catch Me If You Can,' " and Module 13.9 on "Clouds—Bringing It All Together."

1. Run the simulation for fire several times for each of the following situations and discuss the results.
   a. *probBurning* is almost 0; *changeLightning* = *changeImmune* = 0
   b. *probBurning* is 0; *changeImmune* is 0
   c. *probBurning* is 0; *changeLightning* is 0
   d. Devise another situation to consider.

In each of Projects 2–8, revise the fire simulation to incorporate the change indicated in the exercise. Discuss the results.

2. Exercise 1	**3.** Exercise 2	**4.** Exercise 3
5. Exercise 4	**6.** Exercise 5	**7.** Exercise 6
8. Exercise 8		

9. Develop a fire simulation in which every cell in a 17×17 grid has a tree and only the middle cell's tree is on fire initially. Do not consider the possibility of lightning or tree growth. The simulation should have a parameter for *burnProbability*, which is the probability of a tree adjacent to a burning tree catches fire. The function should return the percent of the forest burned. The program should run eight experiments with *burnProbability* = 10%, 20%, 30%, ..., and 90% and should conduct each experiment 10 times. Also, have the code determine the average percent burned for each probability. Plot the data and fit a curve to the data. Discuss the results (Shodor, Fire).

10. a. Develop a fire simulation that considers wind direction and speed. Have an accompanying animation. Do not consider the possibility of lightning.

The simulation should have parameters for the probability (*probTree*) of a grid site being occupied by a tree initially, the probability of immunity from catching fire, the fire direction (value *N*, *E*, *S*, or *W*), wind level (value *NONE* = 0, *LOW* = 1, or *HIGH* = 2), coordinates of a cell that is on fire, and the number of cells along one side of the square forest. The function should return the percent of the forest burned (Shodor, Better Fire).

b. With a wind level of *LOW* (1) and a fixed *probTree*, vary wind direction and through animations observe the affects on the forest burn. Discuss the results.

c. Develop a program to run three experiments with wind levels of *NONE* = 0, *LOW* = 1, and *HIGH* = 2. Have fixed wind direction and *probTree*. The program should conduct each experiment 10 times. Also, have the code determine the average percent burned for each level. Discuss the results.

d. Develop a program to run eight experiments with no wind and *prob-Tree* = 10%, 20%, 30%, . . . , and 90%. The program should conduct each experiment ten times. Also, have the code determine the average percent burned for each probability. Plot the data and fit a curve to the data. Discuss the results.

11. Develop a fire simulation in which a tree once hit by lightning in one time step takes five additional time steps to burn. The fire can spread from the burning tree to a neighboring tree with a certain probability only on the second, third, and fourth time steps after the lightning strike.

12. Develop a fire simulation with accompanying animation in which a section of the forest is damper and, hence, harder to burn. Discuss the results.

Answers to Quick Review Question

From the text's website, download your computational tool's *11_2QRQ.pdf* file for answers to these system-dependent questions.

References

Carter, Phillip. 1996. "Fires decimate Malibu; winds, brush blamed." *Daily Bruin*, October 23. http://www.dailybruin.ucla.edu/db/issues/96/10.23/news.firemalibu.html

Dossel, B., and F. Schwabl. 1994. "Formation of Space-Time Structure in a Forest-Fire Model." *Physica Abstracts*, 204: 212–229.

Gaylord, Richard J., and Kazume Nishidate. 1996. "Contagion in Excitable Media." *Modeling Nature: Cellular Automata Simulations with Mathematica*. New York: TELOS/Springer-Verlag: 155–171.

Shodor Education Foundation. "A Better Fire!!" The Shodor Education Foundation, Inc., 1997–2003. http://www.shodor.org/interactivate/activities/fire2/index.html

Shodor Education Foundation. "Fire!!" The Shodor Education Foundation, Inc., 1997–2003. http://www.shodor.org/interactivate/activities/fire1/index.html

Studebaker, Don. 2003. "Computer Based 3D Wildland Fire Simulation." *The Pacific Southwest NewsLog* website. http://www.fs.fed.us/r5/newslogroundup/photo-sim.html (accessed January 12, 2004; site now discontinued).

Wilderness Society. "The Ecology of Wildland Fire." http://www.wilderness.org/OurIssues/Wildfire/ecology.cfm?TopLevel=Ecology

Wilson, Tracy, Stuart Pfeifer, and Mitchell Landsberg. 2003. "California fires threaten 30,000 more homes." *Pittsburg Post-Gazette*, October 28. http://www.post-gazette.com/pg/pp/03301/234849.stm

MODULE 11.3

Movement of Ants

Downloads

For several computational tools, the text's website has available for download an *Ants* file, which contains the simulation this module develops, and an *11_3QRQ.pdf* file, which contains system-dependent Quick Review Questions and answers. The simulation in this module is based on the material of (Gaylord and Nishidate 1996).

Introduction

> Everyone says stay away from ants. They have no lessons for us; they are
> crazy little instruments, inhuman, incapable of controlling themselves,
> lacking manners, lacking souls. When they are massed together, all
> touching, exchanging bits of information held in their jaws like
> memoranda, they become a single animal. Look out for that. It is a
> debasement, a loss of individuality, a violation of human nature, an
> unnatural act.
>
> *(Thomas 1979)*

Ants are extremely successful constituents of the earth's fauna, but they seem so different from human beings and are generally regarded as pests. So, what can human beings learn from such lowly creatures?

Ants have occupied a variety of ecological niches for millions of years. They are the epitome of "social insects," living in colonies of varying size. These colonies are generally made up of one or more queens, many workers, and various immature stages (egg, larvae, pupae). All of the adults are female, and all but the queen are sterile. Seasonally, a few winged males and females (fertile) are produced, but normally most of the adults are sister workers.

The queen's responsibilities are fairly uncomplicated: she lays eggs. Workers have a variety of chores: tending to young, nest construction, foraging, protecting the nest. Their entire life is dedicated to sustaining the colony.

A nest of ants typically begins with only one individual, the queen. New, mature queens fly from the nest and search for mates from groups of males that have been produced during the same time. In selected meeting places, the queen mates with one or a few males, storing the sperm in special sacs until needed. Then she flies off to find suitable nest sites. Few of these queens successfully establish a new colony, and the males die right after their big moment.

Besides keeping herself alive, a queen must find a suitable site for the new colony, excavate the site, lay the eggs, and care for the developing young. She may also have to forage for food. Some queens live off of stored food reserves and some of her laid eggs, until her young grow up. Once the first workers are produced, they take over all the queen's chores, except laying eggs. The queen can now concentrate on her major role, although she also has some control over the sex ratios and new queen production in the colony. The workers take care of everything else.

Gradually the colony grows as more and more young mature into workers. In many species, worker ants themselves become specialized for all the roles necessary to sustain the queen and the colony. Some remain in the nest, caring for the queen or the young. Others guard the nest, and others forage for food.

There is quite a bit of variability in feeding strategies and food sources used by different species of ants, and many employ more than one type of feeding behavior. Ants may prey on small insects or eat dead insects. Others rely on seeds or raid other ant nests. One of the most interesting strategies is used by the leafcutter ants, which farm nutritious fungus.

Analysis of Problem

Most species of ants communicate their movements when carrying food by leaving trails with a chemical **pheromone**. Thus, by following the scent, other ants can locate the food source. In this module, we simulate the movements of such ants in the presence of a chemical trail.

For the simulation, we use a cellular automaton similar to that in the Module 11.2 on "Spreading of Fire." We hope to observe over time that the simulated ants tend to follow a chemical trail. Thus, the simulation should help us reflect on how behavior on the local level can lead to global behavior, which we can observe in some ants. Through the interactions of many separate individuals, a group of ants as a whole can exhibit **self-organizing** behavior that makes the group appear to have a single consciousness.

Formulating a Model: Gather Data

For the model we develop in this module, we employ empirical observations of ant species that leave pheromone trails. With each step, such an ant tends to turn to and move in the direction of the greatest amount of chemical. As time passes with no ant in a location, the amount of pheromone diminishes there. For a professional model, we should obtain more exact data, such as the average amount of pheromone an ant deposits and the rate at which the chemical decreases.

Formulating a Model: Make Simplifying Assumptions

In formulating a model, we assume that an ant tends to move in the direction of the greatest amount of chemical and that with movement the ant deposits additional pheromone. However, the chemical dissipates with time. For this problem, we start with a straight trail of increasing amounts of pheromone, perhaps lain by ants heading for food. We do not consider food or a nest, although various projects do.

Formulating a Model: Determine Variables

In the module "Spreading of Fire," each cell of the grid contains an integer indicating the state of the cell—empty, tree, or burning tree. In the grid for ant movement, the value in the cell needs to indicate more information—the amount of pheromone in the cell, the presence or absence of an ant, and in the case of an ant, the direction that the ant faces. Consequently, we employ an ordered pair at each cell. The first coordinate of the ordered pair is a nonnegative integer representing the amount of chemical at the site. The second coordinate is a constant (perhaps implemented as an integer 0–4 or as a character) indicating no ant or the direction in which the ant faces, as follows:

- *EMPTY*—cell with no ant
- *NORTH*—north-facing ant
- *EAST*—east-facing ant
- *SOUTH*—south-facing ant
- *WEST*—west-facing ant

Thus, the ordered pair (3, *EMPTY*) indicates a site with level 3 amount of pheromone but no ant, while a cell with no chemical and a south-facing ant has the ordered pair (0, *SOUTH*).

We initialize the grid with zero amount of chemical for each cell except for a trail with gradated amounts of chemical. Moreover, the probability that an ant initially occupies a cell is *probAnt*. For a cell with an ant, we choose a direction at random.

Quick Review Question 1

From the text's website, download your computational tool's *11_3QRQ.pdf* file for this system-dependent question that refers to the initialization of the grid for ant movement.

Formulating a Model: Establish Relationships and Submodels

Ant movement for one time step consists of two actions, sensing and walking. First, the ant tests the neighboring sites and turns to the one with the greatest amount of pheromone. Then, if possible to do so without colliding with another ant, the ant moves to that location. The next three sections develop the *sense* and *walk* functions.

Formulating a Model: Determine Functions—Sensing

As with the fire simulation, for sensing we consider the neighbors to be the cells to the north, east, south, and west, that is, those neighbors in the von Neumann neighborhood. The rules for the function **sense** are as follows:

- An empty cell does not sense.
- An ant turns in the direction of the neighboring cell with the greatest amount of chemical. In the case of more than one neighbor having the maximum amount, the ant picks a direction at random towards one of these cells.

Quick Review Question 2

From the text's website, download your computational tool's *11_3QRQ.pdf* file for this system-dependent question that completes the code for the first *sense* rule.

Quick Review Question 3

From the text's website, download your computational tool's *11_3QRQ.pdf* file for this system-dependent question that refers to the second *sense* rule.

Formulating a Model: Determine Functions—Walking without Concern for Collision

After applying the function *sense* to each cell of the grid, we call the function **walk** for each site to obtain the ordered pair value for a cell at the next time step. The amount of chemical can change in each of the following ways:

- For a cell that remains empty, the amount of chemical decrements by one but does not fall below 0. Thus, the new amount is the maximum of 0 and the current amount minus 1.
- If an ant leaves a cell, the amount of chemical increments by one.
- If an ant stays in a cell, the amount of chemical remains the same.
- If an ant moves into a cell, the cell continues to have its same amount of chemical.

When an ant moves to a new location, the animal faces in the same direction it did before moving.

As we see in the next section, because of concern about ants running into each other, *walk* must have parameters for more neighbors than *sense* has. However, *walk* and *sense* have the same first five parameters, those for the site and the directions to the north, east, south, and west.

Quick Review Question 4

From the text's website, download your computational tool's *11_3QRQ.pdf* file for this system-dependent question that completes the definition of the first *walk* rule.

Quick Review Question 5

From the text's website, download your computational tool's *11_3QRQ.pdf* file for this system-dependent question that completes the definition of one form of the second *walk* bullet.

Quick Review Question 6

From the text's website, download your computational tool's *11_3QRQ.pdf* file for this system-dependent question that completes the definition of the third *walk* rule.

Quick Review Question 7

From the text's website, download your computational tool's *11_3QRQ.pdf* file for this system-dependent question that completes the definition of one form of the fourth *walk* bullet.

Formulating a Model: Determine Functions—Walking with Concern for Collision

An ant stays in its current location if movement in the direction to which it is facing would cause a collision. As Figure 11.3.1 shows, a collision can occur for a north-facing ant at *site* under any of the following circumstances:

- An ant is in the *N* cell.
- An east-facing ant is in the cell to the northwest (*NW*) of *site*.
- A south-facing ant is to the north of the cell to the north (*Nn*) of *site*.
- A west-facing ant is in the cell to the northeast (*NE*) of *site*.

By facing toward the same cell, an ant in any (or all) of *Nn*, *NE*, or *NW* direction prevents the ant in *site* from moving.

Collisions occur for east-, south-, or west-facing ants under similar circumstances. Thus, to determine the return value of the function *walk* we must know the values of a site and its nearest neighbors as well as *their* neighbors (see Figure 11.3.2). The order of the thirteen parameters is as follows: *site, N, E, S, W, NE, SE, SW, NW, Nn, Ee, Ss, Ww*.

Quick Review Question 8

From the text's website, download your computational tool's *11_3QRQ.pdf* file for this system-dependent question that deals with the definition of the *walk* rule for one

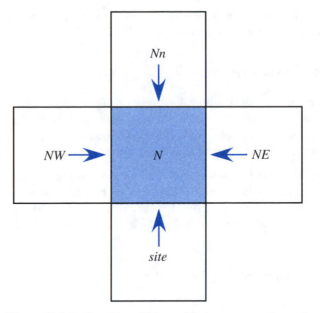

Figure 11.3.1 Possible collisions with movement to the north

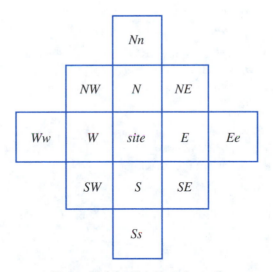

Figure 11.3.2 Neighbors for *walk*

of the situations in Figure 11.3.1 in which a collision should be avoided for the current ant that faces an empty cell to the north and a northeast ant that faces west.

Another collision situation occurs if two ants both want to move into *site*, which is empty. That is, what happens to an empty cell when it is the location of a possible collision? In this case, *site* remains empty. (The projects explore other alternatives, such as allowing one ant to move and the other(s) not.) Also, if the amount of chemical is positive, it decrements by 1. If no chemical is present, the amount remains 0. For the four nearest neighbors of a cell, six possible situations exist for two neighbors wanting to move into *site: N-E, N-S, N-W, E-S, E-W, S-W.*

Quick Review Question 9

From the text's website, download your computational tool's *11_3QRQ.pdf* file for this system-dependent question that deals with the definition of the *walk* rule for the situation in which a collision should be avoided where the current site is empty but ants to the north and east both want to move into the site.

   Because *walk* has parameters of all the cells in Figure 11.3.2, we define a function, *extendLat2*, to extend the boundary by two cells in each direction using periodic boundary conditions. Another function, *applyExtended2*, applies the function *walk* to every cell of the original sublattice in the extended lattice.

Solving the Model—A Simulation

With *grid* being the initial grid and *t* being the number of iterations of the simulation, the algorithm for the ant movement simulation *ants* is as follows:

Algorithm for ants(grid, t)

Function to return a list of grids in a simulation of ant movement, where a cell value of *EMPTY* indicates the cell is empty; *NORTH*, the cell contains a north-facing ant; *EAST*, an east-facing ant; *SOUTH*, an south-facing ant; and *WEST*, an west-facing ant.

Pre

   grid is the initial grid.
   t is the number of time steps

Post

A list of the initial grid and the grid at each time step of the simulation was returned.

Algorithm

   initialize *gridList* be a list containing *grid*
   do the following *t* times:
         elat1 ← call *extendLat1* to get *grid* extended by 1 cell in each direction
         gridSense ← call *applyExtended1* to apply *sense* to each internal cell of *elat1*
         elat2 ← call *extendLat2* to get *gridSense* extended by 2 cells in each direction
         grid ← call *applyExtended2* to apply *walk* to original internal cells of *elat2*
         gridList ← *gridList* with *grid* appended
   return *gridList*

Verifying and Interpreting the Model's Solution—Visualizing the Simulation

We have a number of choices of how to communicate the information in an ant simulation for verification and interpretation of the model's solution. A graphic can indicate the level of chemical attractant with a shade of gray or a level of color. Ants can be represented by a color; an intensity of that color representing the chemical level in that cell, or not at all. Figure 11.3.3 presents a sequence of frames for one such depiction. The ants are represented by color. The level of gray indicates the strength of the chemical at an empty site. For ants in cells, the darker colors indicate greater amounts of chemical. As the sequence shows, ants in contact with a chemical trail move along the path to levels of greater chemical strength. The simulation represents how this social insect can communicate chemically with its sisters for the common good.

Let us consider one coloring scheme. Suppose *maxChem* is the maximum amount of chemical, such as 50, in the cells for all grids in a simulation. For comparisons between grids, we use an overall chemical maximum, although no cell might achieve this maximum in a particular time step grid of the simulation. Suppose *siteChem* is the amount of chemical, such as 10, in a particular cell. Because color amounts range between 0.0 and 1.0, we normalize the latter value by dividing by *maxChem*. For the example values, we have *normalizedChem* = *siteChem*/*maxChem* = 10/50 = 0.2. If the amount of chemical were 40, the normalized value would be 40/50 = 0.8. However, a cell with red, green, and blue values of 0.2, 0.2, and 0.2, respectively, is a darker gray than one with values 0.8, 0.8, and 0.8, respectively. To reverse the situation, we subtract *normalizedChem* from 1.0, so that the amounts of color for each component are 1.0 − 0.2 = 0.8 and 1.0 − 0.8 = 0.2, respectively. If the darker colors are too dark, we can restrict the range to be between 0.5 and 1.0 by multiplying *normalizedChem* by 0.5. In this case, for *siteChem* = 10, 1.0 − *normalizedChem* * 0.5 = 1.0 − (0.2)(0.5) = 0.9, and for *siteChem* = 40, 1.0 − *normalizedChem* * 0.5 = 1.0 − (0.8)(0.5) = 0.6. The minimum amount of chemical, 0, yields a color component of 1.0 − 0.0 = 1.0, while the maximum amount of chemical, here 50, has a value of 1.0 − (50/50)(0.5) = 0.5. A scientific visualization should impart information clearly while not misleading the viewer or suggesting more than is available.

Quick Review Question 10

From the text's website, download your computational tool's *11_3QRQ.pdf* file for this system-dependent question that develops a visualization for the simulation.

Exercises

On the text's website, Ants files for several computational tools contain the code for the simulation of the module. Complete the exercises below using your computational tool.

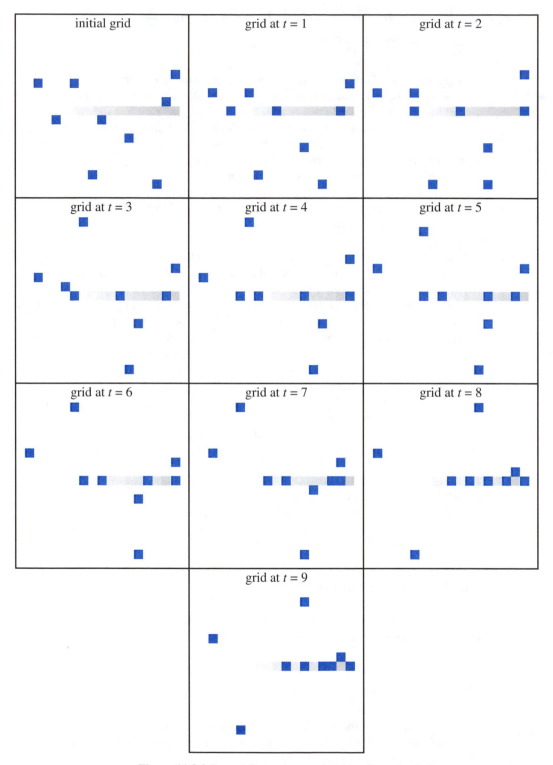

Figure 11.3.3 Several frames in an animation of ant simulation

1. Quick Review Question 5 completes the definition of one form of the second *walk* bullet: If an ant leaves a cell, the amount of chemical increments by one. Give the associated rules.

2. Quick Review Question 7 completes the definition of one form of the fourth *walk* bullet: If an ant moves into a cell, the cell continues to have its same amount of chemical. Give the associated rules.

3. **a.** Quick Review Question 8 completes one of the *walk* rules for a north-facing ant to avoid a collision. Give the other two such *walk* rules.
   b. Give the three *walk* rules for an east-facing ant to avoid collisions.

4. Quick Review Question 9 completes one of the *walk* rules in which a collision should be avoided where the current site is empty. Give the other five such *walk* rules.

5. **a.** In the simulation of this module, an ant can return immediately to a cell from which it just came. Describe the movement of an ant in an area where no other ants are near and initially no chemical deposits exist.
   b. Revise the second *sense* rule so that an ant cannot return immediately to a cell from which it just came in the last time step.

6. Describe how a "stand-off" can occur with two ants facing each other and in close proximity to each other. What, if anything, can cause the stand-off to end?

7. Define a function, *extendLat2*, to extend the boundary by two cells in each direction using periodic boundary conditions.

8. Define a function, *applyExtended2*, to apply a function parameter to every cell of the original sublattice in the lattice extended by *extendLat2*.

9. Define a function, *showGraphs*, to produce the frames for an animation as described in the section "Verifying and Interpreting the Model's Solution—Visualizing the Simulation." Have parameters for a list of square grids representing frames of the simulation at consecutive time steps (*graphList*) and for the maximum amount of chemical (*maxChem*) in the cells of all grids of the list.

Projects

On the text's website, Ants *files for several computational tools contain the code for the simulation of the module. Complete the projects below using your computational tool.*

For additional projects, see Module 13.4 on "Pit Vipers—Hot Bodies, Dead Meat" and Module 13.8 on "Predator-Prey—'Catch Me If You Can.'"

1. **a.** Complete the ant simulation in the file *Ants* of this module.

   Investigate the ant behavior in the following situations:
   b. A path through the center of the grid exists as described in the section "Formulating a Model: Determine Variables."
   c. Initially, no chemical is on the grid.
   d. Two areas of chemical concentrations exist.

2. Revise Project 1 so that for the second *sense* rule an ant cannot return immediately to a cell from which it just came in the last time step.

3. a. Develop a simulation in which a single ant leaves the nest searching for a food source that is unknown to the ant and that is due north of the nest. Initially, the grid does not contain any chemical. The ant only omits chemical once it has found the food. Then, the ant returns directly to the nest in a straight-line fashion, leaving a chemical trail. Because no danger of collision exists, the simulation does not need to have separate sense and walk components. Instead, the simulation can be more like that of Module 10.2 on "Random Walk."

  b. Augment the simulation of Part a so that as soon as one ant returns to the nest, another ant leaves, searching for food.

4. Develop a simulation with a nest, a food source to the north of the nest, and ants that should not collide. An ant emits pheromone only after finding the food and while returning to the nest, and the amount of chemical in a vacant cell decrements by one with each time step. Food diminishes with each ant visit. One possible implementation has an ordered triple for each grid value. The first component is a nonnegative integer representing the level of chemical attractant or amount of food. The second component represents a state: *EMPTY* site, *SEARCHING* ant, *FOOD*, *RETURNING* ant, or *NEST*. The third component is *VOID* in the situation where the site contains food or the nest; otherwise, the value indicates the direction in which the ant faces, *NORTH*, *EAST*, *SOUTH*, or *WEST*. Thus, an ant cannot occupy a food site.

5. An army ant raid can be 20 m wide, 200 m long, and involve hundreds of thousands of ants. The raid is self-organizing, evolving from interactions on the local level into a global pattern. The pattern appears treelike with the forward part of the raid being branchlike. Develop a simulation with no food present that has the following rules, which are based on those of (Franks 2001):

- Every ant deposits pheromone unless the cell is saturated, containing the maximum amount of chemical.
- In new territory, where pheromones are not present, an ant goes randomly to the northeast or to the northwest.
- When pheromone exists, with a certain probability an ant is more likely to follow the peromone trail.
- More than one ant can be in a cell, up to some maximum number of ants.
- Each time step, a constant number of ants leaves the nest, which is one cell.

6. Augment Project 5 to include the following rule: Ants move faster in the presence of more pheromone. For example, you could consider that based on the amount of chemical, an ant makes a move per every one, two, or three time steps. Discuss the effects of varying the speed of the ants.

7. Augment Project 5 to include food and the following rule: Once an ant finds food, it returns to the nest using the same rules as those of Project 5 except it goes to the southeast or southwest (Franks 2001). Discuss the difference in the self-organizing pattern between this simulation and that of Project 5.

8. Augment Project 6 to include food and the following rule: Once an ant finds food, it returns to the nest using the same rules as those of Projects 5 and 6 except it goes to the southeast or southwest (Franks 2001). Discuss the difference in the self-organizing pattern between this simulation and that of Project 6.

For Projects 9–12, repeat the indicated project with the direction being relative to an ant's heading, front right and front left, instead of northeast and northwest, respectively. Have the nest be in one corner of the grid.

9. Project 5 **10.** Project 6 **11.** Project 7 **12.** Project 8

13. Adjust Project 1 or 2 so that the pheromone diffuses to adjacent cells over time.

14. Develop a simulation in which you allow the user to specify the initial locations of ants and the amounts of chemical in various locations. Enable the user to display the grid before continuing with the simulation.

15. Usually, trail following is not completely accurate. Introduce an additional stochastic element in the choice of direction in any of the earlier projects. For example, you might have an ant picking a random direction 25% of the time and face a neighbor with the most chemical 75% of the time. Discuss the advantages and disadvantages of this lack of precision.

16. Adjust the grid on any of the earlier projects to contain obstacles.

17. Develop a cellular automaton simulation to illustate the exploitive competition of Argentine ants versus native ants as described in Project 2 of Module 6.1 on "Competition." Illustrate the competitive factor of discovery time. See Project 15 above for an idea on simulating discovery time (Holway 1999).

18. Develop a cellular automaton simulation to illustrate the exploitive competition of Argentine ants versus native ants as described in Project 2 of Module 6.1 on "Competition." Illustrate the competitive factor of recruitment rate. See Project 6 above for an idea on simulating rate of recruitment.

19. Develop a cellular automaton simulation to illustate the interference exploitive competition of Argentine ants versus native ants as described in Project 2 of Module 6.1 on "Competition." Illustrate the competitive factor of recruitment rate.

20. An alternative form of walking that avoids collision accomplishes movement by testing at each empty cell and allowing exactly one neighboring ant desiring to do so to move into that cell. The test begins as follows:

```
if current site is empty
    if cell to north contains south-facing ant
        move that ant into current cell
        make site to north empty with appropriate amount of chemical
    else if cell to east contains west-facing ant
        move that ant into current cell
        make site to east empty with appropriate amount of chemical
    else if . . . // other cases
    else // no ant to move into cell
        decrement amount of chemical appropriately
```

Note that this algorithm requires extending the lattice by one cell instead of two in each direction. Repeat Project 1 using this version of walking.

21. The walking algorithm in Project 20 has a deterministic and biased order for checking neighbors—north first, then east, etc. Revise the project to pick at random which ant to choose for movement when more than one neighboring ant faces an empty cell.

22. Adjust any of the earlier projects so that the pheromones dissipate at a slower rate than one unit per time step. Describe the effect of this change.

Answers to Quick Review Questions

From the text's website, download your computational tool's *11_3QRQ.pdf* file for answers to these system-dependent questions.

References

Franks, Nigel R. 2001. "Evolution of Mass Transit Systems in Ants: a Tale of Two Societies." *Insect Movement: Mechanisms and Consequences Proceedings of the 20th Symposium of the Royal Entomological Society.* Wallingford, Oxford: CAB International: 281–298.

Gaylord, Richard J., and Kazume Nishidate. 1996. "Chemotaxis." *Modeling Nature: Cellular Automata Simulations with Mathematica.* New York: TELOS/Springer-Verlag: chap. 12, 121–130.

Hölldobler, B., and E. O. Wilson. 1990. *The Ants.* Cambridge, Mass.: Harvard University Press.

Holway, David A. 1999. "Competitive Mechanisms Underlying the Displacement of Native Ants by the Invasive Argentine Ant." *Ecology*, 80(1): 238–251.

Martinoli, Alcherio, Rodney Goodman, and Owen Holland. "Exploration, Exploitation, and Navigation in Ants." EE141: Swarm Intelligence, California Institute of Technology. http://www.coro.caltech.edu/Courses/EE141/Lecture/W3/AM_EE141_W3ExplNav.pdf

Thomas, Lewis. 1979. *The Medusa and the Snail, More Notes of a Biology Watcher.* New York: The Viking Press.

Weimar, Jörg. 2003. "PredatorAgainstPrey." Source code, Technical University of Braunschweig. http://www-public.tu-bs.de:8080/~y0021323/ca/ PredatorAgainst Prey.cdl

12

HIGH PERFORMANCE COMPUTING

MODULE 12.1

Concurrent Processing

Introduction

> We are in the early stage of a revolution in science nearly as profound as
> the one that occurred early in the last century with the birth of quantum
> mechanics. This revolution is caused by two developments: one is the set
> of instruments . . . ; the other is the availability of powerful computing
> and information technology. Together these have brought science finally
> within reach of a new frontier, the frontier of complexity.
> —*John H. Marburger III, Director,*
> *Office of Science and Technology Policy (ITRD 2004)*

Before humankind lies a vast, fascinating, and crucial collection of knowledge—
knowledge that will change our lives. With precision instrumentation, modern laboratory techniques, and ever-increasing computational abilities, we will be able to investigate and understand physical, chemical, and biological systems from the most
fundamental elements of the universe to the largest and most complex systems. The
possibilities for major research breakthroughs, significant technological innovations,
medical and health advances, augmented economic competitiveness, etc. are unfath-
omable.

Enhanced computer technology and power are crucial to progress on this new
frontier. In 2002, the Japanese government began to simulate the earth's climate and
geological activity using what was at that time the world's fastest supercomputer—
the Earth Simulator. This remarkable machine, occupying a building that would hold
four tennis courts, can perform almost 36 trillion calculations per second. This
achievement was the first time the fastest supercomputer had been built outside the
United States (Associated Press 2002). Responding to this challenge and realizing
the associated opportunities, Secretary of Energy Spencer Abraham announced in
2004 a new project called the "Leadership Class Computing Facility for Science,"
which will build the fastest supercomputer in the world. Such computing capability

will permit the development of much more accurate models for chemical, physical, and biological systems (DOE 2004).

The National Oceanic and Atmospheric Administration (NOAA) already has had much success in employing modeling and simulation on supercomputers. For example, NOAA experts predicted that 2004 would be a very active year in the Atlantic for hurricanes, with 6 to 8 hurricanes forming from 12 to 15 tropical storms. The prediction proved accurate as the year had 8 hurricanes and 16 tropical storms. According to retired Navy Vice Admiral Conrad C. Lautenbacher, Ph.D., undersecretary of commerce for oceans and atmosphere and NOAA administrator, this forecast "is the result of thousands of hours of work by NOAA and its partners." He goes on to say, "NOAA investments in high-speed computers, improved weather modeling and extensive Earth observation systems enable our scientists and forecasters to gather and synthesize information and begin the process of preparing the public to take action." In 2003, Hurricane Isabel resulted in 17 deaths and billions of dollars in property damage, so taking the proper action would certainly be prudent. Proper actions are more likely with advance warning from improved hurricane prediction and tracking (NOAA 2004).

Computational biologists at Oak Ridge National Laboratory, a Department of Energy laboratory, are interested in another type of system, the genomes of microbes. One reason for their attention is that some of these microbes have biochemical pathways that permit them to survive some of the most challenging environments—high temperatures, toxic chemicals, radiation, etc. Using supercomputers, scientists may ferret out microbial genes that promote DNA repair from radiation damage or that remediate radioactive wastes. Also, using genes that promote carbon dioxide absorption, we might be able to counteract the buildup of CO_2 in the atmosphere and to slow global warming (Associated Press 2002).

As an engineering example, the Computational Aerosciences (CAS) is a project joining the National Aeronautics and Space Administration (NASA) and other United States governmental agencies with aeronautical industries and academia to reduce the cost and design time for new air- and spacecraft by accelerating the development of **high performance computing** (**HPC**) technology. For instance, the Boeing Corporation has used extensive computational power to design the 7E7, a new, fuel-efficient aircraft (Boeing 2004). At Langley Research Center, scientists are employing cutting edge computational tools to explore unconventional aircraft design, such as the super jumbo blended-wing-body aircraft (Lytle 2001).

On the human side, neuroscientists interested in understanding how the nervous system works are developing models of the molecular activities of neurons and small neural networks, as well as mapping brain systems. These studies require not only sophisticated instruments for imaging, but also high performance computing. Modeling with HPC is an indispensable complement to laboratory experimentation if we are to understand the complexities of the human nervous system, advanced spacecraft design, genomics, weather, and numerous other advanced scientific applications (ITRD 2000).

This chapter is not meant to be an in-depth study of high performance computing but is intended to give an idea of some of the applications, architecture, concepts, challenges, and algorithms.

Analogy

A **processor** or a **central processing unit** (**CPU**) of a computer performs the arithmetic and logic of a computer. It is the brain of the computer. **Concurrent processing** involves having associated, multiple CPUs working concurrently, or simultaneously, on the same or different problems. To achieve the type of high performance for the problems discussed in the introduction, concurrent processing is essential. For an examination of some of the options and problems involved, we consider an analogous situation.

> **Definitions**   A **processor** or a **central processing unit** (**CPU**) of a computer performs the arithmetic and logic of a computer. **Concurrent processing** involves having associated, multiple CPU's working concurrently, or simultaneously, on the same or different problems.

Suppose a scout leader is taking a group of ten scouts in a van to a camping trip. Before the fun begins, they must shop for provisions of about a hundred different items at a grocery store. What are some of the options for shopping? Before reading further, list some your ideas for the task.

1. One option is for the leader to leave all the scouts in the van and to do the shopping alone. This option is analogous to a single processor working on a single program. Meanwhile, ten processors (i.e., scouts), who could be helping, are doing nothing to speed the process.

2. Another alternative is for the leader to tear the grocery list into ten parts and have each scout gather the items on his or her list and meet at a cash register, where the leader is to pay. What difficulties might arise?

   • Initially, each scout must wait for a partial list; and finally, he or she must wait for all the other scouts to finish shopping and for the leader to pay. We have a bottleneck because on part of the overall task only one processor (i.e., the leader) is working.

   • Perhaps only three shopping carts are at the front of the store, so that adequate resources are not immediately available.

   • Moreover, suppose a scout cannot find peanut butter. With this scenario, everyone must wait while the child wanders through the store without assistance from anyone else. A synchronization problem exists. The shopping lists probably could have been divided differently to shorten the wait.

3. A better choice might be that as soon as a scout finishes gathering his or her groceries, he or she helps someone else. However, a speedy scout must know where to go, and both scouts must agree how to divide the work.

4. Another way to help might be to separate scouts into pairs, each consisting of a seeker and a shopper, who have walkie-talkies. The seeker finds an item on the list and tells the location to the shopper. While the shopper is gathering the item into the basket, the seeker searches for the next item. This pipeline

system still has situations in which a processor (i.e., scout) must wait. The shopper must wait for the seeker to find the first item, and the seeker must wait for the shopper to gather the last item. At intermediate stages, the seeker might have difficulty finding an item, causing the shopper to be idle; or the shopper might take a while loading cans of Spam™ into the basket, while the seeker has already found the Vienna sausage, the next item on the list.

5. To avoid the bottleneck at checkout, the leader might give each child money as well as part of the list. However, the leader must have an excess of resources (i.e., money) to distribute to the group.

6. The leader can also do some preprocessing on the grocery list. For example, shopping would be faster if each scout had a list of items on a single aisle. This task, too, could be accelerated with the help of some of the scouts. This scenario would work best if the leader and some of the scouts were familiar with the locations of items in the store.

7. Another alternative consists in the leader emailing a partial grocery list to each scout. Scouts are responsible for buying their parts of the groceries and meeting at the van at a certain time. Upon receiving a receipt, the leader reimburses a scout. The scouts, who might be located at great distances from each other, do not have to shop at the same grocery store. Difficulties can still arise, such as a scout not receiving the message or a scout getting sick and not being able to shop. It would be advisable for the leader to make sure all scouts read their emails and shop. If a scout is sick, the leader can redistribute the workload.

Consider other alternatives along with their advantages and disadvantages.

Types of Processing

Three types of processing exist: sequential, parallel, and distributed. **Sequential processing** involves a single processor working on one program. Such processing is analogous to the leader being the only shopper.

Parallel processing consists in a collection of connected processors in close physical proximity working concurrently. Several examples of a vanload of scouts with the leader shopping together at one grocery store provide analogues to parallel processing.

Distributed processing involves several processors, perhaps at great distances from each other, communicating via a network and working concurrently. The example of the leader emailing the partial lists to the scouts for them to shop at a variety of stores is analogous to distributed processing.

> **Definitions**   **Sequential processing** involves a single processor working on one program. **Parallel processing** consists in a collection of connected processors in close physical proximity, or **tightly coupled**, working concurrently. **Distributed processing** involves several processors, perhaps at great distances from each other, communicating via a network (hence, **loosely coupled**) and working concurrently.

Quick Review Question 1

Indicate the type(s) of processing—sequential, parallel, distributed, or none—for each of the following:

    a. Can involve execution of more than one program at a time
    b. Can involve execution of one program
    c. Can have processors in different countries

Multiprocessor

A **multiprocessor** is a computer system with more than one processor. Communication among the processors is accomplished through shared memory or message passing.

Figure 12.1.1 presents a diagram of a traditional **shared memory multiprocessor**. Although a memory module might be associated with an individual processor, all processors can access all memory modules.

One difficulty with this architecture is maintaining consistency of the data. For example, suppose processor A reads a value, say 2, for a shared variable x; and while A is performing computations with the value, processor B writes a different value, say 3, to x. The values are not consistent. Multiprocessors must provide mechanisms for the programs to ensure consistency of shared data.

A shared memory multiprocessor is specifically designed as a computer system. However, another kind of multiprocessor can be constructed with a network of workstations that communicate with each other through message passing. Figure 12.1.2 gives a diagram of the architecture of a **message-passing multiprocessor** in which each processor has its own associated memory that is inaccessible to other processors.

With a message-passing or a shared memory multiprocessor, programmers must explicitly divide a program into pieces, called **processes**, for concurrent execution. However, in the case of a message-passing system, Computer A cannot access directly a variable, say x, stored in Computer B's memory. Instead, A sends a message to B requesting the value of x; and if acceptable, B sends a message to A with the value of the variable. To handle these operations, programmers write special message-passing calls.

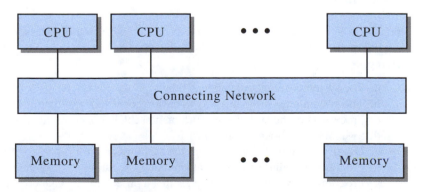

Figure 12.1.1   Architecture of traditional shared memory multiprocessor

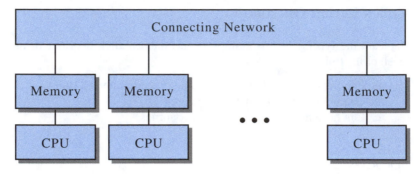

Figure 12.1.2   Architecture of message-passing multiprocessor

A message-passing multiprocessor does require a programmer to code more fairly low-level system calls than does programming a shared memory multiprocessor. However, a shared memory multiprocessor must have its own mechanism to ensure the consistency of shared data, and execution of this mechanism can add significantly to execution time of a program. Additionally, a message-passing multiprocessor with its network of complete workstations has the advantage of **scalability**—with the usually easy addition of more processors to the system, a program's execution speed increases. Also, with advances in technology, a faster, commercially available workstation can easily be swapped for a slower one in a message-passing multiprocessor.

> **Definitions**   A **multiprocessor** is a computer system with more than one processor. In a **shared memory multiprocessor**, processors communicate through shared memory. In a **message-passing multiprocessor**, processors communicate through message passing. A **process** is a task or a piece of a program that executes separately. **Scalability** is the capability of a computer system with expanded hardware resources to exhibit better performance.

Quick Review Question 2

Indicate which type of multiprocessor—shared memory multiprocessor, message-passing multiprocessor, both, neither—exhibits the characteristic for each of the following:

    a. The system is more scalable than others.
    b. The system must provide a way to ensure consistency of data.
    c. Processors of the system can work on a problem concurrently.
    d. System can be upgraded more easily.
    e. Usually, a manufacturer develops the system as a multiprocessor computer.
    f. Programmer splits a program into parts for execution on different processors.
    g. Programmer writes a call to request that a processor send data from its memory to another processor for its memory.

h. Processors can execute several independent programs at the same time.
i. A processor can write directly to the memory of another processor.

Classification of Computer Architectures

Michael Flynn developed a useful classification of parallel computer architectures based on the number of independent concurrent instruction streams, or processes, and data streams (Dowd and Severance 1998). A parallel computer simultaneously might execute a single instruction (SI) stream or multiple instruction (MI) streams. Moreover, a single data (SD) stream or multiple data (MD) streams might move between memory and a processor. Thus, the **Flynn Classification of Computer Architectures** consists of the following categories:

SISD—Single instruction stream, single data stream.
Sequential computers that can only execute one program at a time are in this category.
MIMD—Multiple instruction streams, multiple data streams.
This designation includes multiprocessors that can execute separate programs concurrently.
SIMD—Single instruction stream, multiple data streams.
In this situation, the system executes one program. A front-end processor controls an array of small processors, each member of which simultaneously executes the same instruction on different data.
MISD—Multiple instruction streams, single data stream.
Disagreement exists on whether or not this category has any systems.

Although overlap exists between the categories, the Flynn Classification is helpful in characterizing HPC systems.

Quick Review Question 3

Characterize each of the following computer systems using the Flynn Classification.

a. A collection of 1024 processors executing the same instruction at exactly the same time
b. A network of workstations that communicate through message passing and that collectively can execute a program
c. A shared memory multiprocessor that can execute a program with concurrent processing

Metrics

We can employ several metrics, or measures, to indicate the improvement achieved using various configurations of a MIMD multiprocessor instead of a SISD sequential machine.

For execution on a MIMD machine, a program is divided into separate processes, or tasks, to be executed in parallel on various processors. The **granularity** of parallelism refers to the number of components. We say that a machine has **fine granularity** if it contains many processors, such as a system with thousands of very simple processors, each executing relatively few instructions. A machine with **coarse granularity** contains a small number of processors, such as a system with a dozen very fast and complex processors, each executing many instructions simultaneously. Most commercial multiprocessors are coarse-grained MIMD systems. A measure of the granularity is the ratio of computation to communication:

$$\textbf{Ratio of computation to communication} = \frac{\textbf{Computation time}}{\textbf{Communication time}}$$

This ratio is large in the case of coarse granularity and small for fine granularity. Fine granularity has the advantage that many processors can execute the program simultaneously, but the larger number of processes has the disadvantage of requiring greater communication time. Coarse granularity reduces communication—an advantage—but reduces concurrency—a disadvantage. Thus, the programmer seeks a balance between the extremes of granularity by achieving a larger ratio of computation to communication along with suitable parallelism.

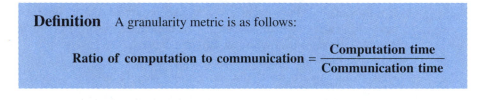

Definition   A granularity metric is as follows:

$$\textbf{Ratio of computation to communication} = \frac{\textbf{Computation time}}{\textbf{Communication time}}$$

Quick Review Question 4

Suppose communication consumes 10% of the time for execution of a concurrent program on a multiprocessor. Determine the ratio of computation to communication.

A commonly used metric for multiprocessor performance is the speedup factor. For a system with n processors, the **speedup factor $S(n)$** is as follows:

$$S(n) = \frac{\textbf{Execution time on sequential computer}}{\textbf{Execution time on system with } n \textbf{ processors}}$$

or

$$S(n) = \frac{\textbf{Number of computational steps on sequential computer}}{\textbf{Number of computational steps in parallel with } n \textbf{ processors}}$$

Often algorithms to accomplish some computation are different on a sequential computer and a multiprocessor, and we employ the times for the best algorithms available on each system in measuring speedup. For example, suppose the best sequential algorithm for a particular task takes 100 milliseconds (ms) while the corresponding work with a 200-processor system requires 2.5 ms. In this case, the speedup is $S(200) = (100 \text{ ms})/(2.5 \text{ ms}) = 40$.

Definition   For a system with n processors, the **speedup factor $S(n)$** is as follows:

$$S(n) = \frac{\text{Execution time on sequential computer}}{\text{Execution time on system with } n \text{ processors}}$$

or

$$S(n) = \frac{\text{Number of computational steps on sequential computer}}{\text{Number of computational steps in parallel with } n \text{ processors}}$$

Quick Review Question 5

Suppose a sequential algorithm takes 24 milliseconds (ms), while the speedup on a multiprocessor with 8 processors is $S(8) = 4$. Determine the execution time on the multiprocessor.

Usually, the maximum speedup possible with n processors is $S(n) = n$, which we call **linear speedup** because the graph is a straight line.[*] This situation is achieved when the time required for execution with n processors is $1/n$ of time for execution on a sequential computer. For example, suppose the time for a sequential algorithm is 1 ms. With linear speedup and 2 processors, the execution time is $1/2$ ms $= 0.5$ ms, so that $S(2) = 1/0.5 = 2$. For 3 processors, the execution time is $1/3$ ms; with 4 processors, $1/4$ ms; etc.

Quick Review Question 6

Suppose maximum speedup is achieved with an 8-processor system for an algorithm that executes in 24 ms on a sequential computer. Determine the execution time on the multiprocessor.

Linear speedup is rarely achieved because of several overhead factors, including the following:

1. Communication time between processors
2. Times when some of the processors are idle
3. Additional computations necessary in the parallel version and unnecessary in the sequential version

In algorithms in the next module, we consider such overheads and speedup factors.

---

[*] Occasionally, a speedup better than $S(n) = n$ can be achieved through comparison with an inferior sequential algorithm or through a special multiprocessor architectural feature, such as a very large amount of memory.

Exercises

1. Complete the matching related to the shopping scouts example in the "Analogy" section. An answer may be used more than once.

   a grocery list   **A** data item
   b item on grocery list   **B** distributed processing
   c leader goes through checkout alone   **C** memory
   d peanut butter   **D** message passing
   e scout   **E** MISD
   f scout leader   **F** parallel processing
   g scouts shopping in same store   **G** processor
   h seeker talks to shopper   **H** sequential processing

2. Give the advantages and disadvantages of a shared memory multiprocessor and of a message-passing multiprocessor.
3. The best sequential sorting algorithm that compares elements requires $n \log n$ computational steps. Suppose a sorting algorithm on a parallel system requires $4n$ computational steps. Determine the speedup factor (Wilkinson and Allen 1999).
4. For each of the overhead factors inhibiting linear speedup, give an analogous example using the shopping scouts.
5. Draw the graph for linear speedup.

Project

1. Write a paper with references on a scientific application that is advanced by high performance computing.

Answers to Quick Review Questions

1. **a.** parallel processing and distributed processing
   b. sequential processing, parallel processing, and distributed processing
   c. distributed processing
2. **a.** message-passing multiprocessor
   b. both. Although the problem is more obvious in shared memory multiprocessors, the problem exists with both architectures.
   c. both
   d. message-passing multiprocessor
   e. shared memory multiprocessor
   f. message-passing multiprocessor
   g. message-passing multiprocessor
   h. both
   i. shared memory multiprocessor
3. **a**. SIMD
   b. MIMD
   c. MIMD

4. ratio of computation to communication = 0.9/0.1 = 9

5. 6 ms, because $S(8) = 4 = (24 \text{ ms})/t$; the parallel algorithm on an 8-processor system is 4 times faster than the sequential one. Thus, $t = (24 \text{ ms})/4 = 6$ ms.

6. 3 ms, because for linear speedup, $S(8) = 8 = 24 \text{ ms}/(\text{Execution time on multi-processor})$. Thus, execution time on multiprocessor = 24 ms/8 = 3 ms

References

Andrews, Gregory R. 2000. *Foundations of Multithreaded, Parallel, and Distributed Programming*. Reading, Mass.: Addison-Wesley-Longman: 664.

Associated Press. 2002. "World's fastest computer simulates Earth." CNN.com/ Technology, November 16. http://archives.cnn.com/2002/TECH/biztech/11/15/ fastest.computer.ap/

Boeing. 2004. "Boeing Launches 7E7 Dreamliner." News Release, April 26. http: //www.boeing.com/news/releases/2004/q2/nr_040426g.html

DOE (Department of Energy). 2004. "DOE Leadership-Class Computing Capability for Science Will Be Developed at Oak Ridge National Laboratory." Press Release, May 12. http://www.energy.gov/engine/content.do?PUBLIC_ID=15871&BT_CODE=PR_PRESSRELEASE&TT_CODE=PRESSRELEASE

Dowd, Kevin, and Charles Severance. 1998. *High Performance Computing*. 2nd ed. Beijing, China: O'Reilly: 446.

ITRD. National Coordination Office for Information Technology Research and Development. 2000. Subcommittee on Computing, Information, and Communications R&D. "Information Technology Frontiers for a New Millennium, High End Computing and Computation." *FY 2000 Blue Book*. http://www.itrd.gov/pubs/ blue00/hecc.html

————. National Coordination Office for Information Technology Research and Development. 2004. "High-End Computing: New Technologies to Explore the Frontier of Complexity." National Science and Technology Council. http://www .itrd.gov/ pubs/blue03/frontier_of_complexity_01.html

Lytle, John K. 2001. Project Manager. "Computational Aerospace Sciences Project." NASA's Glenn Research Project, September 17. http://hpcc.grc.nasa.gov/cas.shtml (accessed July 3, 2004).

NOAA. 2004. National Oceanographic and Atmospheric Administration. "Above-Normal 2004 Atlantic Hurricane Season Predicted." *NOAA News*, May 17. http:// www.noaanews.noaa.gov/stories2004/s2225.htm

Wilkinson, Barry and Michael Allen. 1999. *Parallel Programming*. Upper Saddle River, N.J.: Prentice-Hall: 431.

MODULE 12.2

Parallel Algorithms

Introduction

In this module, we examine some of the algorithms for solving classical problems with scientific applications in parallel. In doing so, we investigate speedups and some of the challenges in programming multiprocessors. While not considering actual code, we can gain an appreciation for some of the aspects of designing parallel algorithms.

Embarrassingly Parallel Algorithm: Adding Two Vectors

Some algorithms are so easy to partition onto non-communicating processors that we call them embarrassingly so. An **embarrassingly parallel algorithm** can divide computation into many completely independent parts with virtually no communication.

> **Definition** An **embarrassingly parallel algorithm** can divide computation into many completely independent parts with virtually no communication.

Addition of two vectors is an example of such an algorithm. On the Cartesian plane to add two vectors, which ordered pairs such as (1, 3) and (2, 5) represent, we add **componentwise**. That is, to obtain the result's first component, or coordinate, we add the first components of the ordered pairs; and the sum of the second components yields the second coordinate of the result. Thus, the sum of (1, 3) and (2, 5) is as follows:

$$(1, 3) + (2, 5) = (1 + 2, 3 + 5) = (3, 8)$$

An ordered pair is a special case of a **vector**, which is an ordered n-tuple of numbers, $v = (v_1, v_2, \ldots, v_n)$. To obtain the sum of two vectors $v = (v_1, v_2, \ldots, v_n)$ and

$u = (u_1, u_2, \ldots, u_n)$, with n elements each, we also compute the sum component-wise, as follows:

$$v + u = (v_1, v_2, \ldots, v_n) + (u_1, u_2, \ldots, u_n)$$
$$= (v_1 + u_1, v_2 + u_2, \ldots, v_n + u_n)$$

With n processors, numbered 1 to n, each processor can perform the sum of two corresponding coordinates without communication with other processors. Thus, the algorithm for Processor i is as follows:

> **Algorithm for Processor i in Calculation of Vector Sum $w = v + u$**
>
> $w_i = v_i + u_i$

> **Definitions**   A **vector** is an ordered n-tuple of numbers, v
> $= (v_1, v_2, \ldots, v_n)$. A **componentwise** vector operation is
> performed component by component, or coordinate by
> coordinate, For n-tuples $v = (v_1, v_2, \ldots, v_n)$ and $u = (u_1, u_2, \ldots, u_n)$, their sum is $v + u = (v_1 + u_1, v_2 + u_2, \ldots, v_n + u_n)$.

In this case, assuming an "ideal" multiprocessor in which communication is not a consideration, the speedup is linear with $S(n) = n$. Of course, if n processors are not available, some processors must calculate more than one coordinate of the result, and speedup is less.

Quick Review Question 1

Suppose we wish to perform the sum of two vectors of 24 elements each. Compute the most number of coordinate sums per processor and the speedup if the following number of processors are available:

> **a.** 24   **b.** 6   **c.** 5   **d.** 4   **e.** 40

Data Partitioning: Adding Numbers

Many applications exist that must compute the sum of a sequence of numbers, x_0, $x_1, \ldots, x_{n-1}$. A sequential algorithm mirrors how we people usually add a column of numbers using a calculator. Initially, we enter the first number (x_0) into the calculator. Correspondingly, with the sequential algorithm, we have a variable, say *sum*, that accumulates the ongoing sum and has an initial value of x_0. On the calculator, we repeatedly press the "+" key and enter the next number from the sequence. With the algorithm, we also add one element at a time to the old value of *sum*, obtaining a new value for *sum*. On the calculator, we complete the process by pressing the "=" key; and in the algorithm we return *sum*. The sequential algorithm is as follows.

Sequential Algorithm to Calculate the Sum of a Sequence of Numbers, x_0, $x_1, \ldots, x_{n-1}$
> $sum \leftarrow x_0$
> for i going from 1 through $n - 1$
> > $sum \leftarrow sum + x_i$
> return sum

One parallel technique of adding a set of numbers uses **partitioning**. A processor, called the **master**, splits the list of numbers into nonoverlapping subsets and sends the subsets to different processors, called **slaves**. Each slave computes the sum of its subset using a sequential algorithm, such as the one above. The slaves send their partial sums to the master, which adds these values to obtain the overall sum. For example, if there are 256 numbers in the sequence and 8 slave processors, each processor computes the sum of $256/8 = 32$ numbers, and the master calculates the final sum involving the 8 partial sums from the slaves. This process is analogous to the scout leader splitting the grocery list and giving each scout a sublist. The scouts work individually to gather their parts of the groceries, which they bring to the leader to purchase. Figure 12.2.1 presents a diagram of the partitioning process, and the following algorithms designate the duties of master and slaves.

Parallel Algorithm for Addition Using Data Partitioning on Message-Passing Multiprocessor

Master's Algorithm

Partition set of n numbers and send n/p numbers to each of p slave processors*
Receive p partial sums from slaves
Compute sum of these p values

Slave's Algorithm

Receive set of numbers from master
Compute sum of these numbers
Send sum to master

* Some processors might get a slightly larger list if p does not divide into n evenly.

Definitions   A parallel **data partition algorithm** uses one processor, the **master**, to partition the data into subsets and to send the subsets to other processors, the **slaves**. Each slave performs the appropriate computations with its subset and sends the result to the master for final processing.

Let us perform a rough analysis of the time involved in this parallel addition by partitioning algorithm. As Table 12.2.1 illustrates, the total time has four phases: initial communication of data n data items from master to p slaves, calculation of the

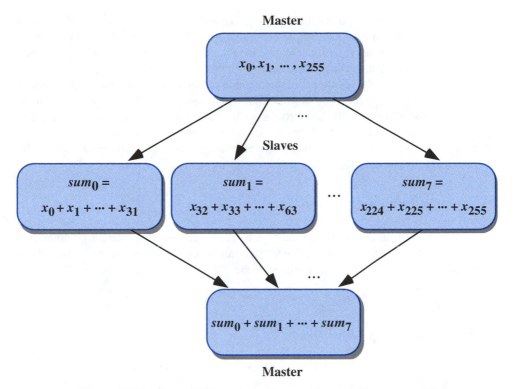

Figure 12.2.1   Sum of 256 numbers using partitioning with master and 8 slaves

Table 12.2.1
Time for Parallel Addition by Partitioning Algorithm

Time to send	Time to add	Time to send	Time to add
256 numbers	32 numbers	8 numbers	8 numbers

sum of n/p numbers by each slave, communication of p partial sums from the slaves to the master, and sum of p numbers by the master. Ignoring communication time, the speedup factor $S(p)$ for p processors is roughly as follows:

$$\text{speedup without communication} = \frac{n}{n/p + p}$$

As an exercise shows, this speedup tends to p for large n. In our example, the speedup is as follows:

$$\frac{256}{\dfrac{256}{8} + 8} = \frac{256}{40} = 6.4$$

We could achieve additional speedup by having the master perform additions as partial sums arrive from the slaves.

In the above computation of speedup, we are ignoring the time for communication. For a worst-case analysis of communication time, assume the multiprocessors do not share memory; communication is sequential; messages cannot overlap; and a message can contain at most one number. In this case, we must move n numbers one at a time before the parallel computation and p numbers afterwards. This communication time might consume as much time as adding the numbers sequentially in one processor. Moreover, all slave processors are idle while the master performs the final addition of p partial sums. If at all possible, we seek to avoid such idle times by so many processors. The divide-and-conquer approach, which the next section discusses, provides an alternative that is useful for many applications.

Quick Review Question 2

Suppose we need to compute the sum of $1024 = 2^{10}$ numbers, and all communication is sequential. Determine how many values are transferred to and from master and slaves with partitioning for each of the following number of slaves:

$$\textbf{a.} \ 2 \qquad\qquad \textbf{b.} \ 8 \qquad\qquad \textbf{c.} \ 256$$

Quick Review Question 3

For the situations in Quick Review Question 2, determine how many addition operations occur at the same time. Consider the parallel computations by the slaves and the computations by the master.

Divide and Conquer: Adding Numbers

Divide-and-conquer algorithms are widely used in computer science, particularly in parallel processing. With such an algorithm, the problem is divided into subproblems of the same form. We continue dividing the problems into smaller and smaller problems. Then we solve the small problems and reassemble the solutions.

Figures 12.2.2 and 12.2.3 diagram a divide-and-conquer solution of adding 256 numbers on 8 processors, $p_0, p_1, \ldots, p_7$. Processor p_0, which initially has all of the numbers, transmits half of the numbers to Processor p_4, so that each is in charge of 128 numbers. Concurrently, p_0 and p_4 send half their numbers (64 numbers each) to p_2 and p_6, respectively. Then, these four processors (p_0, p_2, p_4, p_6) pass half the values (32 numbers each) to the remaining processors (p_1, p_3, p_5, p_7). In all, this tree of divisions to 8 processors in Figure 12.2.2 has $\log_2 8 = 3$ levels of divisions. In Figure 12.2.2, each arrow indicates a message containing half a processor's values, and the relative thickness of the arrow represents the amount of data. At the next lower level of the figure, the new content of a processor's node shows that this message-sending processor has become responsible for the remaining half of its data.

After the "divide" phase comes the "conquer" phase (see Figure 12.2.3). Each processor calculates the sum of its 32 numbers. The odd-numbered processors, p_1, p_3, p_5, p_7, send their results to the even-numbered processors, p_0, p_2, p_4, p_6, respectively. Each of the even-numbered processors adds its answer to the result that an

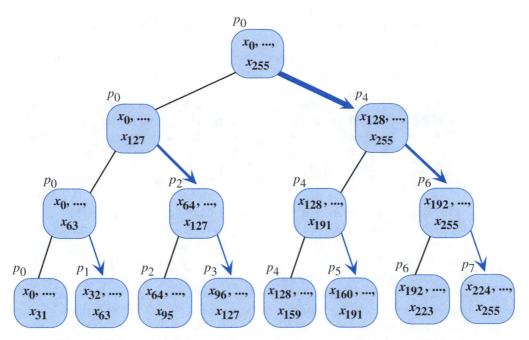

Figure 12.2.2 "Divide" phase of divide-and-conquer algorithm for sum of 256 numbers on 8 processors. Each arrow represents a message containing half the values from the sender.

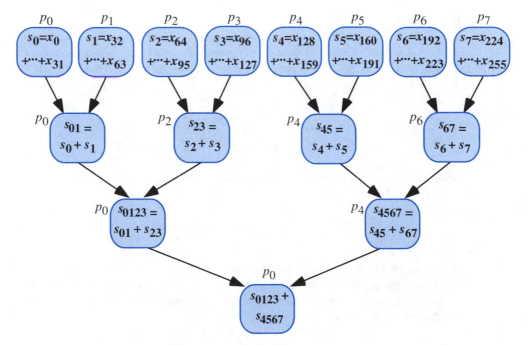

Figure 12.2.3 "Conquer" phase of divide-and-conquer algorithm for sum of 256 numbers on 8 processors. Each arrow represents a single sum value.

odd-numbered processor communicated. Retracing the path when the processors were dividing the data set, p_2 and p_6 send their answers to p_0 and p_4, respectively. Each of Processors p_0 and p_4 adds its two numbers, and finally, p_0 computes the sum of its and p_4's results. In Figure 12.2.3, each arrow represents a single sum value.

> **Definition**   A **divide-and-conquer algorithm** divides a problem into
> subproblems of the same form, and then divides these into
> subproblems of the same form, etc. The small problems are
> solved, and the final solution is assembled.

To analyze the time involve, let us initially ignore communication. At the first step of the "conquer" phase, each processor is adding $n/p = 256/8 = 32$ numbers, or performing $(n/p) - 1 = 31$ addition operations. However, at each level of the tree thereafter, we only have simultaneous sums of pairs of numbers. Thus, after time for the initial additions in the 8 processors, we only need the time to compute 3 more sums. The number of these sums (3) is the same as the number of levels of divisions. In general, a system with p number of processors has $\log_2 p$ of these division levels. Thus, in all we have $(n/p) - 1 + \log_2 p$ sums; and without communication, the speedup factor $S(p)$ for p processors is roughly as follows:

$$\text{speedup without communication} = \frac{n}{n/p - 1 + \log_2 p}$$

This speedup tends to p for large n.

For 256 numbers and 8 processors, during the "divide" phase, communication time includes time to move half the numbers initially, and because of concurrency, one-fourth and one-eighth the numbers, respectively, at the next two levels. Thus, in all, the process must have time to communicate $(256)(7)/8 = 224$ numbers, as the following illustrates:

$$\frac{256}{2} + \frac{256}{4} + \frac{256}{8} = \frac{(256)(7)}{8}$$

In general, for p being a power of 2, the "divide" phases communicates the following number of values:

$$\frac{n}{2} + \frac{n}{4} + \frac{n}{8} + \cdots + \frac{n}{p} = \frac{n(p-1)}{p}$$

For the processors sending their results, communication time is small and approximately proportional to the number of levels, $\log_2 p$. Thus, ignoring startup times for processors, the total communication time is approximately proportional to the following expression:

$$\frac{n(p-1)}{p} + \log_2 p$$

This value is smaller than the communication time for the partitioning algorithm, which is approximately proportional to $n + p$. Moreover, the time in which processors are idle is smaller.

Quick Review Question 4

Suppose we need to compute the sum of $1024 = 2^{10}$ numbers, and all communication is sequential. Determine how many values are transferred concurrently in the divide-and-conquer algorithm of this section for each of the following number of processors:

      a. 2   **b.** 8   **c.** 256

Quick Review Question 5

For the situations in Quick Review Question 4, determine how many addition operations occur at the same time.

Quick Review Question 6

Compare the results of Quick Review Questions 2 and 3 with Quick Review Questions 4 and 5, respectively. For the given situations, determine the better summation algorithm, partitioning or divide-and-conquer:

    a. For a small number of processors, which algorithm uses less communication?
    b. For a large number of processors, which algorithm performs fewer concurrent additions?

Parallel Random Number Generator

Another example of a nearly embarrassingly parallel algorithm is the Monte Carlo estimation of area under a curve (see Module 9.3). For n darts and n processors, each can compute the coordinates of one "dart" hit and increment a shared counter if the "dart" hits below the curve. In general, for n darts and p processors, each processor can compute the number of hits for n/p darts. One processor performs the final step of estimating the area using this count.

However, a problem exists. Each processor must generate random numbers, but generation of the same pseudorandom number sequence by two processors would skew the result. For example, consider the following simple random number generator, which produces the sequence 1, 7, 5, 2, 3, 10, 4, 6, 9, 8 before cycling back to 1:

$$r_0 = 1$$
$$r_n = (7\,r_{n-1})\ mod\ 11, \quad \text{for } n > 0$$

Suppose for simplicity that two processors are available ($n = 2$), and each must use pseudorandom numbers. Using the number of processors as the exponent of the coefficient of r_{n-1}, we have the following:

$$7^2 \bmod 11 = 5$$

Instead of using 7 as the coefficient in the generating function, we employ 5 as follows:

$$r_n = (5\, r_{n-1})\, mod\ 11$$

For one computer, we use the seed $r_0 = 1$, while for the other computer we use $r_0 = 7$, the second number in the original sequence. Thus, one processor generates the sequence 1, 5, 3, 4, 9 with the following random number generator:

$$r_0 = 1$$
$$r_n = (5\, r_{n-1})\, mod\ 11$$

The second processor generates alternate random numbers from the original sequence 7, 2, 10, 6, 8 by using the following random number generator:

$$r_0 = 7$$
$$r_n = (5\, r_{n-1})\, mod\ 11$$

Figure 12.2.4 shows the sequence developed with the generating function $r_n = (7\, r_{n-1})$ mod 11 and the sequences for the two processors with the generating function $r_n = (5\, r_{n-1})\, mod$ 11.

In general, suppose we have a pseudorandom number generator of the following form:

$$r_0 = 1$$
$$r_n = (a\, r_{n-1})\, mod\ m$$

For k processors, we compute the new coefficient as $A = a^k\ mod\ m$. With the processors numbered 0, 1, 2, . . . , $k - 1$, the seed for Processor i is $a^i\ mod\ m$. In the above example, for Processor 0, the seed is $7^0 = 1$; and for Processor 1, the seed is $7^1 = 7$. The algorithm is below, and Figure 12.2.5 shows the sequences for this general case.

Parallel Random Number Generator Algorithm for Processor i of k Processors from Sequential Generator $r_0 = 1$, $r_n = (a\, r_{n-1})\, mod\ m$

$r_0 = a^i\ mod\ m$
$r_n = (A\, r_{n-1})\ mod\ m$, where $A = a^k\ mod\ m$

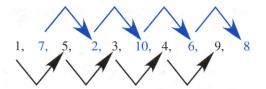

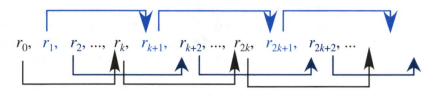

Figure 12.2.4   Parallel random number sequences for two processors

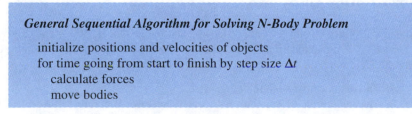

Figure 12.2.5   Parallel random number sequences for k processors

Quick Review Question 7

Consider the following very small, sequential random number generator:

$$r_0 = 1$$
$$r_n = (7\, r_{n-1})\, mod\ 11, \quad \text{for } n > 0$$

Suppose we wish to have the corresponding parallel version for 5 processors. Determine the following:

   a. The generating function
   b. Seeds for the five processors
   c. The sequence for the processor with seed 1

Sequential Algorithm for *N*-Body Problem

The ***N*-Body Problem**, which concerns simulations of the interactions and movements of a number of objects, or bodies, in space, has many applications, including fluid dynamics, evolution of the galaxy, and molecular dynamics. A sequential algorithm for such a simulation has the following general design:

> ***General Sequential Algorithm for Solving N-Body Problem***
>
>   initialize positions and velocities of objects
>   for time going from start to finish by step size Δt
>   calculate forces
>   move bodies

Gravity causes acceleration and movement. The **magnitude of the gravitational force between two bodies** with masses m_1 and m_2 at a distance r apart is

$$F = \frac{Gm_1m_2}{r^2}$$

where $G = 6.67 \times 10^{-11} \text{ m}^3 \text{ kg}^{-1} \text{ s}^{-2}$ is **Newton's gravitational constant**. For example, suppose Body 1 with mass $m_1 = 2$ kg is located at position (1, 3, 0) and Body 2 with mass $m_2 = 4$ kg is at (2, 0, 3), where distances are in meters. We calculate the distance between the objects in a similar fashion to how we compute the distance between points on the plane; we take the square root of the sum of the squares of the differences in corresponding coordinates, as in the following example:

$$r = \sqrt{(2-1)^2 + (0-3)^2 + (3-0)^2} = \sqrt{19} \text{ m}$$

Thus, the following expression gives the magnitude of the gravitational force attracting the two bodies:

$$F = \frac{(6.67 \times 10^{-11} \text{ m}^3 \text{ kg}^{-1}\text{s}^{-2})(2 \text{ kg})(4 \text{ kg})}{19 \text{ m}^2} = 2.8 \times 10^{-11} \text{N}$$

Definitions   The **distance between two points**, (x_1, y_1, z_1) and (x_2, y_2, z_2), is

$$r = \sqrt{(x_2 - x_1)^2 + (y_2 - y_1)^2 + (z_2 - z_1)^2}$$

The **magnitude of the gravitational force between two bodies** with masses m_1 and m_2 at a distance r apart is

$$F = \frac{Gm_1m_2}{r^2}$$

where $G = 6.67 \times 10^{-11} \text{ m}^3 \text{ kg}^{-1} \text{ s}^{-2}$ is **Newton's gravitational constant**.

Quick Review Question 8

Consider Body 1 of mass 35×10^9 kg at location (4000, 0, 5000) and Body 2 of mass 14×10^9 kg at location (2000, 3000, −1000) with distances in meters. Compute the following:

a. The distance between the two bodies
b. The magnitude of gravitational force between two bodies

The **direction of force** on Body 1 at (x_1, y_1, z_1) by Body 2 at (x_2, y_2, z_2) is the unit vector, or vector of length 1, from Body 1 to Body 2, namely,

$$d = \left(\frac{x_2 - x_1}{r}, \frac{y_2 - y_1}{r}, \frac{z_2 - z_1}{r} \right),$$

where r is the distance between the bodies. For example, with bodies at positions (1, 3, 0) and (2, 0, 3), the direction of force on Body 1 by Body 2 is

$$d = \left(\frac{2-1}{\sqrt{19}}, \frac{0-3}{\sqrt{19}}, \frac{3-0}{\sqrt{19}} \right) = \left(\frac{1}{\sqrt{19}}, \frac{-3}{\sqrt{19}}, \frac{3}{\sqrt{19}} \right)$$

Notice that we boldface the name of a vector, such as d, but not the name of a scalar, such as r.

Definition The **unit direction vector from** (x_1, y_1, z_1) **to** (x_2, y_2, z_2), or **direction of the force** on Body 1 by Body 2 at those points, respectively, is as follows:

$$d = \left(\frac{x_2 - x_1}{r}, \frac{y_2 - y_1}{r}, \frac{z_2 - z_1}{r} \right),$$

where $r = \sqrt{(x_2 - x_1)^2 + (y_2 - y_1)^2 + (z_2 - z_1)^2}$ is the distance between the points.

Quick Review Question 9

For the bodies from Quick Review Question 8, determine the direction of force on Body 1 by Body 2.

With magnitude and direction of the force on Body 1 by Body 2, we can compute the force vector as the magnitude of the force (F) times the direction of force. Thus, the force vector F is as follows:

$$F = (F_x, F_y, F_z) = F \left(\frac{x_2 - x_1}{r}, \frac{y_2 - y_1}{r}, \frac{z_2 - z_1}{r} \right)$$

$$= \left(\frac{F(x_2 - x_1)}{r}, \frac{F(y_2 - y_1)}{r}, \frac{F(z_2 - z_1)}{r} \right)$$

This vector indicates that the force that Body 2 exerts on Body 1 in the x-direction is $F_x = F(x_2 - x_1)/r$, in the y-direction is $F_y = F(y_2 - y_1)/r$, and in the z-direction is $F_z = F(z_2 - z_1)/r$. Thus, for the above example with magnitude of the gravitational force between them being $F = 2.8 \times 10^{-11}$ N and the direction vector being

$$d = \left(\frac{-1}{\sqrt{19}}, \frac{3}{\sqrt{19}}, \frac{-3}{\sqrt{19}} \right),$$ the force exerted by Body 2 on Body 1 is the vector $F =$

$$\left(\frac{-2.8 \times 10^{-11}}{\sqrt{19}}, \frac{8.4 \times 10^{-11}}{\sqrt{19}}, \frac{-8.4 \times 10^{-11}}{\sqrt{19}} \right) \approx (0.64 \times 10^{-11}, 1.9 \times 10^{-11}, 1.9 \times 10^{-11}).$$

Quick Review Question 10

For the bodies from Quick Review Questions 8 and 9, determine the force vector indicating the force that Body 2 exerts on Body 1.

The total force on a body is the vector sum of all forces on the body. Below is the sequential algorithm for computing the total force on each of the N bodies, Body 1, Body 2, . . . , Body N. For each Body i except the last, we compute the force vector from it to each Body j, where $j > i$, and vice versa, and add them to the appropriate ongoing sums. Thus, when the algorithm is complete, for each i, F_i is the total accumulated sum of all force vectors on Body i.

> ### Sequential Algorithm to Determine the Total Force on Each of the N Bodies
>
> assume F_i is (0, 0, 0) for i from 1 to N
> for i going from 1 to $N - 1$
>   for j going from $i + 1$ to N
>   calculate distance ($r_{i,j}$) between Body i and Body j
>   calculate direction vector ($d_{i,j}$) from Body i to Body j
>   calculate magnitude ($F_{i,j}$) of force between them
>   add force Body j exerts on Body i ($-F_{i,j}\, d_{i,j}$) to F_i
>   add force Body i exerts on Body j ($F_{i,j}\, d_{i,j}$) to F_j

Quick Review Question 11

In the sequential algorithm to determine the total force on each of the N bodies, suppose the number of bodies is 15. Give the following:

a. The range of numbers for the bodies
b. The sequence of values for j when i is 5

The next major part of the simulation is to move all bodies. To do so, we must first calculate their velocities. Recall that by Newton's Second Law of Motion (see Module 4.1, "Modeling Falling and Skydiving," section on "Physics Background"), a force vector F on a body of mass m creates an acceleration vector a on that body, and

$$F = ma$$

or

$$a = \frac{F}{m}$$

However, acceleration is the derivative of velocity,

$$a = dv/dt$$

Thus, for small change in time Δt, we have the following approximation of the small change in the velocity vector, Δv:

$$\Delta v \approx a\Delta t = \frac{F\Delta t}{m}$$

The rate of change of velocity is approximately a for one time unit, while the change in velocity is about $a\Delta t$ for Δt. We estimate the velocity vector ($v_{t+\Delta t}$) at time step $t + \Delta t$ as the sum of the velocity vector at the previous time step (v_t) and the change in velocity (Δv), as follows:

$$v_{t+\Delta t} \approx v_t + \Delta v \approx v_t + \frac{F\Delta t}{m}$$

We use this velocity vector to calculate the new position of a body. Recall that velocity is the derivative of the change in position with respect to time,

$$v = ds/dt$$

Thus, for a small change in time Δt the small change in position is as follows:

$$\Delta s \approx v\Delta t$$

In one time unit, the change in position is v, while in Δt time units the change in position is $v\Delta t$. As with the new velocity, the new position ($s_{t + \Delta t}$) for the body at time $t + \Delta t$ is approximately the sum of the position at the previous time step (s_t) and the change in position (Δs), as follows:

$$s_{t + \Delta t} \approx s_t + \Delta s$$

$$\approx s_t + v_t\Delta t$$

That is, as Figure 12.2.6 illustrates, we estimate the new position as the old position plus the product of the old velocity and the change in time. We put these various aspects together in the sequential algorithm for computing the new positions of all the bodies.

Sequential Algorithm to Move N Bodies

for i going from 1 to N
    calculate change in velocity vector, Δv, as $F_i(\Delta t/m)$
    calculate change in position vector, Δs, as $v\Delta t$
    add Δv to v
    add Δs to s
    assign (0, 0, 0) to F_i

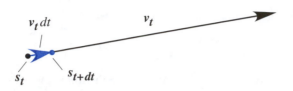

Figure 12.2.6   New position from old

Quick Review Question 12

Consider Body 1 of mass 35×10^9 kg at location $s = (4000, 0, 5000)$ with velocity $v = (500, 300, -100)$. Suppose $\Delta t = 0.1$ sec and $F_1 = (3500 \times 10^9, -1400 \times 10^9, -7000 \times 10^9)$. Evaluate the following:

    a. Δv
    b. Δs
    c. The new v
    d. The new s

Barnes-Hut Algorithm for *N*-Body Problem

Suppose N processors are available so that each processor can be responsible for exactly one body. After computation of the force vector for a body on a time step, the movement of a body is completely independent of the other bodies. Thus, this phase of the simulation can be embarrassingly parallel.

Overall this simulation consumes a great deal of time, particularly if the number of bodies N is large and the time step Δt is small. For each time step, the computation of the total force uses a nested loop whose body executes approximately N^2 times. Concurrency can help in the movement phase. However, if we have one processor responsible for one body's computations, for the force phase, communication times would be approximately proportional to N^2 because of the information required about interactions with all other bodies. A simplification that can reduce communication and speed the process is **clustering** in which we approximate several bodies as a **cluster** that we consider to be one body.

The **Barnes-Hut Algorithm**, another divide-and-conquer parallel algorithm, performs a simulation of the *N*-Body Problem employing clustering. For each time step, the algorithm divides space, which we can consider as a cube, into eight subcubes. Any subcube that does not contain a body is eliminated from further consideration. The algorithm continues the partitioning process on any subcube that contains more than one body. Eventually, we have a collection of cubes of varying sizes that each contains one or no body. Figure 12.2.7 presents an example of the 2D counterpart to this process in which squares are divided into subsquares.

Quick Review Question 13

Partition the square in Figure 12.2.8 into subsquares using the Barnes-Hut Algorithm.

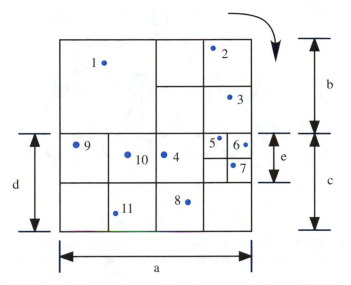

Figure 12.2.7 Partitioning of square into subsquares using Barnes-Hut Algorithm

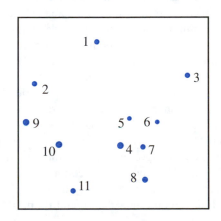

Figure 12.2.8 Square of bodies for Quick Review Question 13

While performing the partitioning process, the algorithm generates an **octtree** with each node corresponding to a cube and with branches going to at most 8 nodes representing subcubes of that cube. Figure 12.2.9 illustrates the 2D counterpart, a **quadtree**, to accompany the partitioned square from Figure 12.2.7. To develop the quadtree, for each subdivided square, we start with the top left square and travel in a clockwise fashion, generating a child node for each square.

Quick Review Question 14

Generate a quadtree for the partition of Figure 12.2.8.

In each node, we store the total mass and center of mass for all the bodies in that subdivision (cube for 3D or square for 2D). As we are computing the total force on a

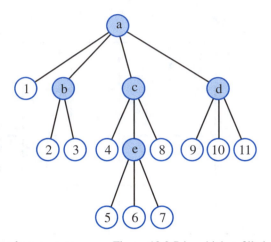

Figure 12.2.9 Quadtree to accompany Figure 12.2.7 in which a filled circle with a letter represents a square and an unfilled circle with a number represents a body

body, say Body 1 in Figure 12.2.9, we **traverse**, or travel through, the tree, starting at the top node, called the **root**, and accumulate the force exerted by the other bodies on Body 1. In determining interactions, if a node representing several bodies is sufficiently far, we do not consider the bodies individually but as a clustered body. For example, in Figure 12.2.9, the cluster that Node e represents might be at a distance greater than some predetermined distance from Body 1. Thus, we do not consider Body 1's interactions with Bodies 5, 6, and 7 individually but perform the computations with the information in Node e as if it were one object interacting with Body 1. Thus, a small enough threshold distance can significantly decrease the amount of communication between processors during the force computation phase.

As with the sequential algorithm, after computing the forces, we move the bodies. Because this phase does not require communication with other processors, this part of the simulation can run very quickly in parallel. Periodically, we reformulate the octtree. The following presents the general Barnes-Hut Algorithm.

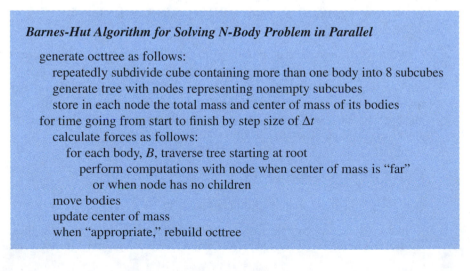

Barnes-Hut Algorithm for Solving N-Body Problem in Parallel

    generate octtree as follows:
        repeatedly subdivide cube containing more than one body into 8 subcubes
        generate tree with nodes representing nonempty subcubes
        store in each node the total mass and center of mass of its bodies
    for time going from start to finish by step size of Δt
        calculate forces as follows:
            for each body, B, traverse tree starting at root
                perform computations with node when center of mass is "far"
                    or when node has no children
        move bodies
        update center of mass
        when "appropriate," rebuild octtree

To calculate the total force on a body requires time on the order of log N for a fairly bushy tree. Because we perform this process for each of the N bodies, the total time is approximately proportional to $N \log N$, a significant improvement over the sequential force computation algorithm, which takes about N^2 steps. The periodic reformulation of an octtree also takes time on the order of $N \log N$. Moreover, the movement phase can easily be preformed in parallel without communication. Thus, overall the Barnes-Hut Algorithm is usually an improvement over the sequential version of simulation of the N-Body Problem. Difficulties do exist, however, in attempts to parallelize. For example, the distribution of the bodies is usually nonuniform, which leads to an unbalance tree and, consequently, longer traversal times.

Quick Review Question 15

Consider the bodies in Figure 12.2.7 with quadtree in Figure 12.2.9.

a. In the sequential algorithm to compute the total force vector for each body, determine the number of times the body inside the nested loop is executed.

b. Suppose in execution of the Barnes-Hut Algorithm the threshold is such that a body only interacts with nodes as follows: Bodies 1, 2, 9, and 11–Node a; Body 3–Nodes b, c, and d; Bodies 4, 5, and 6–Nodes b, d, and e; Body 7– Nodes b and d; Body 8–Nodes b, c, and d; body 10–Nodes b and c. When not interacting with a higher-level node, a body interacts with appropriate descendent bodies. For example, because Body 4 interacts with Nodes b, d, and e but does not interact with Nodes a or c, it interacts with Body 3 immediately below Node a and Body 8, which is a child of Node c. Determine the total number of interactions for the force computation.

Exercises

1. Give an embarrassingly parallel algorithm to compute a scalar times a vector. For example, $3(4, 2, -1) = (3 \cdot 4, 3 \cdot 2, 3 \cdot -1) = (12, 6, -3)$.

2. Suppose a computer graphics screen has resolution 1024×768, that is, 1024 **pixels** (dots on screen) wide and 768 pixels high. Each pixel has levels of red, green, and blue, where each value is between 0.0 and 1.0. Suppose 16 processors are available for parallel computation. Give an algorithm for a nearly embarrassingly parallel algorithm to add 0.05 to the red component up to a maximum of 1.0 for every pixel.

3. As the section "Data Partitioning: Adding Numbers" indicates, ignoring communication, the speedup factor, $S(p)$, for p processors is roughly $\dfrac{n}{n/p + p}$. Show that for large n, this speedup tends to p. One way to do so is first to simplify the expression by obtaining a common denominator for n/p and p and by inverting the resulting fraction and multiplying. Use the fact that for very large n and relatively small p, $n + p^2 \approx n$.

4. Similar to Exercise 3, show that the speedup without communication for the divide-and-conquer addition algorithm, $\dfrac{n}{n/p - 1 + \log_2 p}$, of the section

"Divide and Conquer: Adding Numbers" tends to p for large n.

5. **a.** Give a parallel partitioning algorithm to compute the maximum of n numbers on p processors.
   b. Analyze the communication cost and the speedup.

6. Repeat Exercise 5 for a divide-and-conquer algorithm.

7. Develop a divide-and-conquer algorithm for finding the number of occurrences of a particular element in an array, or vector.

8. **a.** Develop a divide-and-conquer algorithm to perform a parallel merge sort of an array (vector). Hint: After division, each processor sorts its part of the array using an efficient algorithm. Then, the subarrays are merged into larger sorted subarrays.
   b. Analyze the communication and computation times if the number of processors is equal to the number of array elements, n.
   c. Repeat Part b if the number of processors is less than the number of array elements. Assume that the computation time for the sequential sorting algorithm employed is proportional to $m \log m$, where m is the number of elements being sorted.

9. Consider the following random number generator:

$$r_0 = 1$$
$$r_n = (59 \, r_{n-1}) \bmod 349, \quad \text{for } n > 0$$

   a. Suppose 4 processors working concurrently need to generate random numbers. Give the corresponding generating function.
   b. Give the seeds for the four processors.
   c. Give the most number of pseudorandom values the original function generates.
   d. Give the most number of pseudorandom values the parallel function generates.

10. Repeat Exercise 9 for the following random number generator and 3 processors:

$$r_0 = 1$$
$$r_n = (523 \, r_{n-1}) \bmod 1021, \quad \text{for } n > 0$$

11. Suppose a pseudorandom number generator of the following form generates the most number of values possible:

$$r_0 = 1$$
$$r_n = (a \, r_{n-1}) \bmod m$$

   a. Give the number of pseudorandom numbers this function generates.
   b. Suppose $m - 1 = pq$, where p and q are positive integers. Give the number of pseudorandom numbers the following function generates:

$$r_0 = 1$$
$$r_n = (a^p \, r_{n-1}) \bmod m$$

12. Consider the following generating function for a pseudorandom number generator:

$$r_n = (a\, r_{n-1} + c)\, mod\ m$$

For k processors, the parallel version is as follows:

$$r_n = (A\, r_{n-1} + C)\, mod\ m$$

where $A = a^k\ mod\ m$ and $C = c(1 + a + a^2 + \cdots + a^{k-1})\ mod\ m$. Notice that the value of the coefficient is the same as with the version in the text, where $c = 0$. (Wilkinson and Allen 1999)

a. Determine the parallel version for two processors and the generating function $r_n = (7\, r_{n-1} + 4)\ mod\ 11$, for $n > 0$.
b. Repeat Part a for 5 processors.

13. Repeat Exercise 12a for three processors and the generating function $r_n = (229\, r_{n-1} + 1)\ mod\ 10{,}000$, for $n > 0$.

14. Suppose in a Barnes-Hut Algorithm, the measure of "far" is relative. Instead of visiting its children, we use the information from a node if the following is true for some number, *threshold*:

(width of subsquare for node)/(distance to body) < *threshold*

Suppose in Figure 12.2.7 that the smallest square, such as the one containing Body 6, is 1 unit wide; so that the square with Body 4 has width 2 units; and the square with Body 1 is 4 units wide. Suppose *threshold* is 0.5. Estimate (width of subsquare for node)/(distance to body) and if the node information would be used in the force computation for the following situations:

a. Body 4 and Node e
b. Body 1 and Node e
c. Body 1 and Node d

Projects

Complete the projects below using your computational tool.

1. Develop a sequential program to simulate the Parallel Algorithm for Addition Using Data Partitioning by having separate functions representing each processor. Display communications, such as "Slave 3 sending partial sum 537 to Master."
2. Develop a sequential program to simulate a parallel divide-and-conquer algorithm for addition of a list of numbers by having separate functions representing each processor. Display communications, such as "Processor 3 sending partial sum 537 to Processor 2."
3. Develop a sequential program to simulate a parallel divide-and-conquer algorithm for finding the maximum of a list of numbers by having separate functions representing each processor. Display communications, such as "Processor 3 sending its maximum, 537, to Processor 2."

4. Develop a sequential program to simulate a parallel divide-and-conquer algorithm to determine the number of occurrences of a value in a list of numbers by having separate functions representing each processor. Display communications, such as "Processor 3 sending to Processor 2 that it found the value 4 times."

5. For the situation in Exercise 9, develop the sequential random number generator and the random number generators for each processor. Display the complete sets of random numbers generated.

6. Develop a simulation in 2D or 3D of the N-Body Problem using a sequential algorithm. Generate an animation of the simulation.

7. Suppose a pipeline of p processors operates on a stream of integers, 2, 3, 4, . . . , passed from one processor to the next. Each processor remembers the first number, N, it receives and passes to the next processor all remaining numbers in the sequence that are not multiples of N. When the last processor receives a number, the algorithm stops. (This algorithm is a parallel version of the sequential Sieve of Eratosthenes Algorithm.)

a. Determine the task of this algorithm.

b. Develop a program for a sequential version of this algorithm.

c. Develop a program to simulate the pipeline version by having separate functions representing each processor.

d. Write an analysis of the amounts of computation and communication for the sequential and pipeline versions and of the speedup.

Answers to Quick Review Questions

1. a. 1, 24
   b. 4, 6
   c. 5, 5
   e. 6, 4
   e. 1, 24
2. a. $1026 = 1024 + 2$
   b. $1032 = 1024 + 8$
   c. $1280 = 1024 + 256$
   Notice that as the number of processors increases, the communication cost increases.
3. a. $512 = 511 + 1$
   b. $134 = 127 + 7$
   c. $258 = 3 + 255$
4. a. $513 = 512 + 1$
   b. $899 = 512 + 256 + 128 + 3 = 1024(7)/8 + \log_2 8$
   c. $1028 = 512 + 256 + 128 + 64 + 32 + 16 + 8 + 4 + 8 = 1024(255)/256 + \log_2 256$
5. a. $512 = 511 + 1 = (1024/2) - 1 + \log_2 2$
   b. $130 = 127 + 3 = (1024/8) - 1 + \log_2 8$
   c. $11 = 3 + 8 = (1024/256) - 1 + \log_2 256$
   Notice the improvement using 256 processors for the divide-and-conquer algorithm over the partitioning algorithm.

6. **a.** divide-and-conquer
    b. divide-and-conquer
7. **a.** $r_n = (10\, r_{n-1})\, mod\, 11$, for $n > 0$ because $7^5\, mod\, 11 = 10$
    b. 1, 7, 5, 2, 3
    c. 1, 10
8. **a.** 7000 m because

$$r = \sqrt{(2000 - 4000)^2 + (3000 - 0)^2 + (-1000 - 5000)^2}$$

$$= \sqrt{4 \times 10^6 + 9 \times 10^6 + 36 \times 10^6} = 7000$$

   b. $667.0N = F = \dfrac{(6.67 \times 10^{-11})(35 \times 10^9)(14 \times 10^9)}{49 \times 10^6} = 6.67 \times 10^2$

9. $(-2/7, 3/7, -6/7) = ((2000 - 4000)/7000, (3000 - 0)/7000, (-1000 - 5000)/7000)$
10. Approximately $(-190.6, 285.9, -571.7) = 667.0(-2/7, 3/7, -6/7)$
11. **a.** Body 1 through Body 15
    b. integers 6 through 15
12. **a.** $(10, -4, -20)$ because $\Delta v = F_1(\Delta t/m) = ((3500 \times 10^9)(0.1)/(35 \times 10^9),$ $(-1400 \times 10^9)(0.1)/(35 \times 10^9), (-7000 \times 10^9)(0.1)/(35 \times 10^9))$
    b. $(50, 30, -10)$ because $\Delta s = (\text{old } v)\, \Delta t = (500, 300, -100)\, 0.1$
    c. $(510, 296, -120)$ because new $v = \text{old } v + \Delta v = (500, 300, -100) + (10, -4, -20)$
    d. $(4050, 30, 4990)$ because new $s = \text{old } s + \Delta s = (4000, 0, 5000) + (50, 30, -10)$
13. See Figure 12.2.10 for the answer.
14. See Figure 12.2.11 for the answer.

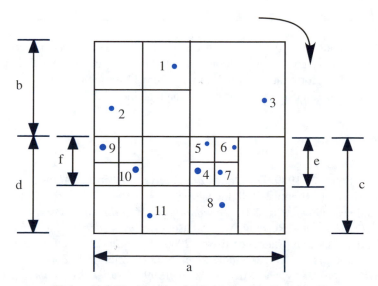

Figure 12.2.10   Partition for Quick Review Question 13

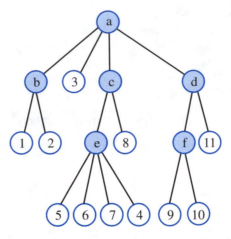

Figure 12.2.11   Quadtree for Quick Review Question 14

Table 12.2.2
Table of Interactions for Figure 12.2.9 and Quick Review Question 15

Body	Interaction with Body	Interaction with Node	Number of Interactions
1		a	1
2		a	1
3		b, c, d	3
4	3, 8	b, d, e	5
5	3, 8	b, d, e	5
6	3, 8	b, d, e	5
7	3, 4, 5, 6, 8	b, d	7
8	3	b, c, d	4
9		a	1
10	3, 9, 11	b, c	5
11		a	1

15. **a.** $55 = 10 + 9 + 8 + 7 + 6 + 5 + 4 + 3 + 2 + 1 = 10(11)/2$ because for $i = 1$, j goes from 2 through 11; for $i = 2$, j goes from 3 through 11; and so forth until for $i = 10$, j goes from 11 through 11.
    b. 38 because of the evaluation in Table 12.2.2

References

Aik, Selim G. 1989. *The Design and Analysis of Parallel Algorithms*. Upper Saddle River, NJ: Prentice-Hall, Inc.: 401.

Andrews, Gregory R. 2000. *Foundations of Multithreaded, Parallel, and Distributed Programming*. Reading, MA: Addison-Wesley-Longman, Inc.: 664.

Miller, Russ, and Laurence Boxer. 2000. *Algorithms Sequential & Parallel, A Unified Approach*. Upper Saddle River, N.J.: Prentice-Hall: 330.

Wilkinson, Barry, and Michael Allen. 1999. *Parallel Programming*. Upper Saddle River, NJ: Prentice-Hall, Inc.: 431.

13

ADDITIONAL CELLULAR AUTOMATA PROJECTS

Overview

In Chapters 8–12, we studied techniques, issues, and applications of computational science empirical models, random walk and cellular automaton simulations, and high performance computing. Projects usually built on or were closely related to the examples discussed and developed in the modules.

As with Chapter 7, Chapter 13 provides opportunities for students to enhance their computational science problem-solving abilities through completion of additional extensive projects. Although not containing examples, each module in the chapter does have sufficient scientific background for students to complete projects in the application area. At the beginning of a module, the prerequisite material is listed. Correspondingly, project sections in previous chapters suggest appropriate Chapter 13 modules for additional projects. Thus, students can work with projects in the current chapter at any time after covering the prerequisites.

As with earlier projects, those of Chapter 13 are appropriate for teamwork. Interdisciplinary teams perform most of the research and development in computational science. Thus, teamwork experiences developing models and simulations in a variety of application areas are important for students studying computational science.

Chapter 13's applications with projects are in a variety of scientific areas, including the following: polymers, solidification, foraging behavior, pit vipers and heat diffusion, growth of mushroom fairy rings, spread of disease, HIV and the immune system, predator-prey interactions, clouds, and fish schooling.

MODULE 13.1

Polymers—Strings of Pearls

Prerequisites: Module 8.3 on "Empirical Models" and Module 10.2 on "Random Walk."

Introduction

The Mesoamerican Civilization of the Maya extended for about 3500 years, between 2000 B.C. and the 1500s A.D. When the Spanish found them during the sixteenth century, the Mayans were playing a very interesting ballgame with political and religious significance that transcended the game as a sport. These games were played on large courts; and the oldest known is one located in Chiapas, Mexico, which dates from about 1400 B.C. The balls they used were large, between 12 and 18 inches in diameter, and were made of solid rubber, likely weighing 8 to 40 lbs. Most Westerners think that rubber originated with Charles Goodyear's vulcanization process that cured natural rubber into a commercially useful product. As it turns out, people of Mesoamerica were processing rubber by 1600 B.C., 3500 years before the first rubber patent in England. The Mayans used rubber not only for making balls, but also for paint, medicines, waterproofing, etc. Interestingly, they took the latex from a local tree, *Castilla elastica*, and mixed it with the liquid from morning glory, *Ipomoea alba*. With stirring and the heat of the region, they produced rubber in a process very similar to Goodyear's (Armstrong; Greider 1999; Hosler et al. 1999).

Rubber is a good example of what scientists term a "natural polymer." But, what is a polymer? The word is from the Greek *polumeres*, which means "having many parts" (poly—"many"; merous—"parts") (Marko). Thus, we use the term **polymer** to describe a class of chemical compounds composed of repeating chemical building blocks. These building blocks are identical or closely related chemical structures, each referred to as a **monomer**. Rubber is a polymer made up of repeating subunits (monomers) of *isoprene* (2-methyl-1,3-butadiene), as in Figures 13.1.1 and 13.1.2 (Michalovic 2000). In Figure 13.1.1, n is a chemist's shorthand to indicate that similar types of bonds link n of these subunits. So rubber is *polyisoprene*. The monomer *isoprene* is also used to generate a number of biologically significant molecules,

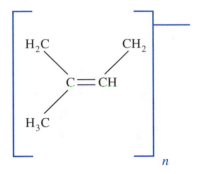

Figure 13.1.1   Rubber polymer with similar types of bonds linking *n isoprenes*

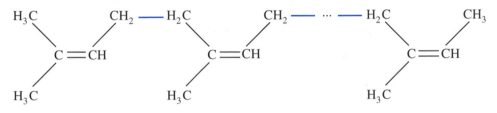

Figure 13.1.2   Rubber polymer

such as lanosterol (precursor to various animal sterols), vitamins A and E, and carotene pigments (Case Western; Garrett and Grisham 2005).

> **Definitions**   A **polymer** is a chemical compound composed of repeating building blocks, which are identical or closely related chemical structures, called **monomers**.

Rubber is not the only natural polymer. Natural polymers abound in nature and include cellulose, starch, chitin, nucleic acids (DNA, RNA), and proteins. Chains of monomers make up each of these polymers. Proteins, for instance, are strings of amino acids. Because each monomer in the protein chain can be any one of the 20 different naturally occurring amino acids, proteins are quite diverse in composition (Garrett and Grisham 2005).

Scientists have learned to duplicate the process of polymer synthesis, **polymerization**, so that quite a variety of synthetic polymers, such as polyethylene and polystyrene, exist. We can now even synthesize rubber. These polymers are found in almost all manufactured products and their packaging (Case Western).

Some polymers, like rubber, are **elastomers**. Elastomers are made up of molecules that are loosely cross-linked (1 in 100 molecules linked to other molecules). By contrast, **plastics** have 1 in 30 cross-linked molecules; and we can shape or mold these stiffer, non-elastic materials (Case Western).

Some of the most interesting developments in polymer chemistry are medically applicable polymers. **Controlled drug delivery**, for instance, combines polymers with medicines to release medicines into the body in predetermined manners. We can manage release to be constant or cyclic over a prolonged time period, or we can cause the medicine to disperse in response to an environmental cue. These combinations improve effectiveness of chemical therapies, reduce the number of administrations, and insure patient compliance. As an example, Nifedipine, classified as a calcium-channel blocker, may be prescribed for treatment of angina and hypertension and can be administered in an **extended release tablet** (**ERT**). This tablet looks like a normal tablet but consists of two layers, an inner "active" drug layer surrounded by an external, inert polymer. This outer layer of polymer is osmotically sensitive. In the digestive tract, osmotic forces cause this outer layer to swell, pressing against the inner layer and forcing small amounts of the medicine out of a previously drilled hole in the tablet. Small, but constantly released doses are absorbed into the circulatory system. This mechanism provides a controlled release of the medication over a 24-hour period. Patients only have to take the medicine once a day, and levels of the drug in the blood remain high enough to be effective. Other drugs, like some for birth control, can be implanted for up to five years of delivery (Brannon-Peppas 1997; Tenanbaum et al. 2003; Pfizer 2003).

Scientists, who study polymers, and those, who study biology and medicine, are teaming up to develop a wide variety of medical applications. For instance, polymer scaffolds have been developed, based on textile industry techniques, that have proven most useful for cell and tissue growth. Such scaffolds have been employed to generate artificial skin to treat burn patients. The "skin" helps to prevent infection and produces chemical factors that promote faster healing. In the future, scientists may use such scaffolds to grow nerve cells for spinal cord injuries or to grow insulin-secreting cells for diabetes patients (Case Western).

Simulations

Scientists are very interested in predicting a protein's native structure from its amino acid sequence because the folding of a protein polymer with its resulting geometric shape determines its function. An understanding of the mechanism would be a major breakthrough in studying the basic science of the cell. Computational scientists are developing computer simulations of polymers to gain insight in solving the protein-folding problem and other polymer questions (Bastolla et al. 1998).

Random walks are commonly used to generate two-and three-dimensional models of polymers. However, two different monomers of a polymer cannot occupy the same space at the same time. Thus, such a simulation generally involves a **self-avoiding walk** (**SAW**), which is a random walk that does not cross itself, that is, a walk that does not travel through the same cell twice. For example, Figure 10.2.1 of Module 10.2 on "Random Walk" displays a walk that is not self-avoiding because at least once (and in this walk, more than once) the walker cycled back to an earlier position. In an algorithm to generate a self-avoiding walk, at each time step eliminating the direction from which the walker comes, the walker selects at random one of the remaining directions. The projects develop several such simulations (Gould and Tobochnik 1988).

> **Definition**   A **self-avoiding walk (SAW)** is a random walk that does not
> cross itself, that is, a walk that does not travel through the same
> cell more than once.

Projects

1. **a.** Develop a simulation and visualization to generate a two-dimensional
   model of a polymer with a self-avoiding walk. Terminate the simulation
   when the random direction would cause the developing path to cross itself. The visualization should show the entire model of the polymer from
   the beginning of the random walk until the end.
   b. Run the simulation a number of times recording the fraction, $f(n)$, of
   times the simulation generates a polymer of length (number of steps) at
   least n. Using empirical modeling, derive an equation for f as a function
   of n (Gould and Tobochnik 1988).
   c. The end-to-end distance is an important geometric property of a polymer
   that influences the texture and other physical properties of the polymer.
   Run the simulation a number of times evaluating the **root-mean-square
   displacement R_n** as follows: For each walk of n steps, evaluate the
   square of the displacement (end-to-end distance) of the nth step (particle)
   from the initial position, the origin; then, over all such n-step walks,
   compute the average of the squares of the displacements; take the square
   root of this average to calculate R_n. If the nth step for trial i is to location
   $(x_{n,i}, y_{n,i})$, the square of the displacement from the origin is $(s_{n,i})^2 =$
   $(x_{n,i})^2 + (y_{n,i})^2$. The formula for the root-mean-square displacement
   follows:

$$R_n = \sqrt{\frac{\sum_{i=1}^{m}(s_{n,i})^2}{m}}$$

   where $s_{n,i}$ is the displacement of the nth particle from the starting position
   during trial i of m successful trials. Using empirical modeling, derive an
   equation for R_n as a function of n (Gould and Tobochnik 1988).
2. Repeat Project 1 in three dimensions, where at each step the walker selects at
   random one of six directions.
3. Repeat Project 1 where bond angles are all 90°, so that the random walker
   turns to the right or left with each step (Gould and Tobochnik 1988).
4. Repeat Project 2 where bond angles are all 90°, so that the random walker
   turns to the right, left, up, or down with each step.
5. The technique in Project 1 is inefficient in calculating R_n (see Part c) for
   large n because of the number of aborted trials. An alternative method uses a

weight W_n associated with each walk of n steps to skew the importance of underrepresented large chains. The weights in a SAW are as follows:

- $W_1 = 1$.
- If no step is possible at step n, $W_n = 0$. In this case, the walk terminates, and the program generates a new walk starting at the origin.
- If the walker can go in any of the three directions at step n, $W_n = W_{n-1}$.
- If the walker can go in exactly two directions at step n, $W_n = (2/3)W_{n-1}$.
- If the walker can go in exactly one direction at step n, $W_n = (1/3)W_{n-1}$.

We estimate the root-mean-square displacement follows:

$$R_n = \sqrt{\frac{\sum_{i=1}^{m}(W_{n,i})(s_{n,i})^2}{\sum_{i=1}^{m} W_{n,i}}}$$

where $W_{n,i}$ is the value of W_n and $s_{n,i}$ is the displacement of the nth particle from the starting position during trial i of m successful trials. Using empirical modeling, derive an equation for R_n as a function of n (Gould and Tobochnik 1988).

6. Repeat Project 5 attempting to produce longer polymer models by aborting any one random walk when the weight becomes smaller than some threshold value, such as 0.15.

7. Using the technique of Project 1a, develop a collection of models for polymers of some length n. Write a program that attempts to construct models of polymers of length $2n$ by attaching the head of one "polymer" to the tail of another. The program should reject the walk if it crosses itself. This technique is at the basis of an accelerated method to calculate R_n (see Project 1c) (Gould and Tobochnik 1988).

References

Armstrong, Joseph E. "Rubber Production: Tapping Rubber Trees, Latex Collection And Processing Of Raw Rubber." Illinois State University. http://www.bio.ilstu.edu/armstrong/syllabi/rubber/rubber.htm

Bastolla, Ugo, Helge Frauenkron, Erwin Gerstner, Peter Grassberger, and Walter Nadler. 1997. "Testing a new Monte Carlo Algorithm for Protein Folding." *Proteins: Structure, Function, and Genetics*, 32: 52–66. http://arxiv.org/pdf/cond-mat/9710030

Brannon-Peppas, Lisa. 1997. "Polymers in Controlled Drug Delivery." *Medical Plastics and Biomaterials Magazine*, November. http://www.devicelink.com/mpb/archive/97/11/003.html

Case Western Reserve University. *Virtual Textbook*. Cleveland, Ohio: Department of Physics, the Department of Macromolecular Science and Engineering, and the Center for Advanced Liquid Crystalline Optical Materials. http://plc.cwru.edu/tutorial/enhanced/files/textbook.htm (accessed April 3, 2004).

Garrett, Reginald H., and Charles M. Grisham. 2005. *Biochemistry*. 3rd ed. Belmont, Calif.: Brooks-Cole Publishing: 258–259.

Gould, Harvey, and Jan Tobochnik. 1988. *An Introduction to Computer Simulation Methods, Applications to Physical Systems, Part 2*. Reading, Mass.: Addison-Wesley: 695.

Greider, Brett. 1999. "Ball Game Rules and Equipment," from Indigenous Religions of the Americas course. University of Wisconsin-Eau Claire. http://www.uwec.edu/greider/Indigenous/Meso-America/ballgame/rules.htm

Hosler, Dorothy, Sandra L. Burkett, and Michael J. Tarkanian. 1999. "Prehistoric Polymers: Rubber Processing in Ancient Mesoamerica." *Science*, 284: 1988–1991 [DOI: 10.1126/science.284.5422.1988] (in Reports).

Marko, John. "What Are Polymers And Why Are They Interesting?" Cornell University, Laboratory of Atomic and Solid State Physics. http://www.lassp.cornell.edu/marko/polymers.html

Michalovic, Mark. 2000. "The Story of Rubber—a Self-Guided Polymer Expedition." *Science Sidetrip: Meet Polyisoprene*. Polymer Science Learning Center and the Chemical Heritage Foundation. http://www.pslc.ws/macrog/exp/rubber/sepisode/meet.htm

Pfizer, Inc. 2003. "Procardia (nifedipine)." New York: Pfizer Labs. http://www.pfizer.com/hml/pi's/procardiaxlpi.pdf

Tenanbaum, David J., et al. 2003. "Polymers and People." *Beyond Discovery: The Path from Research to Human Benefit*. Washington, D.C.: The National Academy of Sciences. http://www.beyonddiscovery.org/content/view.article.asp?a=203

MODULE 13.2

Solidification—Let's Make It Crystal Clear!

Prerequisite: Module 10.2 on "Random Walk."

Introduction

What do snowflakes and steel have in common? At first glance, we probably would say, "Not much." However, if we could look closely enough, we would see that they both are crystalline, possessing amazing structural similarities. Each is made of tree-like structures called **dendrites**, which are formed as the substance cools during the process of **solidification**.

Snowflakes are composed of one or more **snow crystals**. Each crystal is built of water molecules arranged in a very specific, hexagonal **lattice**. These crystals form in the clouds by the condensation of water vapor into ice. At first, while very small, the crystals form as hexagonally shaped prisms, following the original, molecular symmetry. The edges of the facets of this prism grow out more rapidly than the facets themselves, leading to the formation of "limbs." These limbs may, and usually do, produce other branches, leading to the dendrite or tree-like forms.

A number of factors determine the precise shape of the crystal, but temperature is the primary influence. As snowflakes blow and fall through the clouds, they encounter significant variations in temperature, humidity, and pressure. Each snowflake tends to have different environmental "experiences," which lead to the development of different shapes. Why snowflake shape is so temperature-dependent is not completely understood (Libbrecht).

The solidification of snowflakes is fascinating, but the process of solidification has an impressive array of manufacturing applications. Despite the increased use of plastics, think of all the things we use everyday that are metal. Used to produce everything from soda cans to car engines, these metals and alloys are formed from liquids that have "frozen" or solidified. Solidification, therefore, is an important process for generating metal products as well as snowflakes.

Dendrites form within the molten metals/alloys as they solidify during the casting process. These dendrites vary greatly in shape, size, and orientation. Furthermore, the individual dendrites interconnect in various ways to generate a series of intricate **microstructures**. These individual and collective variations greatly influence the structural qualities (e.g., strength and flexibility) of the product (NASA; RPI). There are numerous horror stories of castings that have broken apart from internal defects that originated from thermal stresses occurring during solidification (Seetharamu et al. 2001). According to scientists, we would be able to understand (and therefore control) the properties of materials that solidify dendritically better if we could develop effective computational models of the behavior of individual dendrites (RPI).

Under the influence of Earth's gravity, liquid metal is subject to the influence of **convective currents** as it cools. These currents alter significantly the growth of the dendrites, which makes modeling of "normal" dendritic growth and the effects of convective currents on such growth virtually impossible. Confronting this difficulty, the National Aeronautics and Space Administration (NASA) has teamed with scientists at Rensselaer Polytechnic Institute in the Isothermal Dendritic Growth Experiment (IDGE). Experiments in this program, conducted in conditions of low-gravity that earth orbit offers, have already shed tremendous light on dendritic growth. For instance, scientists, using IDGE data, will be able to separate the effects of convection from other factors that impact solidification of metals and alloys. Such information will go far to improve computational models, which should guide us to improved industrial production of various metals/alloys (RPI).

Projects

1. **a.** We can use the technique of **diffusion-limited aggregation** (**DLA**) to build a dendritic structure. In one form of the algorithm, a **seed**, or initial location for the developing dendritic structure, is in the middle of an $m \times m$ **launching rectangle**. This launching rectangle is a region in the middle of an $n \times n$ grid, where $m < n$. For example, m might be 16, and n might be 40. One at a time, "particles" are released from random positions on the launching rectangle boundary to go on random walks. If the walker comes in contact with another particle (i.e., a neighbor to its north, east, south, or west), with a designated sticking probability, the walker adheres to the particle, resulting in a larger structure. If the walker travels too close to the boundary of the grid, the simulation deletes that walker and releases another random walker from the launching rectangle. Use the DLA algorithm to develop a simulation to generate dendritic structures with the number of particles for the structure as a parameter (Joiner and Panoff 2003).

   b. Develop a visualization that shows the simulation one step at a time, including the random walks. Develop another animation that only shows the particles as a new particle attaches to the growing structure. An attractive enhancement is for the color of the particle to be a function of its distance from the seed. (Follow the link "Simple DLA Example" at the Shodor website for an example of such a simulation with animation (Shodor 2002).)

   c. Run the simulation and visualization a number of times for several different sticking probabilities. Discuss the impact of the sticking probabilities on the resulting structures.

2. Repeat Project 1 considering the eight surrounding cells as a walker's nearest neighbors.

3. Repeat Project 1, Parts a and b, where the sticking probability is 0.33 for contact with one particle, 0.67 for simultaneous contact with two particles, and 1.0 for contact with three. Run the simulation a number of times and discuss the results (Joiner and Panoff 2003).

4. a. Repeat Project 2, Parts a and b, where the sticking probability is based on the number of particles the walker contacts simultaneously. Run the simulation a number of times and discuss the results (Joiner and Panoff 2003).

   b. Adjust the situation so that the sticking probability is 0.1 for contact with one particle, zero probability for two particles, and 0.9 for three or more particles. Run the simulation and animation a number of times and discuss the results (Joiner and Panoff 2003).

   c. Adjust the situation so that the sticking probability is 0.01 for contact with one or two particles, 0.03 for three particles, and 1.0 for more than three particles. Run the simulation a number of times and discuss the results (Joiner and Panoff 2003).

5. Repeat Project 1, Parts a and b, where the sticking probability is greater for bonds continuing in a straight line. For example, a walker is more likely to adhere to a north neighbor if that particle is stuck to a particle to its north. Similar situations exist for the other directions. Run the simulation a number of times and discuss the results (Joiner and Panoff 2003).

6. Repeat Project 5 considering the eight surrounding cells as a walker's nearest neighbors (Joiner and Panoff 2003).

7. Changing conditions affect crystalline formation and cause a great variety in the shapes. During a simulation, we can vary the sticking probability to indicate such changing conditions. Do Project 2 starting with sticking probabilities as in Project 4, Part b. After forming an aggregate with a specified number (such as 100) of particles, use sticking probabilities as in Project 4c for a specified number (such as 100) of particles; then, change to a different sticking probability configuration (Joiner and Panoff 2003).

8. Repeat any of Projects 1–6 considering the impact of wind or gravity on dendritic growth by having the walker travel with a greater probability in a particular direction (Shodor 2002).

9. Repeat any of Projects 1–8 using a launching circle of radius m instead of a launching rectangle. (Follow the link "Diffusion Limited Aggregation Calculator" at the Shodor website for such a simulation example (Shodor 2002).)

10. Repeat any of Projects 1–8 using a launching circle, instead of a launching rectangle, of radius $2r_{max}$, where r_{max} is the radius of the structure so far. Delete a walker if it travels too close to the boundary of the grid or beyond a distance of $3r_{max}$ from the seed. Such adjustments should speed the simulation (Gould 1988).

11. Do Project 10 with the following additional adjustment to speed the simulation by having larger step sizes further away from the structure: If a walker is at a distance $r > r_{max} + 4$ from the seed, then have step sizes of length $r - r_{max} - 2$; otherwise, have step sizes of length 1 (Gould and Tobochnik 1988).

12. Repeat any of Projects 1 or 2 considering accumulation on a structure, such as the deposit of snow on a tree. Have the seed be a tree-like structure or other type of structure on the bottom of the grid. Release random walkers from the north end of the grid with a greater likelihood of traveling south (Joiner and Panoff 2003).

References

Gould, Harvey, and Jan Tobochnik. 1988. *An Introduction to Computer Simulation Methods, Applications to Physical Systems, Part 2*. Reading, Mass.: Addison-Wesley: 695.

Joiner, David, and Robert Panoff. 2003. "Diffusion Limited Aggregation." Keck Undergraduate Computational Science Educational Consortium. http://www.capital.edu/acad/as/csac/Keck/modules.html#diffusion

Libbrecht, Kenneth G. "Snowflake Primer—The Basic Facts About Snowflakes and Snow Crystals." California Institute of Technology. http://www.its.caltech.edu/~atomic/snowcrystals/prmimer/primer.htm

NASA. Fourth United States Microgravity Payload. "Isothermal Dendritic Growth Experiment (IDGE): Science Background." Rensselaer Polytechnic Institute and NASA's Lewis Research Center. http://liftoff.msfc.nasa.gov/shuttle/usmp4/science/idge1.html

RPI (Rensselaer Polytechnic Institute). "Isothermal Dendritic Growth Experiment: Introduction." Rensselaer Polytechnic Institute and NASA's Lewis Research Center. http://www.rpi.edu/locker/56/000756/index.html

Seetharamu, K. N., R. Paragasam, Ghulam A. Quadir, Z. A. Zainal, P. Sthaya Prasad, and T. Sundararajan. 2001. "Finite Element Modeling of Solidification Phenomena." *Sadhana*, 26(1 & 2): 103–120.

Shodor Educational Foundation. 2002. "Software—Diffusion Limited Aggregation Calculator." Computational Science Education Reference Desk. http://www.shodor.org/refdesk/Resources/Models/DLA/

MODULE 13.3

Foraging—Finding a Way to Eat

Prerequisite: Module 11.2 on "Spreading of Fire."

Introduction

Some animals must navigate over long distances, such as in the migration of birds, butterflies, whales, or salmon. This **large-scale navigation** typically uses celestial and/or geomagnetic cues. In animals like migratory birds, circadian (endogenous, daily) rhythms not only help to initiate migration but also may influence an animal's spatial course of migration (Gwinner 1996).

Almost all animals must navigate over shorter distances, such as to forage or seek mates and/or nesting sites. Honeybees, for example, are **central place foragers**, often losing visual and auditory contact with the hive as they search for food. These animals must be able to return to the nest and to "remember" where they have been for return foraging. To do this, they rely on **path integration**, which means that they must take all the angles and distances they experience on their foraging trips and integrate them into "mean home vectors." Like the dead reckoning of human pilots, they perform continuous spatial updating to relate their current location to the hive. This process permits them to return home by more direct routes. Bees appear to use the sun's position and patterns of polarized light as components of a compass, but the calculations of distance and direction are internal (egocentric). Once the route is established, landmarks, which are forms of geocentric information, likely supplement the insect's internal calculator (Wehner et al. 1996; Wehner 1996).

Some birds, like the chickadee, store or cache excess food for retrieval in times of shortages. They are capable of remembering perhaps hundreds of locations over fairly large areas (up to 30 hectare (ha)) for at least a month. Evidence suggests that these birds might navigate to these storage sites using spatial relationships among objects (landmarks) short distances from the caches. Features of the cache sites themselves appear to be less significant. This series of markers form geocentric references that are stored in memory, likely in the hippocampus, as a neural representation of the bird's environment. Many scientists refer to this representation as a

cognitive map. The idea of a cognitive map is a reasonably controversial proposal for those who study spatial navigation. It is intriguing, however, that the hippocampus regions of food-storing birds are larger than comparable bird species that do not store food (Doupe 1994; Sherry 1996; Sherry and Duff 1996). Evidence also exists that the regions of the human hippocampus enlarge in those who are highly dependent on navigational skills (Maguire et al. 2000).

Some extend the concept of a cognitive map to insects. Honeybees and ants certainly seem capable of remembering landmark locations, but whether they are capable of integrating these memories into a map seems improbable. It is more likely that they store a series of snapshots of the surroundings as they make their journeys, which they compare to the current landscape at a particular time (Wehner et al. 1996).

Simulations

The projects in this module develop cellular automaton methods for goal-directed spatial searches in which some form of adaptation occurs. Psychologists have used such simulations to elucidate qualitative properties of animal spatial orientation and learning, such as with a pigeon searching for food in an area where it found food previously, a rat learning its way through a changing maze, a badger returning to multiple food sites when food is no longer present, a dog finding a shortcut, and a shrew maneuvering around a detour. Such studies have revealed that a simulated animal can find its way, not through some "insight" of the problem as a whole, but through repeated local decisions based on its immediate surroundings, or values in neighboring cells (Reid and Staddon 1997; Reid and Staddon 1998). Besides applications to cognitive psychology, scientists are studying the use of such simulation algorithms in the guidance systems of autonomous agents (robots) searching for land mines (Staddon and Chelaru).

For the searching, the robot or animal possesses a cognitive map of the area. A grid represents the cognitive map, and each cell contains the following information:

- Whether the cell contains a barrier, an "animal", or neither
- An expectation value

Usually, only one cell in the grid has an animal, representing the one with the map.

A cognitive map reader is a searching algorithm that is similar to that of a dynamic diffusion model, such as of the diffusion of heat through a metal bar. The change in an empty cell's expectation value, ΔV_t, from time t to time $t + \Delta t$ is a **diffusion rate parameter** r times the sum of the differences in the **expectation value** of the neighbor's $(V_{k,t})$ and the cell's expectation values (V_t), as follows:

$$\Delta V_t = r \sum_{k=1}^{8} (V_{k,t} - V_t), \quad \text{where } 0 < r < 1/8 = 0.125$$

Thus, the expectation value at time $t + \Delta t$ is the following:

$$V_{t+\Delta t} = V_t + \Delta V_t$$

$$= V_t + r \sum_{k=1}^{8} (V_{k,t} - V_t), \tag{1}$$

where $0 < r < 0.125$ and the sum is over the eight neighbors. With subtraction of rV_t occurring eight times, the formula simplifies to the following weighted sum of expectation values of the cell and its neighbors:

$$V_{t + \Delta t} = (1 - 8r)V_t + r \sum_{k=1}^{8} V_{k,t}, \quad \text{where } 0 < r < 0.125$$

If a cell contains an animal and a reward event occurs, such as finding food, then the expectation value becomes some reward value, say 1, usually for only one time step. In contrast, if a nonreward event, such as not finding food, occurs in an animal cell, then the expectation value becomes 0. In this situation, because of diffusion from neighbors, a cell's expectation value increases with time after the animal leaves that cell. The reward/nonreward system assures that the animal does not remain at a reward site. An animal consumes a reward, and the cell's expectation value becomes 1. However, an animal returning to the cell no longer finds the reward, so that the expectation value becomes 0.

In a deterministic version of the algorithm, called the **hill-climbing process**, at each time step, the animal moves to a neighboring cell that has the highest expectation value. In a stochastic version, the move to such a cell occurs with a certain probability.

The simulations involving cognitive maps to illustrate reinforced learning and adaptive behavior contain at least two phases. The first phase often has the animal exploring the grid as the expectation values diffuse. Then, the animal is removed from the grid. Perhaps the diffusion is allowed to continue without the presence of the animal. Finally, the animal is returned to the grid, which might have been altered, to search for its reward (Reid and Staddon 1997, 1998).

Projects

    1. a. Develop a cellular automaton simulation and visualization of open-field foraging with an **area-restricted search**. In an experiment this work is to simulate, a psychologist places a foraging animal into an enclosure that contains a food stash. Once the food has been found, the animal is removed for a while. Then, the animal is returned to the area, which no longer contains any food. Experiments have shown that the animal goes directly to the former location of the food. Not finding the stash, the animal begins searching around the area where the food was. Its path appears erratic and looping and gradually moves further from the former stash site. Simulation results should mimic this behavior.

        For the first phase of the simulation, initialize each cell of a 20×20 grid (cognitive map) to have some very low, uniform expectation value. Assume the food is in the middle of the grid and the diffusion rate parameter (r) is 0.05. Have the animal start at an edge or corner and search with a random walk for the food until found. As discussed in the section on "Simulations," diffusion proceeds; each failure results in zeroing out the cell's expectation value; and success sets the value to 1. The second phase allows diffusion to continue in this grid without the animal, say for 30 time steps. Third, place the animal in the same starting location as in the first phase, and allow the simulation to run for 60 time steps (Reid and Staddon 1997).

b. Run the simulation four times for diffusion rate parameters of 0.001, 0.01, 0.05, and 0.1, respectively. Discuss the search patterns, why they occur according to the algorithm, and how they parallel experimental results.

2. a. Develop a cellular automaton simulation and visualization of a desert ant (genus *Cataglyphis*) conducting an area-restricted search for its nest (see Project 1). The nest is in the ground with an opening less than 1 cm in diameter; and the ant finds the entrance, not with pheromones, but with dead reckoning. In a psychological cognition experiment, an ant that is far from the nest is captured, placed in a different location, and released. By dead reckoning, the ant heads in the compass direction it should have taken from the location where it was captured. Not finding the nest in the expected location, the ant begins searching. The path is in erratic loops of increasing diameters, centered at the supposed nest. However, repeatedly the ant returns to this location before beginning another loop.

For the first phase of the simulation, initialize each cell of a 20×20 grid (cognitive map) to have some very low, uniform expectation value and use a diffusion rate parameter of 0.1. Assume the nest is in the middle of the grid. Have the ant start at a corner and search at random for the nest. As discussed in the section on "Simulations," diffusion proceeds; each failure results in zeroing out the cell's expectation value; and success sets the value to 1. The second phase allows diffusion to continue in this grid without the ant, say for 20 time steps. Third, place the ant at the former nest location, and allow the simulation to run (Reid and Staddon 1997).

b. Graph the ant's distance from the origin as a function of time for several hundred time steps. Discuss the meaning of the graph.

3. Develop a cellular automaton simulation and visualization of a European badger learning to search for multiple sources of food in specific locations. The original psychological experiment was on an open field with three sections. Peanuts were spread out at the same places on the left and right sections in the field for six nights. The badger learned to find the peanuts from these sections, spending very little time in the middle, empty section.

For the first phase of the simulation, initialize each cell of a grid (cognitive map), which is three times longer than wide, to have some very low, uniform expectation value and use a diffusion rate parameter of 0.001. Consider a total of 16 designated food locations in the left and right thirds of the grid. Have the badger start at an edge and search at random for the food. As discussed in the section on "Simulations," diffusion proceeds; each failure results in zeroing out the cell's expectation value; and success sets the value to 1. The second phase allows diffusion to continue in this grid without the badger, say for 300 time steps, simulating the 24-hour period between searches. Third, place the badger again at the edge of the grid, and allow the simulation to run until the badger leaves the grid. Discuss the results (Reid and Staddon 1998).

4. a. Develop a cellular automaton simulation and visualization of a dog with an appropriate cognitive map exhibiting what appears to be spatial "insight" in finding a shortcut to food. The psychological experiment starts at location X and has food at two places, A and B. The three locations form a triangle with X being closer to A than to B. A dog on a leash is lead from X to A, but not allowed to eat, and lead back to X. The same process is repeated taking the dog from X to B to X. Then the dog is released into the field. Although not trained to do so, most dogs travel from X to A, eat the food, and then take the shortcut from A to B instead of following the training path $XAXBX$.

For the first phase of the simulation, initialize each cell of a 20×20 grid (cognitive map) to have some very low, uniform expectation value and use a diffusion rate parameter of 0.05. Simulate the training session. As discussed in the section on "Simulations," diffusion proceeds; each failure results in zeroing out the cell's expectation value; and success sets the value to 1. The second phase allows diffusion to continue in this grid without the dog, say for 50 time steps. Third, simulate the dog's search for both sources of food. Discuss the results (Reid and Staddon 1998).

b. Run the simulation nine times for diffusion rate parameters of 0.01, 0.05, and 0.1 with Phase 2 times of 50, 60, and 60 time steps each. Discuss the search patterns, why they occur according to the algorithm, and how they parallel experimental results.

5. a. Develop a cellular automaton simulation and visualization of a detour problem in which a blind rat adjusts its path to food in a changed landscape. In one psychological experiment, a blind rat is released from a starting gate into an enclosed area. The starting gate is on the left wall, near the northwest corner while the goal, food, is on the right wall, near the opposite corner. The area has a barrier wall parallel to the left and right walls, and the barrier has an opening not far from the south end. For five times, the rat is released and allowed to find the food, which is replenished for each training session. Its path becomes increasingly more efficient as it learns its way. Then, the barrier is removed, and for several more times, the rat searches for food. With some variation, the blind rats quickly adjusted their trails to head straight for the food.

For the first phase of the simulation, initialize each cell of a 20×20 grid (cognitive map) to have some very low, uniform expectation value, and use a diffusion rate parameter between 0.001 and 0.1. Simulate a training session without food in which a rat completely explores the barrier. Second, simulate the rat's search for food for five times. Each new trial begins immediately after success in the previous trial. Finally, have six trials with food but no barrier.

The computational scientists that developed and analyzed simulations of this experiment wrote, "the difference between smart and less smart subjects lies in their ability to change their maps, not in the level of cognitive processing after they've learned the new map." Discuss the results of your simulation as it applies to this statement (Reid and Staddon 1998).

b. Produce other simulated experiments with different configurations of barriers. Discuss your results.

6. Repeat any of Projects 1–5 with a variation of Equation 1, computation of the diffused expectation value in a cell, that has the diagonal neighbors contributing half as much as the north, east, south, and west neighbors.

7. Repeat any of Projects 1–6 using a stochastic instead of a deterministic approach. Thus, with some probability, an animal moves to the neighbor with the largest expectation value. Otherwise, the creature goes to a random neighbor.

References

Doupe, Alliston J. 1994. "Seeds of Instruction: Hippocampus and Memory in Food-Storing Birds." *Proceedings of the National Academy of Sciences USA*, Vol. 91: 7381–7384.

Gwinner, E. 1996. "Circadian and Circannual Programmes in Avian Migration." *Journal of Experimental Biology*, Vol. 199: 39–48.

Maguire, Eleanor A., David G. Gadian, Ingrid S. Johnsrude, Catriona D. Good, John Ashburner, Richard S. J. Frackowiak, and Christopher D. Frith. 2000. "Navigation-Related Structural Change in the Hippocampi of Taxi Drivers." *Proc. Natl. Acad. Sci. USA*, Vol. 97, No. 8: 4398–4403.

Reid, Alliston K., and J. E. R. Staddon. 1998. "A Dynamic Route Finder for the Cognitive Map." *Psychological Review*, Vol. 105, No. 3: 585–601.

———. 1997. "A Reader for the Cognitive Map." *Information Sciences*, Vol. 100. Amsterdam: Elsevier Science Inc.: 217–228.

Sherry, David F. 1996. "Middle-Scale Navigation: the Vertebrate Case." *Journal of Experimental Biology*, Vol. 199: 163–164.

Sherry, David F. and Sarah J. Duff. "Behavioral and Neural Bases of Orientation in Food-Storing Birds." *Journal of Experimental Biology*, Vol. 199: pp. 165–172.

Staddon, J. E. R. and Ioan M. Chelaru. "A Diffusion-based Guidance System for Autonomous Agents." Duke University. www.ee.duke.edu/~lcarin/DeminingMURI/DeminingRouteFinder00.pdf

Wehner, Rudiger. 1996. "Middle-Scale Navigation: The Insect." *Journal of Experimental Biology*, 199: 125–127.

———, Barbara Michel, and Per Antonsen. 1996. "Visual Navigation in Insects: Coupling of Egocentric and Geocentric Information." *Journal of Experimental Biology*, 199: 129–140.

MODULE 13.4

Pit Vipers—Hot Bodies, Dead Meat

Prerequisite: Module 11.2 on "Spreading of Fire;" Module 11.3 on "Movement of Ants" for Projects 3 and 4.

Introduction

Why are engineers like Dr. John Pearce at the University of Texas studying pit viper pits? An important reason is that the United States Air Force believes that a better understanding of these pits will lead to better missile defense. A snake's pits are on each side of its face between the eyes, and the nostrils and are extremely sensitive infrared (heat) detectors. The snake is able to detect heat differences between the background and potential prey that are as small as a thousandth of a degree centigrade. Exactly how they work is not completely understood, but even blind rattlesnakes can strike warm-blooded prey with deadly accuracy. Because visual and heat images are received in the same area of the brain (optic tectum), some suggest that the snake can form a thermal "image" of the size and shape of the prey. The preferred food in the diet of a pit viper is often a small rodent, which is a warm-blooded, or homeothermic, creature. Because such prey usually maintain a body temperature greater than their surroundings, pit vipers have a most effective hunting tool in their pits.

Dr. Pearce and his colleagues are attempting to develop mathematical models of this predator-prey relationship. These models attempt to predict heat emitted from the prey at given distances from the snake and the influence these variations in heat emissions have on the pit receptors. The military is hoping that from these models, they can design more precise missile detection systems (Cantley 2004; Daniel 2001; Powell 2004; Rische 2001).

Simulations of Heat Diffusion

According to **Newton's Law of Heating and Cooling**, the rate of change of the temperature of an object with respect to time is proportional to the difference between the temperatures of the object and of its surroundings. In one dynamic diffusion model, the change in an empty cell's temperature, ΔT_t, from time t to time $t + \Delta t$ is a proportionality constant, r, times the sum of the differences in temperatures of the eight neighbors' temperatures $(T_{k,t})$ and the cell's temperature (T_t), as follows:

$$\Delta T_t = r \sum_{k=1}^{8} (T_{k,t} - T_t), \quad \text{where } 0 < r < 1/8 = 0.125$$

Thus, a cell's estimated temperature at time $t + \Delta t$ is the following:

$$T_{t + \Delta t} = T_t + \Delta T_t$$

$$= T_t + r \sum_{k=1}^{8} (T_{k,t} - T_t),$$

where $0 < r < 0.125$ with the sum being over the eight neighbors. With subtraction of rT_t occurring eight times, the formula simplifies to the following weighted sum of temperatures of the cell and its neighbors:

$$T_{t+\Delta t} = (1 - 8r)T_t + r \sum_{k=1}^{8} T_{k,t} \quad \text{where } 0 < r < 0.125$$

For example, if $r = 0.1$, then 20% of a cell's estimated temperature comes from its temperature at the last time step, while each neighbor contributes 10%.

Alternatively, we can have each of the north, east, south, and west neighbors supplying more than the diagonal neighbors. For example, we could have the site itself providing 25%, each of the four nearest neighbors giving 12.5%, and each diagonal neighbor, 6.25% of the cell's temperature at the next time step. Note that the sum of these percentages, $0.25 + 4(0.125) + 4(0.625) = 1$, is 100%. We can employ similar models using four neighbors instead of eight.

Projects

1. **a.** Develop a simulation and visualization of a pit viper hunting for food in a cool environment. Have one rodent moving and resting at random. Most mammals have an average body temperature of around 37 °C. For example, the normal body temperature of a rabbit is 38.3 °C, and that of a laboratory mouse is 36.9 °C. Have one pit viper that tends to move towards warmth. In the case of the same temperature in all neighboring cells, the snake selects a direction at random. Use periodic boundary conditions along with eight neighbors per site.

   b. Running the simulation repeatedly, determine if the temperature of the environment impacts the viper's hunting ability.

2. Repeat Project 1 assuming that the rodent can, but does not consistently, move twice as fast as the snake.

3. Repeat Project 1 with one pit viper and several rodents. When the snake is hungry, it seeks food; and when full, it rests. After eating, the snake remains full for a while.

4. Repeat Project 1 with one pit viper and several rodents. When the snake is hungry, it searches for food; and when full, it seeks a warm place to rest for a while.

5. Repeat one of Projects 1, 3, or 4 in which a rodent avoids a pit viper if the animal detects the snake. However, the prey does not always see or hear its predator. When avoiding the snake, an animal moves twice as fast as when not eluding its predator.

6. a. For thermal regulation in a hot dessert, a snake seeks cooler places, such as under rocks or in rodent boroughs. Develop a simulation and visualization of a snake seeking such places when its body temperature becomes too hot (say, 42 °C) and tending to stay in the shelter until sufficiently cool. Assume that the rate of heat loss or gain for a snake is proportional to its surface area (Krochmal and Bakkern 2003).

   b. Plot the snake's temperature versus time. Discuss the results

References

Cantley, Caroline. 2004. "Physical Principles of Biological Processes (Part III)." Lectures from Transport Phenomena, Institute for Gravitational Research. http://www.physics.gla.ac.uk/~caroline/IBLS2%20Module%2016b%20Part%20III/2004%20Lectures%20and%20Notes/LECTURE%20COURSE%202004.pdf

Daniel, Peter C. 2001. "How Heat Sensitive Pits Are Used." Notes from Animal Physiology, Hofstra University. http://people.hofstra.edu/faculty/peter_c_daniel/Animal_Physiology/special_topics_fall2001/Senses/SNAKES/SNAKES%20@/PAGE4.htm (July 23, 2004)

Krochmal, Aaron R., and George S. Bakkern. 2003. "Thermoregulation Is the Pits: Use of Thermal Radiation for Retreat Site Selection by Rattlesnakes." *Journal of Experimental Biology*, Vol. 206: 2539–2545.

Powell, Brian J. 2004. "Thermosensory Abilities of Snakes." Notes from Neuroethology, BioNB 424, Cornell University. http://instruct1.cit.cornell.edu/courses/bionb424/students2004/bjp27/physiology.htm

Rische, Becky. 2001. "Heat-Seeking Vipers May Help with U.S. Defense, UT Austin Researcher Finds." Office of Public Affairs, the University of Texas at Austin, May 31. http://www.utexas.edu/opa/news/01newsreleases/nr_200105/nr_vipers010531.html

MODULE 13.5

Mushroom Fairy Rings—Just Going in Circles

Prerequisite: Module 11.2 on "Spreading of Fire."

Introduction

Sometimes in a forest or yard, mushrooms seem magically to grow in circles, which we call "fairy rings" (see Figure 13.5.1). In this module, projects develop simulations along with animations for the expansion and interactions of such mushroom fairy rings.

You might remember exploring your refrigerator and finding an unmarked container that was pushed against the back wall. Without thinking, you open the container to find a disgusting mass of green or black slime. So the word "fungus" probably makes you think of something rather unpleasant, like spoiled food, or reminds you of kicking over "toadstools" during your childhood. In either case, if you are like most people, you probably do not know too much about them.

Figure 13.5.1   Fairy Ring

Fungi are in the business of decay. They depend on nutrients from degradation of the organic matter deposited by or from other organisms. Through their ability to break down rather complex organic molecules, fungi are also responsible for returning to the soil a large quantity of nutrients for plant growth that would be unavailable otherwise. Ecosystems are dependent on this recycling of nutrients from the decay of organic matter. We derive our own direct benefits from fungi also—antibiotics, cheese, beer/wine, truffles, and other edible mushrooms. On the other hand, fungi are responsible for 70% of all major crop diseases and a number of diseases that affect human beings and other animals (Deacon).

Whether we appreciate them or not, we cannot avoid them. In fact, we come into contact with them (or their spores) with every breath we take. Only about 70,000 of the estimated one million species of fungi have even been described (Whitman). Their ecological, economic, and medical significance makes them worthy objects of human curiosity, and scientists worldwide are busily investigating various aspects of their lives.

What Are Fungi?

To consider a model that involves the fungi, we need to understand some of the fundamentals of their biology. Hence, we include a few questions with basic answers. References are provided with more detailed information. We begin by answering, "What are fungi?"

Fungi are multicellular (except for yeasts), spore-producing organisms that depend on absorbing nutrients from their surroundings. The **spores** are reproductive cells that typically are capable of germinating into new fungi. Nutrients are made available by the action of extracellular enzymes secreted by the fungus itself.

What Do Fungi Look Like?

When most people think of a fungus, they think of a mushroom. Mushrooms are clearly one of the more recognizable forms of fungi, but most fungi do not form mushroom-like structures. Amazingly, most of the "body" of a fungus is usually not visible or obvious. Some are spread out over wide areas underground, reaching miles in diameter (Kruszelnicki). The **mushroom** is just the fruiting body of certain types of fungi.

The basic morphological component of all the fungi, except yeasts, is the **hypha**. A hypha is a thin (5–10 μm diameter (Deacon)), branching tubule. Masses of branching hyphae form the body of the fungus, which we usually call the **mycelium**. Many of the hyphae are interconnected within this mass. Much of the fungal mass you may not see. The **mushroom** that you do see is an organized mass of hyphae, which become spore bearing. Hence, mushrooms in mulch or lawn or the shelf-shaped structures on rotting logs are really only the "tip of the iceberg." They produce the spores, but a much larger mycelium underlies the mushrooms. If we examine the undersurface of the mushroom "cap," we will often find the plate-like **gills**, which bear the spores.

How Do Fungi "Feed Themselves"?

Fungi are not chemo- or photosynthetic, and they have no internal "digestive system". So, they must absorb what they need from the environment—water and inorganic and organic nutrients. Many of the organics are too large to be absorbed unaltered. Consequently, fungi produce hosts of extracellular, digestive enzymes. The enzymes degrade the large, and sometimes very complex, organics into molecules they can absorb through their cell wall and plasma membrane.

How Do Fungi Reproduce?

The outgrowth from a spore is by **asexual** cell division. However, most fungi can also reproduce **sexually** through the fusing (mating) of hyphae of opposite mating types. Such sexually produced hyphae can organize into spore-producing structures, such as mushrooms.

How Do Fungi Grow?

Fungi display what is called **apical growth**—that is, they extend or elongate only at the tip of the hyphae without subsequent increase in diameter. Branches form behind the advancing tip, forming more advancing tips (Lepp and Fagg). A new hypha grows out from a spore, and with its continuous branching, yields a circular growth pattern. Growth can be quite rapid, reaching one kilometer of hyphae per day. This rapid growth is made possible because nutrients are rapidly and constantly being delivered toward the tips.

The Problem

Dr. Alan Rayner, an English scientist who has studied forest fungi for years, wrote:

> I have increasingly come to regard the mycelium as a heterogenous army of hyphal troops, variously equipped for different roles and in varying degrees of communication with one another. Without a commander, other than the dictates of their environmental circumstances, these troops organize themselves into a beautifully open-ended or indeterminate dynamic structure that can continually respond to changing demands. Recall that during its potentially indefinite life, a mycelial army may migrate between energy depots; absorb easily assimilable resources such as sugars; digest refractory resources such as lignocellulose; mate, compete and do battle with neighbours; adjust to changing microclimatic conditions; and reproduce. (Rayner 1991)

Circular or radial growth of the mycelium allows the fungus to move into unexploited areas in its search for nutrients. When the innermost part of the mass has exhausted the resources of that area, that portion becomes expendable. The outer mycelium removes and reabsorbs useful nutrients from the central region and transfers them outward. The innermost hyphae die and decay. This ring-shaped growth pattern can be easily seen on fields or on golf course turf as what is termed a **fairy ring** (see Figure 13.5.1). The rate of growth of these rings varies according to the environment (soil texture, moisture, etc.) and species. *Marasmius oreades*, a fairy-ring fungus found extensively in Europe, may increase its radius between 10 and 35 cm per year (Lepp and Fagg). In the Midwest, fairy-ring species, including *M. oreades*, may expand their radii up to 60 cm per year (Illinois Extension Service 1998). It should come as no surprise that property owners and extension agents regard them as diseases.

Fairy rings appear in Shakespeare's *The Tempest* (Act V, Scene I), when Prospero shouts, "you demi-puppets that by moonshine do the green sour ringlets make, whereof the ewe not bites; and you whose pastime is to make midnight mushrumps, that rejoice to hear the solemn curfew . . . ." It was not until Dr. William Withering, who also gave the world the heart medication digitalis from the foxglove plant (Kruszelnicki), dug up the buried mycelia of fairy rings in 1792 that we knew that a fungus was the cause of these phenomena. Before this discovery, one of the most common beliefs was that the rings represented a circular path of fairies dancing— hence, the name "fairy ring."

Although there are more than 50 species of fungi that can produce fairy rings, the most common causes are one of three species: *Marasmius oreades*, *Agaricus campestris*, *Chlorophyllum molybdites*. There is some variation in the appearance, depending on which fungus is present. One common type is characterized by three distinct rings—an inner ring of stimulated grass growth, a middle ring of dead or dying grass, and an outer ring of stimulated grass growth. Both of the stimulated zones are probably the result of release of nitrogen from the decaying organic matter. The central ring contains the dense mass of "feeding" hyphae that prevents penetration of water and depletes nutrients (Illinois Extension Service 1998). The fruiting bodies form within this ring or on the margin next to the outer stimulated zone. Some fungi do not produce this "dead zone."

Because of dead zones, two intersecting fairy rings often merge into a figure 8 pattern (see Figure 13.5.2). Also, as Figure 13.5.3 depicts, a barrier can induce the mushrooms to grow into an arc.

How Do Fairy Rings Get Started?

A fairy ring grows out from spores or transported soil that contains fragments of mycelium. The circular mass extends from this "nucleus." Soil quality and weather conditions determine the rate of growth.

Initializing the System

In many simulations, we model a grid area under consideration with an $n \times n$ lattice, or a two-dimensional square array of numbers. Each cell in the lattice contains a value representing a characteristic of a corresponding location and a state in the

Figure 13.5.2   Intersecting fairy rings

Figure 13.5.3   Arc of mushrooms near a barrier

cell's life cycle. For example, in a simulation for the spread of fairy rings, a cell can contain an integer value, 0–9, representing such states as being empty, containing a spore, or having maturing hyphae. Table 13.5.1 gives a list of possible values with associated constants and their meanings.

To initialize this discrete stochastic system, we can employ the following probability:

probSpore—The probability that a site initially has a spore, or that the grid site is $SPORE = 1$. Thus, *probSpore* is the initial spore density.

Table 13.5.1
Possible Cell Values with Associated Constants and Their Meanings

Value	Constant	Meaning: The Cell Contains
0	EMPTY	**empty** ground containing no spore or hyphae
1	SPORE	at least one **spore**
2	YOUNG	**young hyphae** that cannot form mushrooms yet
3	MATURING	**maturing hyphae** that cannot form mushrooms yet
4	MUSHROOMS	older hyphae with **mushrooms**
5	OLDER	older hyphae with **no mushrooms**
6	DECAYING	**decaying hyphae** with exhausted nutrients
7	DEAD1	**newly dead hyphae** with exhausted nutrients
8	DEAD2	hyphae that have been **dead for a while**
9	INERT	**inert area** where plants cannot grow

Updating Rules

Updating rules apply to different situations. An inert site never can change states; its cell value remains *INERT*. At the next time step, an empty cell (site value *EMPTY*) may or may not have young hyphae (*YOUNG* or *EMPTY*, respectively) growing into it from a neighboring site with young hyphae (*YOUNG*). If a cell contains a spore (*SPORE*), at the next time step the spore may or may not germinate to become young hyphae (*YOUNG* or *SPORE*, respectively). Young hyphae (*YOUNG*) always age to become maturing hyphae (*MATURING*). Maturing hyphae (*MATURING*) may or may not produce mushrooms (*MUSHROOMS* or *OLDER*, respectively); but regardless, these hyphae (*MUSHROOMS* or *OLDER*) eventually exhaust the resources and begin decaying (*DECAYING*). The decaying hyphae (*DECAYING*) die (*DEAD1*). A site with newly dead hyphae (*DEAD1*) cannot support new growth for a while (*DEAD2*) but eventually can (*EMPTY*).

To develop this dynamic, discrete stochastic system, we employ the following probabilities:

> ***probSporeToHyphae***—The probability of a spore (site value of *SPORE*) germinating to form young hyphae (*YOUNG*) at the next time step
>
> ***probMushroom***—The probability that maturing hyphae (cell value *MATURING*) produce mushrooms (*MUSHROOMS*) at the next time step
>
> ***probSpread***—The probability that an empty site (cell value *EMPTY*) gets young hyphae (*YOUNG*) at the next time step from a neighbor that has young hyphae

Figure 13.5.4 summarizes the states and transitions of the model in a state diagram. The ten states, such as *SPORE* and *YOUNG*, give the possible values for a cell. The arrows with labels, such as *probSporeToHyphae*, show the probability with which a cell changes from one state to another in subsequent time steps. The arrow from *EMPTY* to *YOUNG* indicates that the situation is more complicated; if a neighbor has the value *YOUNG*, the empty cell becomes *YOUNG* with a probability of *probSpread*.

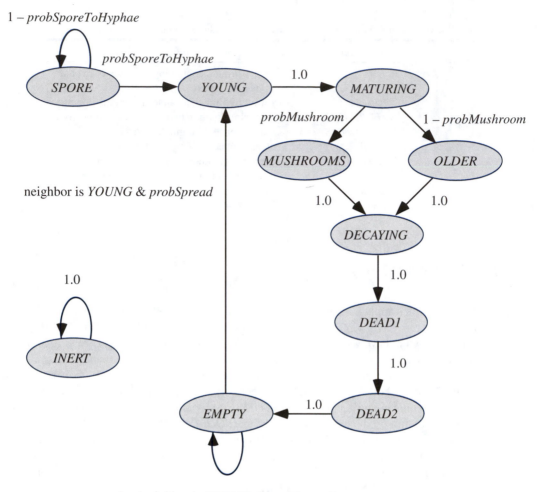

Figure 13.5.4   States and Transitions of the Model in a State Diagram

Displaying the Simulation

For each lattice in the list that a cellular automaton simulation returns, we generate a graphics of a rectangular grid with colors representing the states of the cells, as in Table 13.5.2. Animation of the resulting frames helps us to verify the model and interpret the results.

Projects

1. Develop the simulation of this module using absorbing boundary conditions. Include a function to show the situation above ground. Run the simulation employing various initial grids, as follows:

Table 13.5.2
Possible Cell Values with Associated Constants and Colors

Value	Constant	Colors
0	*EMPTY*	light green
1	*SPORE*	black
2	*YOUNG*	dark grey
3	*MATURING*	light grey
4	*MUSHROOMS*	white
5	*OLDER*	light grey
6	*DECAYING*	tan
7	*DEAD1*	brown
8	*DEAD2*	dark green
9	*INERT*	yellow

    a. As described in the module with various values of *probSpore*. Describe the results.
    b. With exactly one spore in the middle. Verify that the figure seems to agree with the picture in Figure 13.5.1.
    c. With exactly two spores that are several cells apart towards the middle. Verify that the rings merge into the figure-eight pattern observed in nature as in Figure 13.5.2.
    d. With exactly one spore and a barrier. Verify that the results appear to agree with the growth pattern in Figure 13.5.3.
2. Do Project 1 where the probability of young hyphae spreading into a site is proportional to the number of neighbors that contain young hyphae.
3. Adjust the simulation of this module so that new spores can form when mushrooms are present. Consider the following two possibilities:
    a. The probability that a cell can obtain a spore at the next time period is equal to the percentage of mushrooms in the grid.
    b. A cell can obtain a spore at the next time period with a specified probability provided one of its neighbors contains a mushroom.
4. Do Project 1 so that the length of time the hyphae are dead is probabilistic; and on the average, they are dead for two time steps.
5. Do Project 1 using periodic boundary conditions.
6. Do Project 1 so that every extended boundary cell has the value of its immediate neighbor in the original grid.
7. Do Project 1 where the neighbors of a cell include those cells to the northeast, southeast, southwest, and northwest.

References

Deacon, Jim. "The Microbial World—The Fungal Web." Institute of Cell and Molecular Biology and Biology Teaching Organization, University of Edinburgh. http://helios.bto.ed.ac.uk/bto/microbes/fungalwe.htm

Gaylor, Richard J., and Kazume Nishidate. 1996. "Contagion in Excitable Media." *Modeling Nature: Cellular Automata Simulations with Mathematica*. New York: TELOS/Springer-Verlag: 155–171.

Illinois Extension Service. 1998. "Fairy Rings, Mushrooms and Puffballs." *Report on Plant Disease No. 403*. Department of Crop Sciences, University of Illinois at Urbana-Champaign.

Kruszelnicki, Karl S. "Great Moments in Science—Fairy Rings." Karl S. Kruszelnicki Pty, Ltd. http://www.abc.net.au/science/k2/moments/s297489.htm

Lepp, Heino, and Murray Fagg. "The Mycelium." Australian National Botanic Gardens. http://www.anbg.gov.au/fungi/mycelium

Rayner, Alan D. M. 1991. "Conflicting Flows: The Dynamics of Mycelial Territoriality." *McIlvainea*, 10: 24-3557-62.

Spatafora, Joey. "Overview of the Fungi." Oregon State University. http://www.nacse.org/ocid/bot461/lectures/lecture1.htm (accessed May 22 2003).

Whitman, Ross. "Chapter 21 The Origin and Diversity of Life—Part 2." Mississippi University for Women. http://www.muw.edu/~rwhitwam/102S03_Ch21b.html (accessed May 25 2003).

MODULE 13.6

Spread of Disease—"Gesundheit!"

Prerequisite: Module 11.2 on "Spreading of Fire."

Introduction

The "SIR Model" section of Module 6.2 on "Spread of SARS" considers a model for the spread of disease. The **SIR Model** involves the following population groups: **Susceptibles** (*S*) that have no immunity from the disease, **Infecteds** (*I*) that have the disease and can spread it to others, and **Recovereds** (*R*) have that recovered from the disease and are immune to further infection.

In that module, we considered the spread of disease from a systems dynamics point of view. In this module, projects deal with the spread of disease using the approach of cellular automata.

Exercise

1. Suppose an individual is at each grid point. An individual can be well and susceptible (value *SUSCEPTIBLE* = 0) to a disease, sick with the disease that has two phases (values *PHASE1* = 1 and *PHASE2* = 2), or immune (value *IMMUNE* = 3). Let *probSick* be the probability that the individual initially is sick with a disease. Let *probPhase1* be the probability that initially a sick individual is in Phase 1 of the disease. Suppose initially, no individual is immune. Write code in a computational tool to initialize the grid.

Projects

1. Develop a simulation with animation for the contagious spreading of the disease described below. Run the simulation for several probabilities and discuss the results.

Suppose in a population an individual can be susceptible, infectious, or immune to a stomach virus. The infection lasts two days, and immunity only lasts five days before the individual becomes susceptible again. Assume an individual is at each grid point. In the simulation, the value at a grid point can be one of the following:

- 0—susceptible individual
- 1, 2—infectious individual, where the value indicates the day of infection
- 3, 4, 5, 6, 7—immune individual, where the day of immunity is the cell value minus 2. For example, on day 1 of immunity, the cell value is 3.

In the simulation, initialize the grid using the following probabilities:

- *probSusceptible*—the probability the individual is initially susceptible
- *probInfectious*—the probability that an individual that is not susceptible is infectious initially

Uniformly distribute the infected individuals between day 1 and day 2 of the infection. In the initialization, uniformly distribute immune individuals with values 3 through 7. Use constants for the cell values 0 through 7, such as *SUSCEPTIBLE* = 0.

The following rules apply, where the term "neighbor" applies to the cell to the north, east, south, or west:

- If an individual is susceptible and a neighbor is infected, the individual becomes infected.
- The infection lasts for two days.
- Immunity lasts for five days, after which time the individual again becomes susceptible.

Color the graphic as follows:

- Susceptible—full green
- Infectious—blue; full blue on the first day, pale blue on the second.
- Immune—red; full red on the first day with successively paler shades of red on subsequent days

After grid initialization, this model is deterministic, because the next state is always determined by the situation. If a susceptible individual is exposed to the virus, that person will definitely get sick for exactly two days and be immune for exactly five days.

Systematically, run the model for various values of *probSusceptible* and *probInfectious* and discuss the results.

2. Develop a nondeterministic (stochastic) simulation for a situation similar to that in Project 1. In this case, *probCatch* is the probability that a susceptible individual who has a sick neighbor will get sick; and *probBeSusceptible* is the probability that an individual who has been immune for five days will become susceptible. Thus, someone who is exposed to the virus might not become sick, and a person might have longer immunity than five days. Run the simulation for several probabilities and discuss the results. Try to discover a situation in which an epidemic does not occur, that is, in which the disease does not spread to many people over a short period of time.

3. Develop a nondeterministic (stochastic) simulation for a situation similar to that in Project 1. In this case, the probability that a susceptible individual will get sick is the percentage of sick neighbors.

4. Develop a nondeterministic (stochastic) simulation for a situation similar to that in Project 1. In this case, the probability that a susceptible individual will get sick is the average level of infection of the neighbors. For example, suppose the neighbor to the north is susceptible (level of infection = 0); west is immune (level = 0); south is in the first day of infection and very contagious (level = 1); and west is in the second day of infection and less contagious (level = 0.5). Thus, the probability of the individual becoming sick is $(0 + 0 + 1 + 0.5)/4 = 0.375$. The maximum possible total level is 4 and occurs when all neighbors are in the first day of infection. In this case, the probability of the individual catching the virus is $(1 + 1 + 1 + 1)/4 = 1 = 100\%$.

5. Develop a simulation where initially no individuals are sick; but one individual, "Typhoid Mary," is a carrier who never gets sick. Mary walks at random through the grid, and at each step she changes places with the individual in whose cell she steps. Use a contagion situation as in Project 1, 2, 3, or 4. Color Mary as yellow.

MODULE 13.7

HIV—The Enemy Within

Prerequisite: Module 11.2 on "Spreading of Fire."

The Developing Epidemic

"As every month went by, I became more convinced that we were dealing
with something that was going to be a disaster for society."

Anthony S. Fauci, M. D., Director,
National Institute of Allergy and Infectious Diseases (Fauci 1982)

If you had been an attending physician at New York's Bellvue Hospital during the
late 1970s, you might have admitted several patients suffering from a fairly rare
medical problem—*Pneumocystis* pneumonia. At about the same time, doctors in
California were seeing similar patients. Often these patients also were infected op-
portunistically with cytomegalovirus and/or *Candida albicans* (yeast). Furthermore,
there were relatively high occurrences of a fairly rare cancer, Kaposi's sarcoma. The
pneumonia and this cancer were almost never seen except in patients with sup-
pressed immune systems, which apparently characterized each of these patients. The
first papers reporting on these cases appeared in 1981. So it began—the AIDS epidemic in the United States (NIAID 1995).

The 1981 reports did not gain the public's attention. After all there were lots of
other things going on in the United States and the world. Ronald Reagan replaced
Jimmy Carter as president; the Iranian hostages were released; the first shuttle was
successfully launched; MTV first appeared on cable; and IBM introduced its first
PC. During 1981, the Pope and President Reagan survived assassination attempts,
but President Anwar Sadat of Egypt did not (Wikipedia 2004).

Fortunately, these medical reports did alert some in the public health community.
In June, the first patient with the new, unnamed disease was seen at the National In-
stitutes of Health (NIH); and in July, The Centers for Disease Control and Protection
(CDC) formed a task force on "Kaposi's Sarcoma and Opportunistic Infections." In

only a year, there were over 400 cases and 155 deaths in the U.S. from this disease, which now had a name—**Acquired Immune Deficiency Syndrome (AIDS)**. The disease was characterized by a defective cell-mediated immune response. The clustering of the rare opportunistic infections and the Kaposi sarcoma was a result of this weakened immune system, but the cause of the defect was unknown. Substantial evidence, however, pointed to an infectious agent. The unknown agent was apparently acquired sexually, through intravenous drug use, or by transfusion. By the beginning of 1984, 3000 cases of AIDS were reported in the United States, and almost 1300 had died.

In the spring of 1983, scientists at the Pasteur Institute had isolated a new human retrovirus, LAV (lymphadenopathy-associated virus), but they did not claim it to be the cause of AIDS. In 1984, U.S. scientists at the National Cancer Institute showed that a retrovirus (HTLV III—human T cell leukemia virus III) was the apparent cause of AIDS and later concluded that HTLV III and LAV were the same (NIAID 1995).

Despite earlier predictions of a fast cure, this virus, that we now call HIV, has proved to be a very tricky customer and has not yet succumbed to human genius. By March of 1988, more than 84,000 AIDS cases were reported from 136 countries (NIAID 1995).

In 2003, 3 million people died from HIV- and AIDS-associated diseases. By the start of 2004, the world saw an estimated 40 million individuals infected with HIV—more than 37 million adults and 2.5 million children. Each year 40,000 new HIV infections emerge in the United States, half of these cases in young people (under the age of 24) (NIAID, AIDSStat 2004).

Attack on the Immune System

We have learned a lot about HIV and its interactions with the immune system since its discovery. The **human immunodeficiency virus (HIV)** is a type of RNA virus called a **retrovirus**. Retroviruses have some important and unusual characteristics that are significant in certain types of cancer and diseases like AIDS. Retrovirus particles are made up of a **core** (RNA + enzymes) surrounded by a **capsid** (protein). These particles are covered with a lipid **envelope**, which fuses with host cell membranes as the virus enters that cell. This envelope contains glycoproteins that recognize and bind to specific receptors on the host cells. These viruses are able to synthesize DNA from their RNA template, and they carry with them an enzyme that helps to insert the newly synthesized **viral DNA (vDNA)** into a chromosome of the host cell. HIV is actually a special type of retrovirus, called a **lentivirus** ("slow viruses"), associated with slow, degenerative disorders. In such infections, typically a considerable amount of time passes between infection and the appearance of major symptoms (NIAID, HowHIV 2004; Varmus 1998).

HIV infects some incredibly significant cells that play crucial roles in **cell-mediated immunity (CMI)**—CD4+ T-lymphocytes, macrophages, dendritic cells. CMI is one of two major arms of the immune system. It gets its name from its dependence on the effector functions of specific types of immune cells. For example, **CD4+ T-lymphocytes**, also known as **T-helper cells**, help to coordinate immune response through direct interactions with other cells and through the secretion of control chemicals. Some of these chemicals can activate certain cells, which in turn secrete toxins that kill tumor or virally infected cells. Others attract and activate particular white blood cells that engulf invading pathogens. As these cells and lymph tissue become damaged and disabled by HIV, the body becomes progressively more susceptible to various pathogenic agents and cancer. Consequently, death from HIV often occurs from AIDS-associated diseases, rather than from AIDS itself.

Plan of Attack

To enter a cell, the virus must first be able to bind to the cell. Some cells of the body, including certain T-lymphocytes, possess what are termed **CD4** transmembrane receptors. These cells are classified as **CD4+ cells**. Projecting from the HIV envelope is a glycoprotein, **gp120**, which binds to a CD4 receptor like a hand in a glove. Binding induces a conformational change in gp120, which promotes its binding to one of several coreceptor molecules located in the host cell membrane. Once the virus has attached securely to its target, the viral and cell membranes fuse, allowing the virus particle to enter the cell.

Following entry into the cytoplasm of the cell, the capsid is removed from the core of the virus. The core contains RNA and several enzymes, including **reverse transcriptase**. Now activated, reverse transcriptase, using host cell raw materials, synthesizes a double strand of DNA—vDNA. vDNA is transported into the nucleus, where, using an enzyme (integrase), it is inserted into a host cell chromosome. This piece of viral DNA, called a **provirus**, may remain in the chromosome passively (**latent**) or may activate and begin the production of new HIV particles.

The production of new viruses commences with the synthesis of new viral RNA molecules from the vDNA using the host's polymerase. This viral messenger RNA is transported into the cytoplasm, where, using host's ribosomes, enzymes, tRNA, and raw materials, it is translated into HIV **structural proteins** (core, capsid, envelope) and enzymes. The viral messenger RNA, equivalent to viral genomic DNA, is packaged with core proteins and enzymes into new virus particles near the plasma membrane. Envelope proteins are incorporated into the host's membrane, which encases the new virus as it buds from the cell. The last step involves the cleavage of core proteins and enzymes into shorter pieces by a third viral enzyme (protease). Then, the viruses are infectious, capable of invading new host cells.

The primary targets for HIV are these CD4+ T lymphocytes, but the virus also may attack other cells important to an immune response. For instance, cells like **macrophages** and **dendritic cells** normally consume pathogens, presenting essential elements from the microbe on their surfaces. These elements can activate various types of T-cells and also stimulate the production of antibodies. Often, the virus does not destroy macrophages and dendritic cells. In this way, significant quantities of virus are concealed safe from destruction in the very cells that are supposed to help protect the body from infection. Dendritic cells associated with the mucosa (lining) of major virus portals (e.g., vagina, vulva, penis, rectum) pick up viruses and transport them to lymph nodes, which are sites for many types of immune cells (Bugl 2001; Johns Hopkins 2004; NIAID, HIV Life Cycle 2004).

Simulation of the Attack

Computational scientists are employing cellular automaton (CA) simulations to model the immune system and diseases that attack this system, such as AIDS. With CA's stochastic nature, these scientists can use these simulations to estimate the distribution of the system's behaviors as well as the averages; can easily adjust the complex interactions to study the course of an infection and to consider new scenarios, such as new drug therapies; and can express the components and processes in biological language (Kleinstein and Seiden 2000; Sloot et al. 2002). Moreover, in the case of AIDS, CA can model the infection's two time scales over three phases—weeks for the primary response and years for the clinical latency with deterioration of the immune system and for AIDS—much more easily than an approach with differential equations (Sloot et al. 2002).

Projects

1. a. Develop a cellular automata and visualization of an HIV infection. Each site represents one of the following:

- *healthy*—healthy cell
- *infected-A1*—infected cell that can spread the infection
- *infected-A2*—infected cell in its final state before dying due to immune system intervention
- *dead*—infected cell killed by immune system intervention

The system uses the following probabilities:

- *probHIV*—initial probability (fraction) of *infected-A1* cells
- *probReplace*—probability that a *dead* cell will be replaced by a *healthy* cell at the next time step
- *probInfect*—probability that a new *healthy* cell may be replaced by an *infected-A1* cell

The rules for the system are as follows:

- A *healthy* cell with at least one *infected-A1* neighbor becomes *infected-A1* because of infection due to contact before the immune system can respond.
- A *healthy* cell with *numberOfA2* number of *infected-A2* neighbors, where $3 \leq numberOfA2 \leq 8$, becomes *infected-A1* because *infected-A2* cells with concentration above some threshold can contaminate a healthy cell.
- All other *healthy* cells remain *healthy*.
- An *infected-A1* cell becomes an *infected-A2* cell after *responseTime*, the number of time steps for the immune system to generate a response to kill the *infected-A1* cell.
- An *infected-A2* cell becomes a *dead* cell.

- A *dead* cell becomes a *healthy* cell at the next time step with a probability of *probReplace* because the immune system has great ability to recover from an infection's immunosuppressant.
- A new, *healthy* cell may be replaced by an *infected-A1* cell with a probability of *probInfect* because new infected cells can come into the system.

Initialize the grid using *probHIV* = 0.05, indicating that during the primary infection, 1 in 100 to 1 in 1000 T cells harbor viral DNA. Because only 1 in 10^4 to 1 in 10^5 cells in an infected person's peripheral blood expresses viral proteins, use *probInfect* = 10^{-5}. Because the immune system has great ability to replenish dead cells, use *probReplace* = 0.99. Have *responseTime* be 4 weeks, because the time for the immune system to generate a response to kill the *infected-A1* cell is generally between 2 and 6 weeks. Use 8 neighbors for each site and periodic boundary conditions (dos Santos and Coutinho 2001).

b. Plot the numbers of healthy, infected, and dead cells versus time from 0 through 12 weeks and then from 0 through 12 years. To obtain the data, run the simulation a number of times and compute the appropriate average values.

c. Discuss your results. For the visualization in Part a and the graphs in Part b, identify the stages of the infection and explain your results.

2. a. Revise Project 1 to model an HIV infection in the presence of a drug therapy regime, which attempts to block viral replication within the cells. Assume therapy begins at week 300. A therapy has an associated integer rank level, *rankLevel* ($0 \leq rankLevel \leq 8$), indicating the effectiveness of the drug with 0 being the most effective. This rank level models the drug's ability to suppress viral replication and presents a limit to the number of *infected-A1* neighbors that can become infected. At the time of therapy, the first rule in Project 1 changes to be the following (Sloot et al. 2002):

- During drug therapy, a *healthy* cell with *rankLevel* number or more of *infected-A1* neighbors becomes *infected-A1* with a probability of (1 − *probRespond*) ∗ *rankLevel* / 8, where *probRespond* is a response-to-therapy related probability.

b. Consider a *probRespond* function that is constant probability for a set number of time steps and then becomes a significantly smaller constant. Run the simulation for various values of *rankLevel*, and discuss the results. Discuss the impact of the therapy on the simulation visualization and on the graphs of the numbers of healthy, infected, and dead cells versus time.

c. Repeat Part b employing a decreasing linear function for *probRespond*. Because *probRespond(t)* is a probability at time *t* of treatment, its range is between 0.0 and 1.0.

References

Bugl, Paul. 2001. "Cell-Mediated Immunity." From "Immune System," course notes from Epidemics and AIDS. http://uhaweb.hartford.edu/BUGL/immune.htm#cellmed

dos Santos, R.M.Z., and S. Coutinho. 2001. "Dynamics of HIV Infection: A Cellular Automata Approach." *Physical Review Letters*, Vol 87(16): 168102.

Fauci, Anthony S. *Annals of Internal Medicine* editorial (1982).

Johns Hopkins AIDS Service. 2004. "Life Cycle of HIV Infection." http://www.hopkins-aids.edu/hiv_lifecycle/hivcycle_txt.html.

Kleinstein, Steven H., and Philip E. Seiden. 2000. "Simulating The Immune System," *Computer Simulations*, July/August: 69–77.

NIAID (National Institute of Allergies and Infectious Diseases). 1995. "A Brief History of the Emergence of Aids." National Institutes of Health. http://www.niaid.nih.gov/publications/hivaids/4.htm

———. "HIV/AIDS Statistics," National Institutes of Health, 2004. http://www.niaid.nih.gov/factsheets/aidsstat.htm

———. "The HIV Life Cycle." National Institutes of Health. http://www.thebody.com/niaid/hiv_lifecycle/virpage.html

———. 2004. "How HIV Causes AIDS." National Institutes of Health, November. http://www.niaid.nih.gov/factsheets/howhiv.htm

Office of Technology Assessment. "Review of the Public Health Service's Response to AIDS." U.S. Congress, Washington D.C. http://aidshistory.nih.gov/timeline/index.html

Piot, Peter. 2004. "AIDS and the Way Forward" – A World AIDS Day Address Woodrow Wilson International Center for Scholars, Washington, DC, November 30, 2004.

Sloot, Peter, Fan Chen, and Charles Boucher. 2002. "Cellular Automata Model of Drug Therapy for HIV Infection." S. Bandini, B. Chopard, and M. Tomassini, eds. 5th International Conference on Cellular Automata for Research and Industry (ACRI): Geneva, Switzerland. 2002, Lecture Notes in Computer Science 2493. Berlin: Springer-Verlag: 282–293. http://www.science.uva.nl/research/scs/papers/archive/Sloot2002d.pdf

Varmus, Harold. 1988. "Retroviruses." *Science* Vol. 240, No. 4858: 1427–1435.

Wikipedia. "1981." http://en.wikipedia.org/wiki/1981

MODULE 13.8

Predator-Prey—"Catch Me If You Can"

*Prerequisite: Module 11.2 on "Spreading of Fire" for Projects 1 and 7;
Module 11.3 on "Movement of Ants" for Projects 2–6.*

Introduction

We have already dealt with "Predator-Prey Models" in Module 6.4 but from a system dynamics point of view. Projects in this module consider the same subject from the perspective of cellular automata.

About 1970, John Conway developed the *Game of Life*, a 2D cellular automaton with rules for the births and deaths of an imaginary species. This program was the first or one of the first to execute on a parallel computer, a system consisting of multiple processors working together to solve a problem. The rules for the mythical life form in the *Game of Life* are as follows, where the nearest neighbors are the eight surrounding cells:

- A site that is not alive but has exactly three living neighbors has a birth.
- A site that is alive and has exactly two or three living neighbors stays alive.
- All other sites die or stay dead.

Conway carefully developed these rules to enable situations in which patterns grow and evolve over many time steps without becoming stagnant or chaotic. Some patterns, called **life forms**, persist throughout the simulation. **Still-lifes** do not change unless other cells interfere; while some life forms exhibit periodic behavior; and others move across the grid. Project 1 develops the *Game of Life* and explores several life forms. Other projects explore more realistic predator-prey environments (Bedau).

Projects

    1. a. Develop a simulation and animation for Conway's *Game of Life* with periodic boundaries. Run the simulation several times with random initial grids. Then, incorporate the following life forms into grids, and describe their behavior (Bedau):

b. Block: A square of four live cells

c. Traffic light: Three consecutive live cells in a row or a column

d. Glider: Three consecutive live cells in a row with another live cell to the north of the leftmost cell and another live cell to the northeast of the latter cell

e. Other life forms, such as those by Eric Weisstein (Weisstein)

2. a. Develop a simulation with visualization in which a cell can be a predator, a prey, or empty. Use periodic boundaries and eight nearest-neighbor cells. Initialize the grid with a given population density (probability) for a cell being of each type. Each time step of the simulation has two phases: change of state and movement. The rules for change of state are as follows (Weimar and Arab 2003):

- If a prey "meets" (i.e., has as a neighbor) a predator, the predator "eats" the prey (i.e., the prey's site becomes empty). If more than one predator is encountered, a random predator neighbor is selected to "dine."
- A predator dies (i.e., the predator's site becomes empty) when it has gone too long (i.e., a given number of time steps) without food.

During movement, avoid collision as in the text of Module 11.3 on "Movement of Ants" or as in Project 20 of that module.

b. Run the simulation a number of times for various population densities, and discuss the results.

3. a. Develop a simulation with visualization involving predators, which can be in a hungry or full state, and prey. Use periodic boundaries and four nearest-neighbor cells. Each cell can contain up to four hungry predators, four full predators, and four prey individuals, so that a cell can hold from none to twelve individuals. Initialize the grid with given population densities (probabilities) for components of each cell. At each time step of the simulation, the animals undergo predation/reproduction, then direction selection, and then movement. The rules for predation/reproduction are applied in the following order (Alfonseca and Ortega 2000):

- If no prey individuals are in the same cell, a hungry predator dies.
- If at least two prey individuals are in a cell, a hungry predator is in the cell, and less than four full predators are in the cell, then a hungry predator "eats" one prey individual and becomes full.
- If no prey individuals are in the same cell, a full predator becomes hungry.
- If at least two prey individuals are in a cell, a full predator is in the cell, and less than three hungry predators are in the cell, then a full predator "eats" one prey individual, reproduces a hungry predator, and becomes a hungry predator.
- If two or three prey individuals are in a cell, a prey reproduces.

The rule for direction selection is as follows (Alfonseca and Ortega 2000):

- Each individual turns in a random direction (north, east, south, or west) to which no other individual from that category (hungry predator, full predator, or prey) has turned.

Note that the algorithm avoids collisions by having at most four animals in each category moving in different directions.

b. Graph the population densities of predators and prey versus time.

c. Graph the number of predators versus the number of prey.

4. a. Develop a simulation with visualization involving wolves, sheep, and grass on a grid with periodic boundary conditions. A cell is empty or contains one of the following items: a male wolf, a female wolf, a female wolf with cub, a male sheep, a female sheep, a female sheep with lamb, or grass. Associated with each animal is an integer **food ration**, or amount of stored energy from food, up to some maximum value. Assume a population density for each item. The rules are as follows (He et al. 2003):

- A sheep moves into a neighboring empty site, preferring one with grass.
- A lamb leaves its mother and moves into a neighboring empty site. At random this new sheep is a male or female, and its food ration is the same as that of the mother.
- A wolf moves into a neighboring empty site.
- A cub leaves its mother and moves into a neighboring empty site. At random this new wolf is a male or female, and its food ration is the same as that of the mother.
- If its ration of food is less than the maximum, a sheep eats neighboring grass and increases its ration to the maximum amount.
- If a female sheep has at least a designated amount of food ration (such as 2), is of reproduction age (such as 8), and has a male sheep of reproduction age as a neighbor, she becomes a female sheep with lamb.
- If its ration of food is less than the maximum (such as 3), a wolf eats a neighboring sheep and increases its ration to the maximum amount.
- If a female wolf has at least a certain amount of food ration (such as 2), is of reproduction age (such as 8), and has a male wolf of reproduction age as a neighbor, she becomes a female wolf with cub.
- An independent baby matures in a certain number of time steps, such as 8.
- An animal's food ration decreases by 1 at each time step.
- An animal dies when its food ration becomes 0.
- Grass grows in a certain number of time steps, such as 4.

Avoid collisions as in the text of Module 11.3 on "Movement of Ants" or as in Project 20 of that module. Initialize the grid at random with certain densities of each item and with random food rations and ages for each animal. Run the simulation a number of times obtaining situations in which the sheep, wolves, and grass coexist with oscillating densities; in which the sheep become extinct; and in which all animals die.

b. Graph the population densities of sheep, wolves, and grass versus time.

c. Adjust the program to run the simulation a number of times, computing and storing the average number of sheep, wolves, and grass at each time step. Plot these averages versus time. Discuss the results.

5. a. Develop a simulation with visualization involving mobile predators and stationary prey. For example, algae that grow on rocks in inter-tidal areas are a favorite food of some snails. Use periodic boundary conditions, and assume initial population densities for predators and prey. Each predator has a direction to which it turns, a length of time until giving birth (reproduction time), and a length of time until starving (starvation time). The rules are as follows (Gaylord and Nishidate 1996):

- If a predator's reproduction and starvation times are both 0, the predator gives birth and dies. For simplicity, we assume only one child.
- If a predator's starvation time is 0 and reproduction time is positive, the predator dies.
- Prey grows in an empty site with a certain probability.
- If a predator's reproduction time is 0, its starvation time is positive, and a neighboring site with a prey is available, then the predator moves to that site and leaves a child in the old site. Both parent and child get maximum reproduction and starvation times.
- If a predator's reproduction time is 0, its starvation time is positive, and an empty neighboring site is available, then the predator moves to that site with maximum reproduction time and with starvation time decremented by 1 and leaves a child in the old site with maximum reproduction and starvation times.
- If a predator's reproduction and starvation times are positive and a neighboring site with prey is available, then the predator moves to that site with reproduction time decremented by 1 and starvation time set to the maximum.
- If a predator's reproduction and starvation times are positive and an empty neighboring site is available, then the predator moves to that site with its times decremented by 1.
- If a predator does not move, its times decrement by 1.

During movement, avoid collision as in the text of Module 11.3 on "Movement of Ants" or as in Project 20 of that module.

   b. Run the simulation and visualization a number of times for various population densities, and discuss the results.

   c. Adjust the program to run the simulation a number of times, computing and storing the average number of individuals in each species at each time step. Plot these averages versus time, and plot the number of predators versus the number of prey. Adjust the number of predators and prey to obtain graphs that resemble those in Figures 6.4.2 and 6.4.3 of Module 6.4 on "Predator-Prey Model."

6. Adjust Project 5 to allow for a prey population in which each individual is mobile and can give birth.

7. a. Simulations can help illuminate ecosystem problems when a species becomes extinct or varies greatly in size. Consider a hierarchy of species, numbered 1, 2, ..., m, where species i is higher on the food chain than species $i-1$ and $m \geq 2$. Thus, species $i-1$ is food or prey for species i, and species i is predator for species $i-1$. Develop a predator-prey cellular automaton simulation and visualization for an ecosystem. Each cell is empty or contains exactly one animal. Initially, with probability $probSpecies_i$, a cell contains an individual from species i, and the sum of these probabilities is less than 1. Use periodic boundary conditions, eight surrounding cells as neighbors, and the following rules (Yang 2002):

- If a cell has no predator and no prey neighbors, then the cell obeys the rules of Conway's *Game of Life* (see Project 1).
- If a cell contains a live animal and the predator neighbors outnumber the prey neighbors, then the animal in this cell dies.
- If the prey neighbors outnumber the predator neighbors, then the cell stays or becomes alive.
- If the number of prey neighbors equals the number of predator neighbors, a positive number, then the state of the cell does not change.
- With probability $probDie_i$, a cell containing species i dies.

b. For $m = 2$, adjust the program to run the simulation a number of times, computing and storing the average number of individuals in each species at each time step. Plot these averages versus time, and plot the number of predators versus the number of prey. Adjust the number of predators and prey to obtain graphs that resemble those in Figures 6.4.2 and 6.4.3 of Module 6.4 on "Predator-Prey Model."

c. For $m > 2$, run the simulation and visualization several times with $probDie_i = 0$ for all species. Then, after making $probDie_i$ a small, positive number for one species, repeat the experiment. Discuss the results.

d. For $m > 2$, adjust the program to run the simulation a number of times, computing and storing the average number of individuals in each species at each time step. Run the simulation a number of times with $probDie_i = 0$ for all species, and plot the average number of individuals in each species versus time. Then, after making $probDie_i$ a small, positive number for one species, repeat the experiment. Discuss the results.

e. For $m > 2$, adjust the program to run the simulation a number of times, computing and storing the average number of individuals in each species at each time step. Plot the average number of individuals in each species versus time. Then, after increasing $probSpecies_i$ for some species i and adjusting the corresponding probabilities for the other species, repeat the experiment. Discuss the results.

References

Alfonseca, Manuel, and Alfonso Ortega. 2000. "Representation of Some Cellular Automata by Means of Equivalent L Systems." *Complexity International*, Vol. 7. http://journal-ci.csse.monash.edu.au/ci/vol07/alfons01/alfons01.pdf

Bedau, Mark. "The Game of Life." Reed College, Portland, Oregon. http://www.reed.edu/alife/classes/gameoflife.html

Gaylord, Richard J., and Kazume Nishidate. 1996. "Predator-Prey Ecosystems." *Modeling Nature: Cellular Automata Simulations with Mathematica*. New York: TELOS/Springer-Verlag: 143–154.

He, Mingfeng, Hongbo Ruan, and Changliang Yu. 2003. "A Predator-Prey Model Based on the Fully Parallel Cellular Automata." *International Journal of Modern*

Physic C, 14(9): 1237–1249. http://www.worldscinet.com/ijmpc/14/1409/ S0129183103005376.html (abstract)

Weimar, Jörg Richard and Meraa Arab. 2003. "Simulating Predator and Prey in the Forest." Project work by M. Arab in lectures *Simulation with Cellular Automata* by J. R. Weimar, Technical University of Braunschweig. http://www-public.tubs .de:8080/~y0021323/ca/

Weisstein, Eric W. 1995–2002. "Eric Weisstein's Treasure Trove of the Life Cellular Automaton." Wolfram Research, Inc. http://www.ericweisstein.com/encyclopedias/ life/

Yang, Xin-She, 2002. " Characterization of Multispecies Living Ecosystems with Cellular Automata," *Artificial Life VIII*, Standish, Abbass, Bedau (eds) (MIT Press). pp 138–141. Available from: http://parallel.hpc.unsw.edu.au/complex/ alife8/proceedings/sub857.pdf

MODULE 13.9

Clouds—Bringing It All Together

Prerequisite: Module 11.2 on "Spreading of Fire" for Projects 1–4 and 7;
Module 11.3 on "Movement of Ants" for Projects 5 and 6;
Module 12.2 on "Parallel Algorithms,"
Section on "Sequential Algorithm for N-Body Problem," for Project 8.

Introduction

> The Clouds consign their treasures to the fields;
> And, softly shaking on the dimpled pool
> Prelusive drops, let all their moisture flow
> In large effusion, o'er the freshen'd world.
> —*James Thompson*
> *Seasons—Spring* (line 173)

Clouds. Their endless variety and beauty inspire human beings to dream, to imagine, and to write poetry. They help to cool the earth by reflecting sunlight, but they also help to keep it warm by trapping heat radiated from the earth's surface. But, what are clouds? Scientists tell us that they are collections of water droplets and ice, as well as non-aqueous solids and liquids (Alcorn 2003). During the summer, air near the earth is warmed and rises. As the air rises, it expands and cools, generating relative humidity of 100% or saturated air. The moisture required for saturation decreases as the air temperature decreases. After saturation is achieved, further cooling triggers water vapor to condense into small droplets. These droplets, and sometimes ice crystals, may form by condensing on suspended aerosols (salt, dust, pollution, etc.). The products of all this condensation are light, fluffy **cumulus clouds**. Known as "fair weather clouds," they are characterized by flat bases and do not give rise to any rain (Alcorn 2003; UIUC 1997; Odman 2004; SSEC 2004; Strathclyde 2004; Yin 2004).

However, if cumulus clouds are overlain by large quantities of cold, unstable air, the warm air may continue to ascend as mighty updrafts. As a consequence, clouds develop further vertically and are transformed into **taller, cumulonimbus** forms. Cumulonimbus clouds, their anvil-shaped tops reaching altitudes exceeding 12,000 meters, appear as towers and sometimes in lines called **squall lines**. In these "thunderheads," water droplets continue to condense, and updrafts in the cloud carry the smaller ones upward. The countless droplets collide with each other as they move. Some of these collisions result in droplets merging to form larger droplets. This process is referred to as **coalescence**. The droplets that are large enough begin to fall through the cloud, colliding and coalescing as they tumble downward. Some of these are large enough to make it to the earth's surface. If liquid, we term it "rain" (UIUC 1997; Brown 2003).

Projects

1. **a.** Develop a 2D cellular automaton simulation and visualization of **coalescence** of droplets in a cross section of a cloud. Each grid point contains a number representing the droplet's size or indicating that the cell is empty. At a step of the simulation, each droplet moves in a random direction. If more than one droplet moves into a cell, the droplets coalesce. The size of the new droplet is the sum of their sizes. Initialize the grid with droplets according to a probability that corresponds to the **relative humidity**, which is (partial vapor pressure)/(saturation vapor pressure) for a particular temperature. At initialization, give each droplet a normally distributed random size (Gaylor and Nishidate 1996).

   b. Produce a bar chart of the size distribution of droplets at the end of the simulation.

   c. Produce a graph of the average size of droplets versus time. Discuss your findings. What will happen eventually if you allow your simulation to run long enough?

2. The simulation in Project 1 eventually produces one large droplet, which is an unrealistic result (Gaylor and Nishidate 1996).

   a. Repeat Project 1 and achieve a steady-state distribution of droplet sizes by adding small droplets with a designated probability throughout the simulation.

   b. Achieve a steady-state distribution of droplet sizes by removing larger drops with a certain probability.

   c. Achieve a steady-state distribution of droplet sizes using the techniques of Parts a and b.

3. Repeat Project 1 and achieve a steady-state distribution of droplet sizes by breaking each larger droplet into two droplets with a certain probability. The new droplet forms in a random vacant neighbor and is of random size. The sum of the sizes of the two droplets equals the size of the original droplet (Gaylor and Nishidate 1996).

4. Repeat Project 1 including the formation of rain and eight neighbors per cell. Have the ground towards the south. Do not allow medium-sized droplets to move to the north, northeast, or northwest. Medium-large droplets can only "fall" south, southeast, or southwest. Large droplets can only head toward

the south. Remove droplets that travel off the south boundary, indicating that they have fallen to the ground. Continually add small droplets to the grid at random with a certain probability.

5. **a.** Develop a 2D cellular automaton simulation and visualization of cloud evolution, such as one might view from the ground. Thus, the cross section is a horizontal slice of the cloud. Each cell has three Boolean values (Dobashi et al. 2000):

- *humidity*—true if the cell contains enough water for cloud droplets
- *cloud*—true if the cell has cloud droplets
- *act*—true if the vapor in the cell is ready to transition to cloud droplets

The following are probabilities in the simulation:

- *probHumidity*—probability that a non-cloud cell has enough humidity to transition to cloud
- *probExtiction*—probability that a cloud cell becomes a non-cloud cell
- *probAct*—probability that a cell that is not ready to act (i.e., to transition from vapor to cloud) becomes ready

The transition rules are as follows:

- If a cell's value of *act* is false, the cell's *humidity* value remains the same at the next time step.
- If *cloud* or *act* is true in a cell, *cloud* is true at the next time step.
- If *act* is not true in a cell but *humidity* is true and at least one neighbor's value of *act* is true, then the cell's value of *act* becomes true.
- With probability *probHumidity*, a cell's *humidity* value becomes true.
- If *humidity* is true, it remains true.
- If *cloud* is true, then with probability *probExtiction*, *cloud* becomes false.
- With probability *probAct*, a cell's *act* value becomes true.
- If *act* is true, it remains true.

Initialize the grid at random with only the middle cell having *cloud* as true.

b. Add wind to your simulation by letting *velocity* be an integer indicating wind velocity from left to right across the grid. Thus, the state of a cell in column j at one time step is the state of cell in column ($j - velocity$) at the previous time step.

6. Develop a 3D version of Project 5.

7. Develop a 2D cellular automaton simulation and visualization of the first step in the formation of precipitation, **condensation** or **deposition** of vapor on particles, called **condensation nuclei**, to form droplets. The nuclei are typically tiny (from 0.05 to 0.5 micrometers in radius) solid particles, such as dust, smoke, or salt. The process, which occurs in high humidity, is fast at first, but then slows. Specifically, experiments have shown that the rate of change of the radius (r) of a cloud particle is proportional to $1/r$. For a normal nucleus, which neither attracts nor repels water, if the relative humidity (RH; see Project 1 for definition) is less than 100%, evaporation exceeds condensation so that the droplet shrinks. However, some of the nuclei attract water molecules and thus promote condensation at RH less than or equal to

100%. Also, if the RH is greater than 100%, with a normal nucleus, the reverse happens, and the droplet grows. Condensation is important until the cloud particle becomes a cloud droplet with a radius of about 100 μm. At that time, the collision-coalescence process (see Project 1) becomes more significant (Clayson 2003; McCormack 1999).

For the simulation, initially have a cluster of several cells representing the condensation nucleus and other cells picked at random indicating vapor with the number of such vapor particles based on RH. Have vapor particles move at random. If a vapor particle has a neighbor that is part of the growing droplet, the vapor particle condenses on the droplet with a designated probability based on conditions. Repeat the simulation a number of times with various condensation nuclei. Allowing the simulation to run for a while, discuss the ultimate shapes of the droplets.

8. Develop a simulation of the precipitation process in a relatively warm cloud with temperatures above freezing. We call droplets with radii larger than 0.25 mm "**raindrops**", and those with radii larger than 2 mm frequently split in two. Suppose an updraft of air pushes smaller droplets up into the cloud. Larger droplets, which have higher terminal velocities, fall. For instance, the terminal speed of a cloud droplet with radius 0.05 mm is 7 cm/sec; with a radius of 0.1 mm, the terminal speed of a droplet is 70 cm/sec; while a raindrop of radius 1 mm has a terminal speed of 550 cm/sec. Some particles that appear on a collision path, such as a large falling **collector drop** and a much smaller rising droplet, do not collide; but the smaller droplet streamlines past the larger one. **Collision efficiency** is $E = d^2/(r_1 + r_2)^2$, where d is the **critical distance**, or distance between the center lines of the drop and droplet, and r_1 and r_2 are the corresponding radii. For a coordinate system in which the xz-plane is horizontal, suppose the collector drop and smaller droplet are at locations (x_1, y_1, z_1) and (x_2, y_2, z_2), respectively. Assume that the collector drop is going straight down and the droplet is going straight up. Then, the critical distance is the distance between the two points projected onto the xz-plane, namely $\sqrt{(x_2 - x_1)^2 + (z_2 - z_1)^2}$.

Even if collision occurs, coalescence might not because sometimes the droplets bounce off each other. Studies indicate that droplets tend to become charged during thunderstorms, and droplets with opposite charges are more likely to coalesce. Collisions can also cause a drop to break apart. Drops involved in the collision/coalescence process usually have radii no larger than 2.5 mm. **Coalescence efficiency** is the portion of collisions that result in the droplets sticking together. Laboratory experiments indicate that if the radius of a collector drop is less than 0.4 mm or the radius of a droplet is less than 0.2 mm, then coalescence efficiency is about 1.0. If the collector drop's radius is between 1 mm and 2.5 mm, then coalescence efficiency is less than 0.2. When the radius of the droplet is about 60% that of the drop, coalescence efficiency is small, but the collision efficiency is close to 1.0. **Collection efficiency** is the product of the collision and coalescence efficiencies (Brown 2003; Flory 2004).

References

Alcorn, Marion. 2003. "Exercise 10—Clouds." Meteorology 304, Department of Atmospheric Sciences, Texas A&M University. http://www.met.tamu.edu/class/Metr304/Exer10dir/clouds.html

Brown, Derek W. 2003. "The Collision and Coalescence Process." From The Plymouth State University Meteorology Program's tutorial on "Precipitation: Formation to Measurement." http://vortex.plymouth.edu/precip/precip2aaa.html

Clayson, Carol Anne. "Precipitation." Lecture notes from "An Introduction to Atmospheric Sciences." http://www.met.fsu.edu/Classes/Met1010-clayson/lecture13.html

Dobashi, Yoshinori, Kazufumi Kaneda, Hideo Yamashita, Tsuyoshi Okita, and Tomoyuki Nishita. 2000. "A Simple, Efficient, Method for Realistic Animation of Clouds." Proceedings of the 27th International Conference on Computer Graphics and Interactive Techniques: 19–28.

Flory, Dave. "Chapter 4—Atmospheric Aerosol and Cloud Processes." Lecture notes from "General Meteorology." http://www.meteor.iastate.edu/classes/mt301/lectures/mteor301_ch4.pdf

Gaylor, Richard J., and Kazume Nishidate. 1996. "Coalescence." *Modeling Nature: Cellular Automata Simulations with Mathematica*. TELOS/Springer-Verlag: 107–112.

McCormack, John. 1999. "Precipitation." Course notes from Atmospheric Sciences 171: "Introduction to Meteorology and Climate," University of Arizona. http://www.atmo.arizona.edu/students/courselinks/fall99/atmo171-mcc/atmo171_f99_10.html

Odman, Amy. 2004. "Supercooling, Clouds and Precipitation." General Science 109, Meteorology, Portland Community College. http://spot.pcc.edu/~aodman/supercooling-clouds-precipitation.doc (accessed June 12, 2004; site now discontinued).

SSEC (Space Science and Engineering Center). 2004. "Cloud Identification." Satellite Meteorology Course Modules, Cooperative Institute for Meteorological Satellite Studies, University of Wisconsin–Madison. http://cimss.ssec.wisc.edu/satmet/modules/clouds/index.html

Strathclyde (University of Strathclyde). 2004. "Weather." Physics 12-490, "Communicating Physics," Physics Department. http://dutch.phys.strath.ac.uk/CommPhys2002Exam/Rachel_Kennedy/cloudformation.htm

UIUC (University of Illinois–Urbana-Champaign). 1997. Weather World 2010. "Clouds and Precipitation." Department of Atmospheric Sciences. http://ww2010.atmos.uiuc.edu/(Gh)/guides/mtr/cld/home.rxml

Yin, Yan. "Cloud Microphysics." Atmospheric Science Lectures, Physics 28520, 2004, University of Wales Aberystwyth. http://users.aber.ac.uk/yyy/teaching/ph28520/b8.pdf

MODULE 13.10

Fish Schooling—Hanging Together, not Separately

Prerequisites: Module 11.2 on "Spreading of Fire" and Module 12.2 on "Parallel Algorithms," Section on "Sequential Algorithm for N-Body Problem."

Introduction

Imagine yourself suspended in the splendid blue of the Caribbean, gazing out over a beautiful, underwater garden. You are diving on one of Nature's treasures—a coral reef. These gardens are the most diverse places in the ocean, home to one in four known marine species of plants and animals. You look out over the massive coral heads, decorated with sea fans and whips, various worms, and sea urchins and teeming with fish. A yellowtail damselfish darts in and out of a crevice to protect its territory from rival species. Small herds of parrotfish are grazing loudly on the coral, converting the algae into energy for themselves and the hard skeletal material into sand. Butterfly fish browse the reef for small invertebrates, and a pair of French angelfish munches on some of the many sponges tucked about the reef.

As you glide along the coral walls, you see a small school of blue chromises, moving as if they were articulated parts of one organism. Suddenly, in unison, they scurry away, and you wonder why. Soon, you know why. To your left you see the reason—a beautiful school of jacks are heading for that area of the reef. The chromises have left to avoid this oncoming mass of predators. You wonder at the precision with which this group of 50 swims, turning left, then right, up and down. They seem almost choreographed.

Fish schooling has fascinated human observers for years. Fish schools are social troupes of fish, frequently of comparable age and size, traveling as units, moving in synchrony in the same direction. We wonder why schooling is so common in various fish species (80%) and also how fish are able to coordinate such behavior (Brooks and Yasukawa; Wells).

There are several hypothesized advantages for schooling behavior. Two of the most common explanations are foraging efficiency and protection from predators. Many "eyes" increase the chances of finding food. Everyone follows those of the group who locate food. They are also able to overwhelm some prey in a group, whereas they would have less chance for success individually. While foraging, there are also many eyes to watch for potential predators. Moreover, while groups of fish are more easily detected by predators, the large group may resemble a single, larger organism and discourage attack (Brooks and Yasukawa; Stout; Wells).

Recently, scientists have discovered that fish in their swimming motions create eddies. Schooling fish exploit the energy of eddies created by their neighbors to push them forward (Liao et al. 2003). So, it seems that schooling also decreases energy expenditure for foraging.

How fish are able to coordinate their movements so that they can respond instantaneously to changes in direction and speed of their schoolmates is complex. Most fish, especially those that school, have eyes on the sides of their heads. This location is advantageous to detecting changes in lateral events. Additionally, fish possess a lateral line system on their flanks that is sensitive to pressure changes. Swimming movements of the school generate water displacement that is detected by this system as changes in pressure (Stout).

Simulations

Scientists are attempting to understand biological aggregations, such as fish schools and bird flocks, and to determine pertinent biological and mechanical features and evolutionary behaviors. However, observing individual and group behavior in the laboratory and nature is quite challenging because of the inherent difficulties in three-dimensional tracking of animals in air and water. What they do know has enabled scientists to devise "traffic rules" of an individual animal's response to its neighbors. Using these, computational scientists have developed mathematical models and computer simulations of a group's dynamics. Typically, such a simulation assigns forces that act upon the direction and speed of an individual, while the environment and actions of close neighbors moderate these forces. With these studies, scientists hope to determine the mechanics of relationships between individual behaviors and group spatial patterns. Also, with more detailed observations of aggregations, such as fish schools, and better simulations of observed behaviors, computational scientists hope to determine the behavioral algorithms that some animals, such as fish, employ (Parrish et al. 2002).

Projects

1. Develop a 2D simulation and visualization of fish schooling (or bird flocking) behavior. Suppose each fish follows three rules (Graham 2003):

- Cohesion Rule: A fish moves towards the mean position of its "closest neighbors." Consider each cell to indicate a position (x, y). For the x coordinate of the mean position, take the integer average of the x-coordinates of the closest neighbors. Perform a similar computation for the y-coordinate.
- Separation Rule: A fish does not get closer than some minimum distance to any neighbor.

- Alignment Rule: A fish heads in the mean direction to which its "closest neighbors" head. Consider each fish to have one of eight directions, represented as integers 1–8, in which it is headed. For the mean direction, take the integer average of the closest neighbor's direction values.

A fish with no close neighbors swims at random. Employ periodic boundaries, and initialize the fish with random positions and orientations.

2. Repeat Project 1 considering the boundaries as walls that the fish should avoid. Thus, when a fish moves "close" to a wall, say within two cells of the wall, it turns in a random direction that does not take it closer to the wall.

3. Repeat Project 1 or 2 taking into account the influence of a shark. When not "close" to fish, the shark moves at random. When close to a fish, the shark moves towards the prey, and the fish moves away from the shark. Have the fish and shark move faster when in close proximity to each other, and have the shark move faster than the fish. If a shark catches a fish, the predator eats the prey (Malone 2002).

4. Repeat Project 1 having the fish move from one wall towards part of the opposite wall, which is an entrance to a cave. Fish can go through the entrance to safety. Once in the cave, a fish is no longer a participant in the simulation. Consider the other boundaries as walls that the fish should avoid. Your simulation should also take into account the influence of a shark as in Project 3 (Malone 2002).

5. Repeat Project 2. Initially, have all fish head in the same direction. After several time steps, turn a certain percentage of the fish in a different direction. Run the simulation a number of times with various percentages. Discuss how a percentage of fish turning in a different direction impacts the behavior of the school (Huse et al. 2000).

6. Repeat any of Projects 1–5 only considering neighbors to the sides of a fish and not taking into account the influence of neighbors to the front and rear, which are out of a fish's field of view.

7. Develop a 2D simulation and visualization of fish schooling behavior in a coral bed. Each fish remains close to its nearest neighbor. Any fish that is beyond some threshold distance from its nearest neighbor is subject to shark attack. In the visualization, use different colors to indicate schooling fish, shark, and loner fish (StarLogo 1997).

References

Brooks, Rebecca L., and Ken Yasukawa. "Schooling Behavior in Fish." Laboratory Exercises in Animal Behavior from Animal Behavior Society. http://www.animalbehavior.org/ABS/Education/Labs/lab_schooling.html

Graham, William C., Jr. 2003. "Virtual Ecosystems—Fish Schooling." A Virtual Sea. http://www.ecovis.org/Schooling.htm

Huse, Geir, Steve Railsback, and Anders Fernø. 2002. "Clupeoids, A Fish Schooling Simulator Based on Boids." Humboldt State University. http://math.humboldt.edu/~simsys/Clupeoids.html

Liao, James C., David N. Beal, George V. Lauder, and Michael S. Triantafyllou. 2003. "Fish Exploiting Vortices Decrease Muscle Activity." *Science*, 302(5650): 1566–1569.

Malone, Greg. "Integrated Science Class Goes to Fish School." Electric Kiva, 2002. http://www.electrickiva.com/lessons/starlogo/classproj/fishschool.htm

Ocean Institute. "Schooling Fish." http://www.ocean-institute.org/edu_programs/materials/P/Org/F_G/School.htm

Parrish, Julia K., Steven V. Viscido, and Daniel Grunbaum. 2002. "Organized Fish Schools: An Examination of Emergent Properties." *Biological Bulletin*. 202: 296–305.

StarLogo Sharing Project. 1997. "Fish School." Massachusetts Institute of Technology. http://education.mit.edu/macstarlogo/community/fish.html (accessed January 14, 2004).

Stout, Prentice K. "Fish Schooling." Rhode Island Sea Grant. http://seagrant.gso.uri.edu/factsheets/schooling.html

Tovey, Craig. "Self-Organizing Social Structure in Fish Groups." http://www-2.cs.cmu.edu/~ACO/dimacs/tovey.html

Wells, Kentwood D. "Respiration, Locomotion, and Schooling in Fishes, Schooling Behavior in Fishes." Lecture notes for Ecology and Evolutionary Biology 214, University of Connecticut. http://www.eeb.uconn.edu/courses/vertbio/lecture6.htm (accessed January 12, 2004).

GLOSSARY OF TERMS

absolute error — Absolute value of the difference between the exact answer and the computer answer.

absorbing boundary conditions — Conditions so that border cells of a cellular automaton grid are considered to have no neighbors off the original lattice.

acceleration — Rate of change of velocity with respect to time.

acceleration due to gravity — Approximately -9.81 m/sec^2, where up is considered the positive direction.

adrenalin — Epinephrine.

aerobic respiration — Cellular respiration, where oxygen serves as the final electron receptor. Typically, this type of respiration uses the Krebs cycle and the electron transport chain, both found in the mitochondria.

Ampere (A) — Unit of current for a charge of one coulomb to pass through a region in one second.

angular acceleration — Rate of change of angular velocity.

angular velocity — Rate of change of an angle with respect to time.

antigen — Substance which the body recognizes as foreign and which stimulates an immune response.

antigen-presenting cell — Specialized cell that can present antigens on its surface for interaction with T cells.

antigen-specific T-cell — T-cell, or T-lymphocyte, that bears receptors that bind specifically to an antigen.

anthropogenic — Of human origin.

antiderivative — Function F if $F'(t) = f(t)$, or the derivative of F is f.

apical growth — Extension or elongation only at the tip of hyphae.

aposematic coloration — Bright coloration often displayed by prey as a warning of their toxic composition.

assignment statement — Statement that causes the computer to store the value of an expression in a memory location associated with a variable.

associative property — Property where grouping in addition or multiplication does not matter: $(a + b) + c = a + (b + c)$ and $(ab)c = a(bc)$.

atm — Abbreviation for atmosphere.

atmosphere — Air surrounding the earth. As a measurement, one atmosphere (atm) is the atmospheric pressure at sea level.

ATP synthase — Enzyme that synthesizes ATP from ADP and phosphate. In the mitochondrion, the enzyme is localized in a complex of proteins that permit the passage of protons from high concentration across the inner membrane into the mitochondrial matrix. As this gradient runs downhill, the energy released is available to help the enzyme make ATP.

atria — In the human heart, the two upper chambers that receive blood returning from the body and lungs.

average velocity — Ratio of the change in position to the change in time.

Barnes-Hut Algorithm — Divide-and-conquer parallel algorithm employing clustering that is a simulation of the N-Body Problem.

binary number system — Number system with base 2; used by most computers.

biomagnification — Cumulative increase in the concentrations of a substance in successively higher levels of the food chain.

biosphere — All living things.

bit — 0 or 1 in the binary number system.

blood pressure — Hydrostatic (fluid) pressure that moves the blood through the circulation.

BMI — Body mass index.

body mass index (BMI) — Anthropometric measure, defined as 703 multiplied by the weight in pounds and divided by the square of height in inches. BMI is used to indicate if someone is at a healthy weight, overweight, or obese.

Boyle's Law — For gas at a particular temperature, $PV = K$, where P is pressure, V is volume, and K is a constant.

CA — Cellular automaton.

calculus — Mathematics of change.

capacitance — Ability to store charge.

capacitor — Electronic circuit element for storing charge.

capsid — Protein coating that covers the core of a virus.

carbohydrates — Organic molecules composed of the elements carbon (C), hydrogen (H), and oxygen (O) in the ratio of one C to two H to one O.

carbon cycle — Movement of carbon from one earth subsystem to another.

cardiac output — Product of the stroke volume and the heart rate.

carrying capacity — Maximum population size that an environment can support indefinitely.

CD4+ T-lymphocytes — T-helper cells, bearing CD4 receptors, promote both humoral and cell-mediated immunity.

cell-mediated immunity (CMI) — Part of body defense that is mediated by antigen-specific T-cells and various nonspecific cells of the immune system.

cellular automaton (plural, automata) — Type of computer simulation that is a dynamic computational model and is discrete in space, state, and time, where space is represented as a regular, finite grid and a discrete state is associated with each grid element (cell).

central place forager — Animal that, after traveling in search for food, always returns to a central place.

central processing unit (CPU) — Part of a computer that performs the arithmetic and logic.

Charles' Law — $PV = nRT$, where P is pressure, V is volume, T is temperature in kelvin, n is the number of moles, and R is the constant 0.0832 l atm/(mol K).

child — In radioactivity, substance formed by radioactive decay from another substance.

circulatory system — System of interconnected spaces and tubes that transport fluids in multicellular animals.

clustering — Partitioning data into subsets, or clusters, so that the elements of a cluster have some common trait, such as proximity.

CMI — Cell-mediated immunity.

coalescence — Merging of rain droplets to form larger droplets.

coalescence efficiency — Product of the collision and coalescence efficiencies. The portion of collisions that results in the droplets of rain sticking together.

coarse granularity — Machine with few processors, each executing many instructions simultaneously, so that the ratio of computation time to communication time is large.

coenzymes — Organic cofactors that associate with enzymes and help them catalyze.

cognitive map — Series of markers forming geocentric references that are stored in an animal's memory as a neural representation of the animal's environment.

collision efficiency — In clouds, $E = d^2/(r_1 + r_2)^2$, where d is the distance between the center lines of a drop and a droplet, where r_1 and r_2 are the radii of the droplets.

common logarithm — Logarithm to the base 10, usually written log n; log $n = m$ if and only if 10^m is n.

community — All the organisms living in an area.

competition — Struggle between individuals of a population or between species for the same limited resource.

componentwise — Vector operation performed component by component, or coordinate by coordinate.

computational science — Emerging interdisciplinary field that is at the intersection of the sciences, computer science, and mathematics.

computer simulation — Having a computer program imitate reality in order to study situations and make decisions.

concurrent processing — Having associated, multiple CPUs working concurrently, or simultaneously, on the same or different problems.

condensation — Deposition of vapor on particles (condensation nuclei) to form droplets.

condensation nuclei — Particles upon which water vapor is deposited to form rain droplets.

continuous distribution — Probability distribution with continuous values.

continuous model — Model in which time changes continuously as opposed to discretely.

contractility — Ability to shorten, as in heart muscles.

controlled drug delivery — Combining a polymer with a medicine for release of drug into the body in predetermined manners.

convective current — Current within a medium caused by a difference in temperature.

core — Component of a virus that contains its genetic material (DNA or RNA) and various proteins.

cost function — Function of quantity that returns the total cost in producing the items.

coulomb — Unit of electric charge.

CPU — Central processing unit.

cumulonimbus — Tall, anvil-shaped clouds.

cumulus clouds — Light, fluffy clouds with flat bases that do not give rise to any rain.

current — Rate of change of charge with respect to time having basic unit of measure of Ampere.

Dalton's Law — Partial pressure of a gas is the product of the fraction of the gas in the mixture and the total pressure of all gases, excluding water vapor.

data partitioning — In master-slave parallel processing, technique where the master processor splits data into nonoverlapping subsets and sends each subset to a different slave processor.

decay chain — Sequence of several radioactive decay events, producing in the end a stable product.

decimal number system — Number system using the base 10.

decompression sickness — Painful and life-threatening condition experienced by divers who return to the surface of the water too rapidly after deep-water dives; also known as the bends. This condition results from nitrogen bubbles expanding as pressure decreases during ascent.

defibrillator — Medical device that causes a predetermined amount of current to flow across the heart so that normal electrical patterns are restored.

definite integral — $\int_b^a f(t)dt = \lim_{n \to \infty}(\text{left-hand sum}) = \lim_{n \to \infty}(f(t_0)\Delta t$

$+f(t_1)\Delta t + \cdots + f(t_{n-1})\Delta t)$

and $\int_a^b f(t)dt = \lim_{n \to \infty}(\text{right-hand sum}) = \lim_{n \to \infty}(f(t_1)\Delta t +$

$f(t_2)\Delta t + \cdots + f(t_n)\Delta t)$

where $\Delta t = (b-a)/n$ and $a = t_0 < t_1 < \cdots < t_n = b$.

dehydrogenase — Class of enzymes which catalyze oxidation of substrates by removing hydrogen or electrons, often in energy-producing pathways.

dendrites — Small, tree-like structures that form during the process of solidification of crystalline structures.

dendritic cells — Special types of antigen-presenting cell (APC), located in skin and mucosal membranes that activate T-lymphocytes.

dependent variable — Variable that relies on other variables.

derivative — For a function $f(t)$, the derivative at t is $f'(t) = \lim_{h \to 0} \dfrac{f(t+h) - f(t)}{h}$, provided the limit exists; instantaneous rate of change of a function with respect to an independent variable; geometrically, the slope of the tangent line to the curve f at t.

deterministic behavior — Behavior of systems that is predictable.

deterministic model — Model that is predictable.

diastolic pressure — Pressure in the arteries as the left ventricle relaxes.

differential calculus — One of the two branches of calculus dealing with problems involving the derivative.

differential equation — Equation containing one or more derivatives.

diffusion-limited aggregation (DLA) — Simulation technique to build a dendritic structure by adding one random walking particle at a time.

discrete distribution — Distribution with discrete values.

discrete model — Model in which time changes in incremental steps.

disintegrations per minute (dpm) — A standard expression of radioactive decay.

distributed processing — Several processors, perhaps at great distances from each other, communicating via a network and working concurrently.

distribution of numbers — Description of the portion of times each possible outcome or each possible range of outcomes occurs on the average.

divide-and-conquer algorithm — Algorithm that divides a problem into subproblems of the same form, and then divides theses into subproblems of the same form, etc. The small problems are solved, and the final solution is assembled.

distributive property — $a(b + c) = ab + ac$.

DLA — Diffusion-limited aggregation.

double-precision number — Floating point number using twice as many bits in a computer representation than a single-precision number; typically contains 14 or 15 significant digits and has magnitude between 10^{-308} and 10^{308}.

downwelling Process where currents move ocean surface water to lower depths.

dpm Disintegrations per minute; a measure of radioactive decay.

dynamic model — Model which changes with time.

dynamical disease — Disease in which blood cell counts may oscillate, perhaps in an involved or chaotic manner.

e — Part of exponential notation, where *aen* represents $a \times 10^n$, a is a decimal fraction, and n is the exponent. Symbol for $2.718281 \ldots$. Base for the natural logarithm.

ecological niche — Complete role that a species plays in an ecosystem.

elastomer — Substance made up of molecules that are loosely cross linked.

electron transport system — System that removes and passes along electrons and protons from reduced coenzymes.

electronic potential — Potential energy per unit charge at a point, or the work per unit charge to bring a positive charge from infinity to the point.

embarrassingly parallel algorithm — Algorithm in which computation can be divided into many completely independent parts with virtually no communication.

empirical model — Model employing only data to predict, not explain, a system and consisting of a function that captures the trend of the data.

envelope — Outer coating made up of lipids and proteins that surrounds some virus particles.

environmental subsystem — An interdependent part of the earth's system.

enzyme — Organic catalyst in a chemical reaction for a biological system.

enzyme kinetics — Quantitative study of enzyme activity.

EPC — Euler's Predictor-Corrector Method, or Runge-Kutta 2 Method.

epinephrine (adrenalin) — Chemical messenger of the sympathetic nervous system, including the adrenal medulla, released during times of stress. A major effect is to increase heart rate.

equilibrium solution — Solution of a differential equation where the derivative is always zero; solution of a difference equation where the change of the dependent variable is always zero.

Euler's Method — Method of numeric integration that estimates $P(t)$ as $P(t - \Delta t) + P'(t - \Delta t)\, \Delta t$, where Δt is the change in t.

Euler's Predictor-Corrector Method (EPC) — Runge-Kutta 2 Method.

exploitative competition — Competition where one individual (species) reduces the availability of the resource to the other.

exponential function — Function of the general form $P(t) = P_0 a^{rt}$, where P_0, a, and r are real numbers.

exponential notation — Floating point number represented as a decimal fraction times a power of 10.

F — Farad.

FAD — Flavin adenine dinucleotide.

flavin adenine dinucleotide (FAD) — A major coenzyme involved in oxidation-reduction reactions of metabolism. By taking on two electrons and two protons it is converted to a reduced form, $FADH_2$.

fairy ring — Naturally occurring arc of mushrooms arising at the periphery of a radially spreading underground mycelium.

farad (F) — Measure of capacitance, equivalent to having a capacitor hold a charge of 1 coulomb for a potential difference of 1 volt across its conductors, or 1 farad = 1 coulomb/volt.

fine granularity — Machine with many processors, each executing relatively few instructions, so that the ratio of computation time to communication time is small.

finite difference equation — Discrete approximation to a differential equation of the form (new value) = (old value) + (change in value).

finite geometric series — $a^{n-1} + \cdots + a^2 + a^1 + a^0$ for $a \neq 1$ and positive integer n where a is the base.

Fishing Maximum Economic Yield (FMEY) — Maximum profit of fishing as it relates to economic yield.

Fishing Maximum Sustainable Yield (FMSY) — The cost of fishing effort to produce a maximum yield in the Gordon-Schaefer Fishery Production Model.

fitting data — Obtaining a function that roughly goes through a plot of data points and captures the trend of the data.

floating point number — Real number expressed with a decimal expansion and stored in a computer in a fixed number of bits.

flux — Transfer of carbon or other element from one reservoir to another.

Flynn Classification of Computer Architectures — Classification of parallel computing architectures based on the number of independent concurrent instruction streams, or processes, and data streams; contains categories SISD, MIMD, SIMD, and MISD.

FMEY — Fishing Maximum Economic Yield.

FMSY — Fishing Maximum Sustainable Yield.

free energy — Energy available for cellular work.

functional response — Predator's behavioral reaction to changes in prey density.

Fundamental Theorem of Calculus — If f is continuous (unbroken) on the interval from a to b and $f(t) = F'(t)$ is the derivative, or rate of change, of F with respect to t, then $\int_a^b f(t)dt = F(b) - F(a)$ or $\int_a^b F'(t)dt = F(b) - F(a)$.

fungus — (plural, **fungi**) Multicellular (except for yeast), spore-producing organism that depends on absorbing nutrients from its surroundings.

gametocyte — Sexual stage of the malarian parasite that circulates freely in a host's blood and that the female mosquito obtains in her blood meal. Male and female gametocytes fuse in the mosquito's gut to form an oocyst, which divides to produce sporozoites.

generalist — Predator that may use alternative prey as densities of their primary prey decline.

global warming — Gradual increase in average temperature of the earth's atmosphere and oceans.

glycogen — Branched polymer of glucose used as a storage carbohydrate by animals.

glycolysis — Initial sequence of chemical reactions of glucose oxidation that results in the production of two molecules of pyruvate, ATP, and NADH.

Gompertz differential equation — Model for predicting the growth of malignant tumors: $dN/dt = kN \ln(M/N)$, $N(0) = N_0$, where N is the number of cancer cells and k and M are constants.

Gordon-Schaefer Fishery Production Model — Model relating fishing biological and economic yields; assumes a quadratic yield function and linear cost-of-effort function.

gp 120 — Glycoprotein that protrudes from the surface of HIV and binds to CD4+ T-cells.

granularity — In a parallel computer system, the ratio of computation time to communication time, which is related to the number of processors. See "coarse granularity" and "fine granularity."

greenhouse effect — Warming of the atmosphere caused by the trapping of infrared radiation by "greenhouse gases."

greenhouse gas — Atmospheric gas that absorbs infrared radiation, preventing the radiation's loss to space.

half-life — In radioactivity, the period of time that it takes for a radioactive substance to decay to half of its original amount. In drug dosage, amount of time for a body to eliminate half of the drug.

henry (H) — Unit of measure of inductance, where one henry = V s/A (volt sec/amp).

Henry's Law — For the amount of any gas in a liquid at a particular temperature, $V_g/V_L = sP_g$, where V_g is gas volume, V_L is liquid volume, s is the solubility coefficient for the gas in that liquid, and P_g is the pressure of the gas.

HIV — Human immunodeficiency virus.

homeothermic — Term used to describe an animal that can maintain its core body temperature at a nearly constant level regardless of the environmental temperature.

Hooke's Law — Within the elastic limit of a spring, $F = -ks$, where F is the applied force, k is the spring constant, and s is the displacement (distance) from the spring's equilibrium position.

host Organism whose body supplies nourishment and shelter for another.

human immunodeficiency virus (HIV) — Type of RNA virus belonging to the retroviruses that is the causative agent of AIDS (Acquired Immune Deficiency Syndrome).

hydrosphere — All water of the earth, including bodies of water, ice, and water vapor in the atmosphere.

hypha — (plural **hyphae**) Fungal filament that is the basic morphological component of all fungi, except yeasts.

hypnozoite — Dormant life stage of some species of malarian parasites that develop from merozoites. Once out of dormancy, they may reinvade other liver cells, where they produce more merozoites.

ideal gas — Gas in which the volume of its atoms is insignificant in comparison to the total volume of the gas and in which atom interactions are negligible except for the energy and momentum exchanged during collisions.

ideal gas laws — Laws that describe the behaviors of an ideal gas.

impulse — Product of the thrust and the length of time of force application.

indefinite integral — $\int f(t)dt = f(t) + C$, where $F'(t) = f(t)$ and C is an arbitrary constant.

independent variable — Variable on which other variables depend.

Individual Fishing Quota (IFQ) — Part of the total allowable catch that must be owned by a fisher to participate in fishing operations. These quotas are properties, which may be bought and sold.

inductance — Ability, measured in henrys, of a circuit element to store energy and oppose changes in current flowing through it.

inductor — Circuit element, such as a coil of wire, that dampens sudden changes in current.

infected — Individual within a population that has a disease and can spread it to others.

initial condition — Value of the dependent variable when the independent variable is zero.

instantaneous velocity — For position $s(t)$ at time t and $t = a$, the limit of the average velocity from $t = a$ to $t = a + h$ as h approaches 0, or $s'(t) = \lim_{h \to 0} \dfrac{s(t + h) - s(t)}{h}$, provided the limit exists.

integral calculus — One of the two branches of calculus dealing with problems involving the integral.

interference competition — Direct interaction between individuals (species), where one interferes with or denies access to a resource.

interpolation — Computing intermediate data values between existing data values.

isolation — Separation of a person who has a communicable disease (e.g., a SARS patient) from those who are healthy and susceptible.

keystone predator — Dominant predator; predator that has a major influence on the community structure.

kinetic friction — Force that tends to slow a body in motion.

Kirchhoff's Current Law — Sum of the currents into a junction equals the sum of the currents out of that junction.

Kirchhoff's Voltage Law — In a closed loop, the sum of the changes in voltage is zero.

Krebs' Cycle — Pathway in energy metabolism in which electrons are removed from pyruvate and placed onto oxidized coenzymes.

lactate fermentation — Following glycolysis, the conversion of pyruvate into lactate, which reoxidizes coenzyme NADH to NAD^+ for glycolysis.

large-scale navigation — Guided movement of an animal over long distances.

launching rectangle Surrounding a developing simulated dendritic structure in diffusion-limited aggregation, a rectangle from which new particles are released.

left-hand sum — $f(t_0)\Delta t + f(t_1)\Delta t + \cdots + f(t_{n-1})\Delta t$, where f is continuous (unbroken) function for $a = t_0 < t_1 < \cdots < t_n = b$ and $\Delta t = (b - a)/n$.

limit — Generally, number approached by $f(x)$ as x approaches some number c.

linear combination — Linear combination of $x_1, x_2, \ldots$, and x_n is the sum $a_1 x_1 + a_2 x_2 + \cdots + a_n x_n$, where $a_1, a_2, \ldots$, and a_n are constants.

linear congruential method — Method to generate pseudorandom integers from 0 up to, but not including, the *modulus* using $r_0 = seed$ and $r_n = (multiplier \times r_{n-1} + increment)$ mod *modulus*, for $n > 0$, where *seed*, *modulus*, and *multiplier* are positive integers and *increment* is a nonnegative integer.

linear function — Function, whose graph is a nonvertical straight line, that has the form $y = mx + b$, where m is the slope and b is the y-intercept.

linear least-squares regression — Line that "best captures" the trend of the data, $(x_1, y_1), (x_2, y_2), \ldots, (x_n, y_n)$; Line $y = mx + b$, where

$$b = \frac{\sum x_i^2 \sum y_i - \sum x_i y_i \sum x_i}{n \sum x_i^2 - (\sum x_i)^2} \quad \text{and} \quad m = \frac{n \sum x_i y_i - \sum x_i \sum y_i}{n \sum x_i^2 - (\sum x_i)^2}.$$

linear regression — linear least-squares regression.

lipophilic — Term used to describe a substance that can combine with or dissolve in fat (lipid).

lithosphere — Outer solid part of the earth, including the crust and topmost mantle.

loading dose — Initial dosage of a drug that is much higher than the maintenance dosage.

logarithm — Logarithm to the base b of n, written $\log_b n$, is m if and only if b^m is n.

logarithmic function — Function, $\log_b(n) = m$, that is the inverse of the exponential function $g(x) = b^x$, so that $b^m = n$.

logical operator — Symbol (such as *AND*, *OR*, and *NOT*) used to combine or negate expressions that are true or false.

logistic function — Function to model population, $P(t) = \dfrac{MP_0}{(M - P_0)e^{-rt} + P_0}$,

where M is the carrying capacity, P_0 is the initial population, r is the continuous growth rate, and t is time.

loop — Segment of an algorithm that is executed repeatedly.

lymphocyte — White blood cell, such as a T-cell, that mediates the immune response.

macrophage — Specialized white blood cell that ingests bacteria and foreign substances and processes them, often displaying products on its surface for T-cells.

magnitude — 10 to the power when the number is expressed in normalized exponential notation.

mantissa — Significand.

marginal cost — Instantaneous rate of change of the cost with respect to quantity.

marginal revenue — Instantaneous rate of change of revenue with respect to quantity.

master — In high performance computing, a processor that commands all other processors, the slaves.

Maximum Therapeutic Concentration (MTC) — Largest amount of a drug that is helpful without having dangerous or intolerable side effects.

mean arterial pressure — Average pressure during an aortic pulse cycle.

MEC Minimum Effective Concentration.

merozoite — Life stage of malarian parasite produced asexually from sporozoites or other merozoites, released into the blood, where it may infect other host cells.

message-passing multiprocessor — System in which the processors communicate through message passing.

methylation — Chemical addition of a methyl group (CH_3) to another atom or molecule.

methylmercury — Organic form of mercury, synthesized from metallic or elemental mercury by sulfate-reducing bacteria in sediments. Aquatic organisms absorb this form of mercury, which tends to accumulate in the top components of a food chain (fish).

Michaelis-Menten constant — Measure of the affinity of an enzyme for a particular substrate; the concentration of substrate, where the velocity of the reaction is equal to half of the maximum velocity (v_{max}).

Michaelis-Menten equation — Equation that describes the relationship in an enzymatic reaction between substrate concentration [S] and reaction velocity (v), as follows:

$$v = \frac{v_{max}[S]}{K_m + [S]}.$$

microstructure — During solidification of crystalline structures, complex formed by interconnections of dendrites.

MIMD — Multiple instruction streams, multiple data streams computer architecture; a category in the Flynn Classification of Computer Architectures.

Minimum Effective Concentration (MEC) — Least amount of drug that is helpful.

MISD — Multiple instruction streams, single data stream computer architecture; a category in the Flynn Classification of Computer Architectures.

mitochondria — Membrane-bound cellular compartments that are the sites of most energy production in aerobic (using oxygen) cells.

mod — Function to return the remainder in integer division.

modeling — Application of methods to analyze complex, real-world problems in order to make predictions about what might happen with various actions.

modulus — Divisor when calculating the remainder in integer division for the *mod* function.

mole — Molecular weight of a substance expressed in grams.

monomer — Chemical building block of a polymer.

monosaccharide — Carbohydrate composed of a single sugar unit, which is often made up of a five- or six-carbon skeleton.

Monte Carlo simulation — Simulation model involving an element of chance.

most significant digit — Leftmost of the significant digits of a number.

MTC Maximum Therapeutic Concentration.

multicompartment model — In modeling drug dosage, representation of the body with more than one compartment.

multiprocessor — Computer system with more than one processor.

mushroom — Fruiting body of certain types of fungi.

mycelium — Body of a fungus.

myocardial infarction — Medical term for a heart attack.

NAD⁺ (nicotinamide adenine dinucleotide) — Major coenzyme involved in oxidation-reduction reactions of metabolism. NAD^+ accepts 2 electrons and 1 proton to become NADH.

natural logarithm — Logarithm to the base e, usually written $\ln n$; $\ln n = m$ if and only if e^m is n.

N-Body Problem — Problem concerning the interactions and movements of a number of objects, or bodies, in space.

neighbor — In a cellular automaton simulation, one of the cells that surrounds a lattice site.

newton (N) — Measure of force, where $1 \text{ N} = 1 \text{ kg m/sec}^2$.

Newtonian friction — Drag on a larger object moving through a fluid which is expressed as $F = 0.5CDAv^2$, where C is a constant of proportionality (the drag coefficient) related to the shape of the object, D is the density of the fluid, A is the object's projected area in direction of movement, and v is velocity.

Newton's Law of Heating and Cooling — Rate of change of the temperature of an object with respect to time is proportional to the difference between the temperatures of the object and of its surroundings.

Newton's Second Law of Motion — Force F acting a body of mass m gives the body acceleration a according to the formula $F = ma$.

Newton's Third Law of Motion — For every action, there is an equal and opposite reaction.

nitrogen narcosis — Sudden feeling of judgment-impairing euphoria experienced by divers resulting from increased residual nitrogen in the blood.

normalized number — Number in exponential notation with the decimal point immediately preceding the first nonzero digit.

octtree — Tree with each node having branches descending to at most eight nodes; in the Barnes-Hut Algorithm, each node corresponds to a subcube in the 3D partitioning process.

ohms (Ω) — Measure of the resistance of a resistor. $1 \, \Omega = 1$ V/A (V = volts; A = amps); the resistance of a circuit in which a potential difference of 1 volt produces a current of 1 ampere.

one-compartment model — In modeling drug dosage, simplified representation of the body as one homogenous compartment, where distribution is instantaneous.

one-term model — In empirical modeling, a function with one independent variable that can capture the trend of data whose plot is always concave up or always concave down.

optic tectum — Roof of the midbrain that receives visual images. It also receives "heat images" in pit vipers.

overflow — Error condition that occurs when there are not enough bits to express a value in a computer.

oxidation — Removal of electrons or hydrogens from a molecule.

oxidative phosphorylation — Production of ATP using the proton gradient established by the electron transport system.

pacemaker — Group of cells (sinoatrial node) located in the right atrium of the heart that conducts impulses through the right and left atria, signaling these chambers to contract and pump blood into the ventricles.

parallel processing — Collection of connected processors in close physical proximity working concurrently.

parasympathetic nervous system — Subdivision of the autonomic nervous system that generally opposes the effects of the sympathetic nervous system.

Major effects of parasympathetic activity include increasing digestive activities and decreasing heart rate.

parent — In radioactivity, substance from which a second substance forms by radioactive decay.

path integration — When an animal takes all the angles and distances it experiences on a foraging trip and integrates them into a "mean home vector."

period Length of time to complete a full cycle of an oscillating function, such as a function describing the motion of a spring or pendulum.

periodic boundary conditions — Conditions so that a border cell of a cellular automaton grid is considered to have the corresponding cell on the opposite border as a neighbor. Thus, we consider the grid to wrap as a torus, so that on a grid row, the leftmost site has the rightmost cell as its west neighbor and vice versa; similarly, on a column, the topmost site has the bottommost cell as its north neighbor, and vice versa.

pheromone — Chemical produced by animals that sends specific signals to other members of the same species.

pits — Pair of surface concavities, located between the eyes and nostrils of certain snakes (pit vipers), that house extremely sensitive innervated "heat detectors" for hunting.

pixel — Picture element; dot on a computer monitor's screen.

plasma — Fluid portion of the blood, containing suspended blood cells and clotting factors.

plastic — Long-chain carbon polymer that has a high degree of cross linking.

platelet — Blood cell that helps the blood to clot.

Poiseuille's Equation — Model for blood flow in an arteriole: $Q = \pi r^4 \Delta P/(8\eta L)$, where Q is the blood flow through a vessel over time, r is the cross-sectional area of the vessel, ΔP is the pressure gradient, η is the viscosity of the blood, and L is the vessel length.

polymer — Class of chemical compounds composed of repeating chemical building blocks.

polymerization — Process of polymer synthesis.

polynomial function of degree n — Function of the form $f(x) = a_n x^n + \cdots + a_1 x + a_0$, where $a_n, \ldots, a_1$, and a_0 are real numbers and n is a nonnegative integer.

potential difference — Difference in electronic potential between two points.

precision — In mathematics, number of significant digits in an number.

predation — When one species (predator) kills and consumes another species (prey).

predator — Organism that kills and consumes another organism (prey).

pressure — Weight of matter per unit area.

prey — Organism that is killed and consumed by another organism (predator).

probabilistic behavior — Behavior of systems with an element of chance.

probabilistic model — Model that exhibits random effects.

probability function — For a discrete distribution, returns the probability of occurrence of a particular argument; for a continuous distribution, indicates the probability that a given outcome falls inside a specific range of values.

process — Task or a piece of a program that executes separately.

processor — Part of a computer that performs the arithmetic and logic.

profit — Total gain from producing and selling a given quantity of items.

provirus — Viral DNA that may remain in the chromosome passively (latent) or may activate and begin the production of new viral particles, as in HIV.

pulmonary circulation — Circulatory loop that transports blood to and from the lungs.

pyruvate — Three-carbon product of glycolysis, resulting from the splitting and oxidation of glucose.

quadratic function — Function of the form $f(x) = a_2x^2 + a_1x + a_0$, where a_2, a_1, and a_0 are real numbers.

quadtree — Tree with each node having branches descending to at most four nodes; in the Barnes-Hut Algorithm, each node corresponds to a subsquare in the 2D partitioning process.

quarantine — Limitation on freedom of movement of an individual for a period of time to prevent spread of a contagious disease to other susceptible members of a population.

radiative forcing — For global warming, increased infrared absorption and warming as a result of increased concentration of greenhouse gases.

random walk — Refers to the apparently random movement of an entity, taking single steps in apparently random directions.

rate of absorption — For gas absorption by tissues, $dP_{tissue}/dt = k(P_{lungs} - P_{tissue})$, where P_{lungs} is the partial pressure of the gas in the lungs, P_{tissue} is the partial pressure of the gas in the tissue, and $k = \ln(2)/t_{half}$, where t_{half} is the time for the tissue to absorb or release half of the partial difference of the gas.

real number — Number that can be expressed with a decimal expansion and used to measure continuous quantities.

recovered — Individual that has recovered from a disease and is immune to further infection.

recursion — Process of a function or task calling itself.

red blood cell — Blood cell used for oxygen transport between the lungs and tissues.

reference dose — Amount of a substance that may be ingested on a daily basis for a lifetime with no adverse effects on health.

relative error — Absolute error divided by the absolute value of the exact answer, provided the exact answer is not zero.

relative humidity — Ratio of the actual water vapor pressure to the vapor pressure that would occur if the air were saturated at the same ambient temperature.

reproductive number — Expected number of secondary infectious cases resulting from an average infectious case once an epidemic is in progress.

reservoir — Portion of the atmosphere, hydrosphere, lithosphere, or biosphere where various forms of an element of a biogeochemical cycle are stored.

resistance — Measurement of the ability of a resistor to reduce the flow of charges.

resistor — Device used to control current in electrical circuits by imparting resistance.

respiratory distress syndrome — Inflammatory disease of the lung, characterized by a sudden onset of edema and respiratory failure.

retrovirus — Type of enveloped, RNA virus that uses reverse transcriptase to convert its RNA into DNA in the host cell.

revenue — Total amount of income from selling a given quantity of items.

reverse transcriptase — Enzyme used by retroviruses to form a complementary DNA sequence (cDNA) from viral RNA.

right-hand sum — $f(t_1)\Delta t + f(t_2)\Delta t + \cdots + f(t_n)\Delta t$, where f is continuous (unbroken) function for $a = t_0 < t_1 < \cdots < t_n = b$ and $\Delta t = (b - a)/n$.

RLC circuit — Electrical circuit with a resistor, an inductor, and a capacitor.

root — In mathematics, unique top node in a (rooted) tree.

round — For a normalized number, to round to precision k is usually to round down if the $k + 1$ significant digit is less than 5 and to round up, otherwise.

round down — For a normalized number, to truncate the significand to the desired number of significant digits.

round-off error — Problem of not having enough bits to store an entire floating point number and approximating the result to the nearest number that can be represented.

round up — For a normalized number, to truncate the significand to the desired number of significant digits, adding one to the last of the remaining significant digits.

Runge-Kutta 2 Method — Method of numeric integration that employs a correction to each Euler's Method estimate.

Runge-Kutta 4 Method — Method of numeric integration, where each approximation is a weighted average of four estimates.

scalability — Capability of a computer system with expanded hardware resources to exhibit better performance.

schooling — Behavior of certain species of aquatic animals to swim in large groups for protection against predators or for foraging.

self-avoiding walk — A random walk that does not cross itself, that is, a walk that does not travel through the same cell twice.

scientific notation — Exponential notation where the decimal point is placed immediately after the first nonzero digit.

seed — In a method for generating pseudorandom numbers, an initial value; in diffusion-limited aggregation, an initial simulated dendritic structure.

serum — Fluid portion of the blood that remains after the blood clots.

sequential processing — Single processor working on one program.

shared memory multiprocessor — System in which two or more processors communicate through shared memory.

significand — For a floating point number expressed in exponential notation as $a \times 10^n$, the integer formed by dropping the decimal point from a.

significant digits — For a floating point number, all digits except leading zeros; for an integer, all digits except leading and trailing zeros.

simple harmonic oscillator — System which satisfies the following properties: The system oscillates around an equilibrium position. The equilibrium position is the point at which no net force exists. The restoring force is proportional to the displacement. The restoring force is in the opposite direction of the displacement. The motion is periodic. All damping effects are neglected.

simple pendulum — Pendulum where the mass for the bob is concentrated at a point, the stiff string has no mass, and friction does not exist.

SIMD — Single instruction stream, multiple data streams computer architecture; a category in the Flynn Classification of Computer Architectures.

sink — In a biogeochemical cycle, the destination of an element coming from a source.

single-precision number — Floating point number using half as many bits in a computer representation as a double-precision number; typically contains 6 or 7 significant digits and has magnitude between 10^{-38} and 10^{38}.

SIR model — Spread of disease model which considers the susceptible, infected, and recovered population groups.

SISD — Single instruction stream, single data stream computer architecture; a category in the Flynn Classification of Computer Architectures.

slave — In high performance computing, one of many processors that is commanded by one particular processor, the master.

slope — For a line, a change in y over the corresponding change in x.

solidification — Becoming solid.

source — In a biogeochemical cycle, the origin of an element as it flows to a sink.

specific impulse — Impulse per pound of burned fuel, or the quotient of impulse and the change in the fuel's weight.

speed — Magnitude of velocity, the change in position with respect to time.

spore — Reproductive cell that typically is capable of germinating into a new fungus.

sporozoite — Life stage of malarian parasite that accumulates in the salivary glands of the female mosquito. A female anopheles mosquito inoculates her human host with this stage, which invades liver cells.

squall line — Line of storms with a well-developed gust front at the leading edge.

stable solution — Solution q to a differential equation dP/dt or a difference equation ΔP, where there is an interval (a, b) containing q, such that if the initial population is in that interval then $P(t)$ is finite for all $t > 0$ and $\lim_{\Delta t \to \infty} p(t) = q$.

static model — Model that does not consider time.

stochastic behavior — Behavior of systems with an element of chance.

stochastic model — Model that exhibits random effects.

Stokes' friction — Friction F on a particle that is approximately proportional to its velocity, $F = kv$, where k is a constant of proportionality for the particular object and fluid and v is the particle's velocity.

stroke volume — Volume of blood that the left ventricle ejects upon contraction.

substrate-level phosphorylation — Synthesis of ATP from ADP and P, where the P used to make ATP has come from an organic compound that has a higher energy level than does ATP.

substrate — Molecule that is acted upon by an enzyme.

susceptible — Individual within a population that has no immunity to a disease.

sympathetic nervous system — Subdivision of the autonomic nervous system responsible for mobilizing the body's energy when stressed or aroused. Major effects of sympathetic activity include raising of blood pressure and increasing heart rate.

systemic circulation — Circulatory loop carrying blood to and from the body (except for the lungs).

systemic vascular resistance — Resistance or impediment of the blood vessels in the systemic circulation to the flow of blood.

systolic pressure — Highest pressure exerted as the left ventricle contracts.

T-cell (T-lymphocyte) — White blood cell that matures in the thymus and bears antigen-specific receptors on its surface.

T-helper cell — White blood cell (lymphocyte) that displays one of two protein structures on the surface of a human cell, allowing HIV to attach, enter, and thus infect the cell. A T-helper cell mediates both cell-mediated immunity and antibody production.

therapeutic range — Drug concentrations between the Minimum Effective Concentrations and the Maximum Therapeutic Concentration.

thrust — Mechanical force caused by the acceleration of a mass of gas and in the opposite direction to gas flow.

T-lymphocyte — T-cell.

transient equilibrium — In radioactive decay from *substanceA* to *substanceB*, when the ratio of the mass of *substanceB* to the mass of *substanceA* is almost constant.

transmission constant — Infection rate indicating the infectiousness of a disease.

tree — In mathematics, a rooted tree, often called a tree, is a connected, hierarchical structure of nodes (points), which contain information, and edges connecting them that has no cycles and has a unique node, the root, at the top.

truncate — For a normalized number, to chop off all digits of the significand beyond the desired number of significant digits.

truncation error — Error that occurs when a truncated, or finite, sum is used as an approximation for the sum of an infinite series.

two-compartment model — In modeling drug dosage, representation of the body as two chambers (e.g., gastrointestinal tract and blood).

type-1 predator functional response — Response where the predator consumes a constant proportion of prey, regardless of prey density.

type-2 predator functional response — Response where the predator consumes less as it nears satiation, which determines the upper limit on consumption.

type-3 predator functional response — Response where predation increases slowly at low prey density, increases rapidly at higher densities, but levels off at satiation, even if prey density continues to increase.

unconstrained growth — Growth with no limiting factors.

underflow — Error condition that occurs when the result of a computation is too small for a computer to represent.

uniform distribution — Discrete distribution in which all possible outcomes have an equal chance of occurring; continuous distribution in which equal-length intervals of outcomes have an equal chance of occurring.

unstable solution — Solution q to a differential equation dP/dt or a difference equation ΔP that is not stable.

upwelling — When deep currents bring cool, nutrient-rich bottom ocean water to the surface.

V — Volt.

validation — Process that establishes if the system satisfies the problem's requirements.

v_{max} — Limit of the rate of an enzymatic reaction.

vasoconstriction — Decrease in blood vessel diameter.

vasodilation — Increase in blood vessel diameter.

vector — In biology, animal that transmits a pathogen, or something that causes a disease, to another animal; in mathematics, ordered n-tuple of numbers, $(v_1, v_2, \ldots, v_n)$.

venous return — Flow of blood to the heart.

ventricle — In the human heart, one of two lower chambers that pump blood to the body and the lungs.

ventricular fibrillation — Chaotic electrical disturbance in the heart ventricle.

verification — Process which determines if a solution works correctly.

volt (V) — Unit of measure of potential difference.

voltage — For a point in an electrical circuit, the voltage difference between a point and a circuit reference point, the ground.

voltage difference — Difference in electronic potential between two points.

von Neumann neighborhood — In a two-dimensional grid of a cellular automaton simulation, a set of cells directly to the north, east, south, and west of a grid site and the site itself.

white blood cell — Blood cell that is part of the body's defense mechanism against infections.

y-intercept — Value of y when $x = 0$ in the equation of a line, $y = mx + b$; where the graph of a function crosses the y-axis.

ANSWERS TO SELECTED EXERCISES

Chapter 2

Module 2.2

1. 0.6385×10^5 **13.** Magnitude $= 10^{25}$, precision $= 6$ **16.** 4, 6
22. 0.1×10^{-5} to 0.999×10^5 **24. a.** 0.001   **b.** 0.0160%   **c.** 0.009
d. 0.144%   **27. a.** $0.36000000094 \times 10^5$ **b.** 0.36000×10^5
c. 0.94×10^{-4} **d.** 0.26×10^{-8} **29.** 6.23; 12.4; 0.625%
34. a. $-1/3 = -0.333$ **b.** -0.4161

Module 2.3

1. a. 11.625 km/hr   **b.** 11.0 km/hr   **c.** -20.9 km/hr
4. a. atoms/day   **b.** negative   **c.** atoms/day^2
7. a. For $f(t) =$ number of students with influenza on day t of the epidemic, $f'(t)$
b. 288 students

Module 2.4

1. d. underestimate $= 7.8$; overestimate $= 15.15$ **f.** $y(t) = 15t - 4.9t^2 + 11$
i. 11.48   **j.** 11.48 m

Chapter 3

Module 3.2

2. a. $15000e^{0.02(20)} = 22377.4$
b. 15024.9; 15049.8; 15074.8   **7. a.** $Q = Q_0\, e^{-0.0239016t}$

Module 3.3

1. b. $P(t) = \dfrac{e^{rt}MP_0}{M - P_0 + e^{rt}P_0}$ **2. a.** $\pi/2 + \pi n$, where n is an integer

Module 3.5

4. b. 0.099021   **5. b.** Assuming *elimination_constant* $= 0.0315$, 50.90 mg

Chapter 4

Module 4.1

1. **a.** $v = ds/dt$, $a = d^2s/dt^2 = dv/dt = -9.81$ m/sec^2, $s_0 = 11$ m. $v_0 = 15$ m/sec
   b. $v = -9.8t + 15$, $s = -4.9t^2 + 15t + 11$

Module 4.2

2. **a.** $m\dfrac{d^2s}{dt^2} = -km$

Module 4.3

1. **a.** 1   **b.** θ **c.** $F = -mg\theta$ **e.** $d^2(\theta)/dt^2 = -g\theta/l$

Module 4.4

2. $dv/dt = -I_{sp}\, g\, (dm/dt)/m + g - 0.5CDAv^2/m$, where C is the drag coefficient, A is the rocket's cross-sectional area, and the density of the atmosphere is $D = 1.225\, e^{-0.1385y}$, where y is altitude < 100 km.

Chapter 5

Module 5.2

1. $P_2 = 324$ at $t = 16$ hr

Module 5.3

1. Starting with $P_1 = 212$ at $t = 8$ hr, $P_2 = 449.44$ at $t = 16$ hr

Module 5.4

1. Starting with $P_1 = 222.24$ at $t = 8$ hr, $P_2 = 493.906$ at $t = 16$ hr

Chapter 6

Module 6.1

1. **a.** $dW/dt = aW - bWB$, $dB/dt = cB - dWB$, $W_0 = 20$, $B_0 = 15$
   b. $a = bB$ and $c = dW$, where b and d are any nonnegative real numbers

Module 6.2

2. $dS_Q/dt = qk(1 - b)I_U S/N_0 - uS_Q$ 11. a. $3^{10} = 59,049$

Module 6.3

1. $d[E]/dt = -d[ES]/dt$ 6. a. $d[S]/dt = -k_1[E][S]^n + nk_2[ES]$

Module 6.4

2. $\Delta s = (k_s * s(t - \Delta t) - k_{hs} * h(t - \Delta t) * s(t - \Delta t)) * \Delta t$
$\Delta h = (k_{sh} * s(t - \Delta t) * h(t - \Delta t) - k_{sh} * s(t - \Delta t) * h(t - \Delta t)^2/M) * \Delta t$

Module 6.5

2. $d(human\_hosts)/dt = (prob\_bit)(prob\_vector)(uninfected\_humans) - (recovery\_rate)(human\_hosts) - (malaria\_induced\_death\_rate)(human\_hosts) - (immunity\_rate)(human\_hosts)$

Chapter 9

Module 9.2

1. 6 4. a. 219, 244, 36   b. 0.627507, 0.69914, 0.103152
10. $20.0 * rand + 6.0$ 14. $int(21.0 * rand + 6.0)$

Chapter 11

Module 11.2

1. Have two constants for burning, such as *BURNING1* and *BURNING2*. Replace the rule for *BURNING* with two rules: If the site is *BURNING1*, then return *BURNING2*. If the site is *BURNING2*, then return *EMPTY*.
4. *spread* has 9 parameters: *site, N, NE, E, SE, S, SW, W, NW*. The rule for determining if a tree will burn at the next time step begins as follows: if *site* is *TREE* and (*N, NE, E, SE, S, SW, W,* or *NW* is *BURNING*)

Module 11.3

5. a. The ant goes back and forth between its first cell and the cell it initially selected at random from its neighbors, none of which had any chemical at the start of the simulation. As the ant leaves a cell, the amount of chemical

increases by 1. Thus, after the first time step, the cell from which it just came has the maximum amount of chemical in the neighborhood.

Chapter 12

Module 12.1

3. $0.25 \log n$

Module 12.2

1. Algorithm for Processor i in Calculation of Scalar Product av: return av_i

7. The divide phase is identical to that for the divide-and-conquer algorithm for adding numbers. The conquer phase is the same except that each processor returns the number of occurrences of a particular element in its subarray, which the receiving processor adds to its number of occurrences.

9. a. $r_n = (81 \, r_{n-1}) \, mod \, 349$, for $n > 0$

INDEX

←(left-facing arrow), 78
ΔP, 88
$\Delta population$, 74
Δt, 39–40, 74

A (ampere or amp), 289, 293
absolute error, 19–20
absorbing boundary condition, 411
acceleration due to gravity, 115
acceleration, 48, 114–116
Acquired Immune Deficiency Syndrome (AIDS), 505
aerobic respiration, 301; and ATP synthase, 301; and electron transport systems, 301; and Krebs' Cycle, 301; and mitochondria, 301; and NADH, 301; and oxidative phosphorylation, 301
AIDS (Acquired Immune Deficiency Syndrome), 505
air: density, 118; pressure at sea level, 265
alkaptonuria, 213
ambient pressure, 268
amp (A), 289, 293
ampere (A), 289, 293
amplitude, 333
analytical solution, 79–82
anchor, 144
angular acceleration, 141–142
Anopheles mosquitoes, 238
ant movement simulation, 423–434; algorithm, 428; constants, 424; modeling, 423–429; neighbors for walk, 427; and pheromones, 423; and self-organizing, 423; and sensing, 425; solving model, 428; and verification and

interpretation, 429; and visualization, 429; and walking, 425–428
anthropogenic, 279, 304
antiderivative, 62–63
apical growth, 494
aposematic coloration, 224
area, Monte Carlo simulation of, 367–373; algorithm, 371; implementation of, 371; and measure of quality, 370–371; throwing darts for, 368–370
area-restricted search, 485
Argentine ants, 195–196
arithmetic errors, 27
array, 392, 393
arrow, left-facing (←), 78
aspirin, 90–101, 108–109
assignment: operator, 22; statement, 21–22
associative properties, 27
atm (atmosphere), 265
atmosphere (atm), 265
atmosphere, 274
ATP, 300; synthase, 301
atria, 283
autonomic nervous system, 284

Barnes-Hut Algorithm, 462–465
base 10, 18
base 2, 18
bends, 271
Berkeley Madonna, 15; diagram, 75
binary, 18
biological aggregations, 522
biomagnification, 303
biosphere, 274
bits, 18
blood cell populations, 259–264; destruction coefficient of, 260; destruction

rate of, 260; Lasota production function of, 261–262; Mackey and Glass production function of, 263; model of, 260–262; model parameters of, 260–262
blood flow, 285–286; mean velocity of, 286; Poiseulle's Equation of, 286; pressure gradient of, 286
blood pressure, 284; and cardiac output, 284; and diastolic pressure, 284; and heart rate, 284; and mean arterial pressure, 284; and stroke volume, 284; and systemic vascular resistance, 284; and systolic pressure, 284
blood vessel: cross-sectional area of, 286; length of, 286
blood: flow, 285–286; nitrogen in, 269; pressure, 284; viscosity of, 286; volume of, 269
bob, 140
body mass index, 305
Box-Muller-Gauss Method, 380–381, 387
Boyle's Law, 267–268
Briggs-Haldane model, 220
Brownian Motion, 393
bungee jump, 135–137
buoyancy, 138
Burk, Dean, 216

C (coulomb), 288, 293
calculus, 36
Candida albicans, 504
capacitance, 291, 293; capacitor, 291; and farads, 291
capacitor, 291
capsid, 505

carbohydrates, 299; and metabolism, 299–302

carbon cycle, 274–277; and downwelling, 275; and flux, 275; and hydrocarbons, 275; and photosynthesis, 274; and reservoirs, 274; and sink, 275; and upwelling, 275

carbon dating, 82–84

carbon dioxide (CO_2), 274, 279

cardiac output, 284

cardiovascular system, 283–287

carrying capacity, 87–88

CD4 transmembrane receptors, 506

CD4+ T-lymphocytes, 505, 506

cell initiation algorithm, 408

cell values in fire simulation, 407

cell, 392, 393

cell-mediated immunity, 505–506; and CD4 transmembrane receptors, 506; and CD4+ T-lymphocytes, 505, 506; and dendritic cells, 506; and gp120, 506; and macrophages, 506

cellular automaton simulations, 15, 392, 393; array, 392, 393; Brownian Motion of, 393; cell of, 392, 393, 407; and cell values, 407; grid of, 392, 393, 407; lattice of, 392, 393, 407; and neighbor, 409; periodic boundary conditions, 411–414; rules of, 392, 393; site of, 392, 393; and von Neumann neighborhood, 409

Celsius temperature, 268, 269

central place foragers, 483

central processing unit (CPU), 439

chain, radioactive, 256

chance, 359; Monte Carlo simulation of, 359; and random numbers, 359

change in population, 74

change in time, 39

chaparral, 407

Charles' Law, 268–269

child in radioactive chain, 256

circulation, 283

classification of computer architectures, 443; and Flynn Classification, 443

clouds, 516–524; and coalescence, 517; and coalescence efficiency, 519; and collection efficiency, 519; collector drop of, 519; and collision efficiency, 519; and condensation, 518; and condensation nuclei, 518; and critical distance, 519; cumulonimbus, 517; cumulus, 516; and deposition, 518; and raindrops, 519

cluster, 462

coalescence, 517; efficiency, 519; and relative humidity, 517

coarse granularity, 444

coefficient of drag, 118

coenzymes, 300

cognitive map, 484; reader, 484

collection efficiency, 519

collector drop, 519

collision efficiency, 519

common logarithm, 328–329; and log n, 328

community, 191

competition, 191–197; exploitive, 192, 196; interference, 192, 196; interspecific, 192; intraspecific, 191

component-wise vector operation, 448–449

computational science, 3–5

computational tools, 15; *Excel*, 15; *Maple*, 15; *Mathematica*, 15; *MATLAB*, 15

computer simulation, 358

concentration, 99

concurrent processing,

437–447, 439; architecture, 443; and central processing unit, 439; and metrics, 443–445; and multiprocessor, 441–443; and processor, 439; types of, 440–441

condensation, 518; nuclei, 518

conductor, 289

constrained growth, 87–96

continuous distribution, 376, 377; and probability density function, 376, 377

continuous growth rate, 74

continuous model, 7

contractility, 284

convective currents, 480

core, 505

coronavirus, 198

corrector, 170–173

cosine, 331, 333–334

cost, 309; fixed, 309; function, 309; marginal, 312–313

coulomb (C), 288, 293

critical distance, 519

Cultural Revolution in China, 227

cumulonimbus, 517; coalescence, 517; squall lines, 517

cumulus clouds, 516

current, 288–289, 293; direction of, 289

cytomegalovirus, 504

d^2y/dt^2, 48

Dalton's Law, 266–267

damped spring, 134–135

data errors, 17

data partitioning, 449–452; master, 450; use of partitioning, 450; slaves, 450

decay constant, 256

decimal, 18

decompression sickness, 271

defibrillators, 288; circuit for, 292–293

defininte integral, 60–61

dehydrogenases, 299

dendrites, 479

dendritic cells, 506
density function, 376, 377
density: of air, 118; of atmos-
    phere, 154; of water, 118
dependent variable, 8
deposition, 518
derivative, 37, 41–42, 46
deterministic behavior, 7
deterministic model, 7
diagram, in *Berkeley
    Madonna, STELLA, Ven-
    sim*, or text's format, 75
diastolic pressure, 284
difference equation, 74–75
differential calculus, 36
differential equation, 47–48,
    64, 74; for enzyme kinet-
    ics, 217–218
differentiation, 53
diffusion model, 484, 490;
    rate parameter, 484
diffusion-limited aggregation
    (DLA), 480
diffusion rate parameter,
    484
Dilantin, 101–105, 107–108
directly proportional, 73
discrete distribution,
    376–380; probability den-
    sity function of, 376, 377
discrete model, 7
disease, spread of, 501–503;
    and infecteds, 501; and re-
    covereds, 501; and the SIR
    Model, 501; and suscepti-
    bles, 501
disintegration constant, 256
disk equation, 233
distributed processing, 440
distribution, 374–388; Box-
    Muller-Gauss Method,
    380–381, 387; continuous,
    376; discrete, 377–380; ex-
    ponential, 382–384;
    Gaussian, 380; Maximum
    Method, 386, 387; normal,
    380–381; rejection
    method, 384–385, 387;
    Root Method, 386, 387;
    statistical, 374–377; uni-
    form, 375
distributive property, 27

divide-and-conquer algo-
    rithms, 452–455
DLA (diffusion-limited ag-
    gregation), 480
double-precision number, 19
downwelling, 275
dP/dt, 73–74, 88
drag, 118; coefficient of, 118
drug delivery, controlled,
    474–475
drug dosage, 98–110; half-
    life, 99; and maximum
    therapeutic concentration,
    99; and minimum effective
    concentration, 99; and
    minimum toxic concentra-
    tion, 99; and one-
    compartment models,
    99–103; and therapeutic
    range, 99; and two-
    compartment model,
    106–107
drum, 144
dy/dt, 41
dynamic model, 7; continu-
    ous, 7; discrete, 7
dynamic systems, 71
dynamical diseases, 259

ecological niche, 191
elastomers, 474
electric charge, 289, 293
electrical circuits, 288–298
electron transport system,
    301; and ATP synthase,
    301; and oxidative phos-
    phorylation, 301
electronic potential, 289
embarrassingly parallel algo-
    rithm, 448–449; sum of
    two vectors, 448–449
empirical model, 335–353;
    multiterm, 349–351; NIST
    datasets, 335; one-term,
    340–348; sequence of
    transformations, 342
endothelial factors, 285
envelope, 505
environmental subsystems,
    274
enzyme concentration, 214
enzyme kinetics, 213–223;

and differential equations,
    217–218; Michaelis-
    Menten, 214–216; model,
    218–219
enzymes, 213
enzyme-substrate complex,
    215
EPC, 173
epinephrine, 286
equilibrium, 92; secular, 257;
    solution, 92; transient, 257
error propagation, 24–27
error, 17–35, 165–166, 174,
    185
escapement gear, 144
Euler's Method, 163–169;
    error, 165–166
Euler's Predictor-Corrector
    (EPC) Method, 170–175;
    algorithm, 173–174; cor-
    rector, 170–173; error, 174
evasion, 305
Excel, 15
expectation value, 484
exploitive competition, 192,
    196
exponent, 18
exponential distributions,
    382–384
exponential function,
    327–328
exponential notation, 18
exposed, 203

F (farads), 291, 293
$FADH_2$, 301
fairy rings, 492–500; con-
    stants, 496; definition, 495;
    display, 498; initialization,
    495; probabilities, 497;
    simulation, 495–498; state
    diagram, 498; updating
    rules, 497–498
falling, 113–128
farads (F), 291, 293
feedback loop, 121
fermentation, 300; lactate, 300
fine granularity, 444
finite difference equation,
    74–75
finite geometric series,
    104–106

fire simulation, 406–421; algorithm, 414, 416; applying function, 414; cell, 407; cell initiation algorithm, 408; cell values, 407; constants, 407; grid, 407; grid extended, 413; lattice, 407; neighbor, 409; periodic boundary conditions, 411–414; probabilities, 408; *spread*, 408; updating rules, 408–411; visualization, 417; von Neumann neighborhood, 409

fish schooling, 521–524; and biological aggregations, 522

fishing economics, 308–317; cost, 309; Fishing Maximum Economic Yield, 314; Fishing Maximum Sustainable Yield, 314; fixed cost, 309; Gordon-Schaefer Fishery Production Curve, 314; marginal cost, 312–313; marginal revenue, 312–313; price per item, 310; profit, 311–312; quantity, 309; revenue, 309–310

Fishing Maximum Economic Yield, 314

Fishing Maximum Sustainable Yield, 314

fitting data, 322

floating point numbers, 18; equality of, 28; fractional part, 18; mantissa, 18; significand, 18

flux, 275

Flynn Classification, 443; MIMD, 443; MISD, 443; SIMD, 443; SISD, 443

food ration, 512

foraging behavior, 483–488; and central place foragers, 483; cognitive map, 484; and large-scale navigation, 483; and path integration, 483; simulations, 484–485

force, direction of, 459

fossil fuels, 276

fractional part, 18

free energy, 300

friction, 118

fructose 1,6-bisphosphate, 300

functional response, 232

functions, 322–334

Fundamental Theorem of Calculus, 62–63

fungi, 492–495; fairy ring, 495; gills, 493; growth, 494; hypha, 493; mushroom, 493; mycelium, 495; nucleus, 495; reproduction, 494; spores, 493

Game of Life, 510; lifeforms, 510; still-lifes, 510

gametocytes, 238–239

Garrod, Archibald, 213

Gaussian distribution, 380

generalist, 233

generating function, 360

geometric series, 104–106

gills, 493

global warming, 278–282; consequences of, 279–280

glycogen, 299

glycolysis, 300; and dehydrogenases, 299; and fructose 1,6 bisphosphate, 300; and NADH, 300

Gompertz differential equation, 93, 168

Gordon-Schaefer Fishery Production Curve, 314

gp120, 506

granularity, 444

greenhouse effect, 278

greenhouse gases, 279; carbon dioxide, 279; methane, 279, 280; nitrous oxide, 279, 281

grid, 392, 393

ground of circuit, 289

growth, 75; rate, 74

H (henry), 292, 293

half-life, 83, 99

harmonic motion, 134

heart rate, 284; and auto-

nomic nervous system, 284; and epinephrine, 286; and pacemaker, 284; and vagus nerve, 284

heat diffusion, 489–490; Newton's Law of Heating and Cooling, 489; simulation, 489–490

heavy metals, 303

henry (H), 292

Henry's Law, 269–270

Hg (mercury), 303

high performance computing (HPC), 438

hill-climbing process, 485

HIV, 504–509; attacking immune system, 505–506; and *Candida albicans*, 504; and cell-mediated immunity, 505–506; and cytomegalovirus, 504; and Kaposi's sarcoma, 504; and *Pneumocystis* pneumonia, 504; retrovirus, 505; simulation, 507

Hooke's Law, 130

hormones, local, 285

host, 238

hydrocarbons, 275

hydrosphere, 274

hypha, 493

hypnozoites, 238

ideal gas laws, 266–273; Boyle's Law, 267–268; Charles' Law, 268–269; Dalton's Law, 266–267; Henry's Law, 269–270

ideal gas, 266

IFQ (Individual Fishing Quota), 308–309

implementation errors, 18

imposed voltage, 289

impulse, 151

indefinite integral, 62–63

independent variable, 8

index, 25

Individual Fishing Quota (IFQ), 308–309

inductance, 292–293; and henry, 292; and inductor, 292

inductor, 292
infecteds, 200, 501
infinite series expansion, 29
initial condition, 74
initial velocities, 214
instantaneous growth rate, 74
instantaneous rate of change, 36, 114
instantaneous velocity, 40
integer part, 363
integral calculus, 36, 53
integral, 62
integrand, 61
integration, 53; numeric, 63
interference competition, 192, 196
interpolation, 349
interspecific competition, 192
intraspecific competition, 191
isolation, 203
I_{sp} (specific impulse), 151

Kaposi's sarcoma, 504
Kelvin temperature, 268, 269
keystone predator, 234
kinetic friction, 118
Kirchhoff 's Current Law, 295–296
Kirchhoff 's Voltage Law, 293–295; and the RLC circuit, 293–294
K_m, 214
Krebs' Cycle, 301; FADH$_2$, 301; NADH, 301; substrate-level phosphorylation, 301

lactate, 300; dehydrogenase, 300; fermentation, 300
large-scale navigation, 483
Lasota production function, 261–262
lattice, 392, 393
launching rectangle, 480
least-squares fit, 338
left-facing arrow (←), 78
left-hand sum, 59, 60
Lehmer, D. J., 360
lentivirus, 505
life forms, 510
lim, 40
limit, 40

limits of integration, 61
linear combination, 337
linear congruential method, 360; generating function, 360; and Lehmer, D. J., 360; mod, 360; modulus, 360; multiplier, 361; seed, 360
linear damping, 144
linear function, 323–324
linear regression, 339–340; predictor variable, 339; response variable, 339
linear speedup, 445
Lineweaver, Haus, 216
lipophilic, 305
Lipsitch model, 203
lithosphere, 274
ln n, 329
loading dose, 108
log n, 328
logarithm: to base 10, 328–329; to base e, 329–330; common, 328–329; function, 328–330
logistic equations, 90–91
logistic function, 330–331
loop, 24–26; variable, 25
loosely coupled, 440
Lotka-Volterra Model, 225–227

Mackey and Glass production function, 263
macrophages, 506
magnitude of gravitational force between two bodies, 458; and Newton's gravitational constant, 458
magnitude, 19
malaria, 237–252; life cycle, 238; model, 239–249
Malthusian model, 73
Manhattan Project, 360
mantissa, 18
Maple, 15
master, 450
Mathematica, 15
MATLAB, 15
Maximum Method, 386, 387

maximum sustainable yield, 95
maximum therapeutic concentration, 99
mean arterial pressure, 284
mean, 370, 380
mercury (Hg), 303; budget, 304–305; evasion, 305; natural emission, 305; oxidized, 304; particulate, 304; particulate removal, 304–305; pollution, 303–307; riverine flow, 305; toxicity, 304, 306, 307
merozoites, 238–239
message-passing multiprocessor, 441, 442; scalability, 442
methane (CH$_4$), 279, 280
methylation, 303
methylmercury, 303, 304, 306; lipophilic, 305; reference doses, 306; toxicity, 306, 307
metrics, 443–445; granularity, 444; ratio of computation to communication, 444; speedup factor, 444–445
Michaelis-Menten, 214–216; constant, 214; equation, 214
microstructures, 480
MIMD, 443
minimum effective concentration, 99
minimum toxic concentration, 99
mitochondria, 301
mod function, 360
model classifications, 7
modeling, 6–11, 238–249, 358, 423–429; analysis, 8, 238–239, 423; assumptions, 8, 240–241, 424; data, 8, 239–240, 423–424; errors, 17–18; equations and functions, 9, 243–244, 425–428; interpretation, 9, 247–249, 429; maintenance, 10; process, 6–11;

modeling (*cont.*)
  relationships, 242–243,
    424–425; report, 9; solu-
    tion, 9, 244–247, 428;
    steps of, 8–10; submodels,
    8, 424–425; variables, 8,
    241, 424; verification, 9,
    247–249, 429; visualiza-
    tion, 429
modulus, 360
moles, number of, 269
monomer, 473
monosaccharides, 299
Monte Carlo simulation, 359,
    367–373; genesis,
    359–360; Manhattan Pro-
    ject, 360; and Ulam,
    Stanislaus, 360; and von
    Neumann, John, 359–360
most significant digit, 19
multicompartment model, 107
multiplier, 361
multiprocessor, 441–443; and
    message-passing, 441, 442;
    processes, 441, 442; shared
    memory, 441, 442
multiterm empirical model,
    349–351
mushroom, 493
mycelium, 493
myocardial infarction, 288
myogenic factors, 285

N (newton), 117
NAD$^+$, 300
NADH, 300, 301
Napier, John, 328
natural logarithm, 329–330;
    ln n, 329–330
N-Body Problem, 457–465;
    Barnes-Hut Algorithm,
    462–465; cluster, 462;
    force direction, 459; oct-
    tree, 463; quadtree, 463;
    equential algorithm,
    457–462; traversal, 464
neighbor, 409
neurohumoral factors, 285
newton (N), 117
Newtonian friction, 118
Newton's Gravitation Law,
    154

Newton's gravitational con-
    stant, 458
Newton's Law of Heating
    and Cooling, 84, 489
Newton's Second Law of
    Motion, 117, 141, 150
Newton's Third Law of
    Motion, 150
nicotinamide adenine dinu-
    cleotide, 300
nitrogen: narcosis, 271; solu-
    bility coefficient for, 269
nitrous oxide (N$_2$O), 279, 281
normal distributions,
    380–381
normalized number, 18
nucleus, 495

O(Δt), 166
octtree, 463
ohm (Ω), 290, 293
Ohm's Law, 290
on the order of Δt, 166
one-compartment model,
    99–103; of repeated doses,
    101–103; of single dose,
    99–101
one-term empirical model,
    340–348
oocyst, 238–239
overestimate, 54, 58
overflow, 22–23
oxidation, 299

pacemaker, 284
parabola, 324
parallel algorithms, 448–470;
    data partition, 449–452;
    divide-and-conquer,
    452–455; embarrassingly
    parallel, 448–449; random
    number generation,
    455–457
parallel processing, 440
parasympathetic nervous sys-
    tems, 284
parent in radioactive chain,
    256
partial vapor pressure, 517
partitioning, 450
parts per billion (ppb), 281
parts per million (ppm), 279

path integration, 483
pendulum clock, 144–145;
    anchor, 144; drum, 144;
    escapement gear, 144; pe-
    riod, 144; stop, 144;
    weight, 144
pendulum, 141–147
period, 144, 332
periodic boundary conditions,
    411–414
pharmacokinetics, 99
pheromone, 423
phosphorylation: oxidative,
    301; substrate-level, 301
photosynthesis, 274
pit vipers, 489–491; and heat
    diffusion, 489–490
pixels, 465
plasma, 99, 259
Plasmodium falciparum,
    238–239
plastics, 474
platelets, 259
Pneumocystis pneumonia,
    504
Poiseuille's Equation, 286
Polymerase Chain Reaction
    (PCR), 393
polymerization, 474
polymers, 473–482; and con-
    trolled drug delivery,
    474–475; and elastomers,
    474; and extended release
    tablet, 475; and monomers,
    473–474; and plastics, 474;
    and polymerization, 474;
    and root-mean-square dis-
    placement, 476; and self-
    avoiding walk, 475; simu-
    lation of, 475–476; and
    weight in SAW, 477
polynomial function,
    325–326; degree of,
    325–326; quadratic func-
    tion, 324–325
position, 114–116
potential difference, 289
potential, 289
ppb (parts per billion), 281
ppm (parts per million), 279
precision, 19, 20
predation, 224

predator, 224
predator functional response, 232–233; Type I, 233; Type II, 233; Type III, 233
predator-prey, 224–236, 510–515; food ration, 512; Game of Life, 510
predictor variable, 339
pressure, 265–266
pressure gradient, 286
prey, 224
price per item, 310
probabilistic behavior, 7
probabilistic model, 7
probability density function, 376
probability function, 376
processes, 441, 442
processing, types of, 440–441; distributed, 440; parallel, 440; sequential, 440
processor, 439
profit, 311; function, 311–312
proportional, 73
proteins, structural, 506
provirus, 506
pseudorandom numbers, 359–366
pulmonary circulation, 283
pyruvate, 300

quadratic function, 324–325; parabola, 324
quadtree, 463
quantity, 309
quarantine, 203

radiative forcing, 279
radioactive chain, 256–258; child, 256; model, 256; parent, 256
raindrops, 519
random numbers, 359–366, 374, 378, 379, 384; and parallel generator, 455–457; and ranges, 361–363
random walk, 392–402; Algorithm 1, 394–395, 397; Algorithm 2, 398; Algo-

rithm 3, 399; animation, 395–398; and average distance covered, 398–399; Brownian Motion, 393
rate of absorption of gas, 270–271
rate of change, 36–52, 74
ratio of computation to communication, 444
reaction rates, 214
recovereds, 200, 501
recursion, 386
red blood cells, 259
refinement, 9
Rejection Method, 384–385, 387
relative error, 19–20
relative humidity, 517; partial vapor, 517; saturation vapor pressure, 517
repeated doses, 101–106; mathematics of, 103–106; finite geometric series, 104–106
reproductive number, 207
reservoirs, 274
resistance, 290–291, 293; ohms, 290, 293; and Ohm's Law, 290; resistor, 290–291
resistor, 290–291
respiratory distress syndrome (RDS), 199
response variable, 339
retrovirus, 505; capsid, 505; core, 505; envelope, 505; and lentivirus, 505; and provirus, 506; reverse transcriptase, 506; structural proteins, 506
revenue, 309; function, 309–310; marginal, 312–313
reverse transcriptase, 506
Revised Euler's Method, 164
right-hand sum, 59, 60
RLC circuit, 293–294
rockets, 149–158
Root Method, 386, 387
root, 464
root-mean-square displacement, R_n, 476

round down, 21
round-off error, 22
round up, 21
rules, 392, 393
Runge-Kutta 2, 170–175; algorithm, 173–174; corrector, 170–173; error, 174
Runge-Kutta 4, 176–187; algorithm, 184–185; error, 185; estimate, 183–184; first estimate (∂_1), 176–177; fourth estimate (∂_4), 181–183; second estimate (∂_2), 177–179; third estimate (∂_3), 179–181

$s'(t)$, 41
$s''(t)$, 48
SARS, 198–212; model, 202–207
satiation, 233
saturation vapor pressure, 517
SAW (self-avoiding walk), 475–477
scalability, 442
scientific notation, 18
SCUBA, 265–273
second derivative, 48
secular equilibrium, 257
seed in DLA, 480
seed in random number generation, 360
SEIR Model, 203
self-avoiding walk (SAW), 475–477
self-organizing, 423
separation of variables, 79
sequential processing, 440
serum, 99
Severe Acute Respiratory Syndrome (SARS), 198–212
shared memory multiprocessor, 441, 442
sigma ($\sum_{i=1}^{n}$), 339
significand, 18
significant digits, 19
simple harmonic motion, 134
simple harmonic oscillator, 134
simple pendulum, 141
simplification, 9

simulation, 358–388; disadvantages, 359; of fairy ring, 495–498; of foraging behavior, 484–485; of heat diffusion, 489–490; of HIV, 507; modeling, 358; of polymers, 475–476; program, 78–79; when to use, 358

sine, 331–334; period, 332

single dose, 99–101

single-precision number, 19

sink, 275

SIR model, 199–202, 501; infecteds, 200; recovereds, 200; susceptibles, 200

site, 392, 393

skydiving, 122–124

slope, 44, 323; of curve, 46; of tangent line, 42–47

snow crystals, 479

solidification, 479–482; convective currents, 480; dendrites, 479; diffusion-limited aggregation, 480; launching rectangle, 480; microstructures, 480; seed in DLA, 480; snow crystals, 479

solubility coefficient for nitrogen, 269

source, 275

specific impulse (I_{sp}), 151

speed, 116

speedup factor, $S(n)$, 444–445; and linear speedup, 445

spores, 493

sporozoites, 238–239

spring, 130–135; damped, 134–135; overdamped, 138; undamped, 132–133; underdamped, 138

squall lines, 517

square root function, 326–327, 329–330

stable, 92

standard deviation, 370, 380

static model, 7

statistical distributions, 374–377

steady-state level, 260

STELLA, 15; diagram, 75

still-lifes, 510

stochastic behavior, 7

stochastic model, 7

Stokes' friction, 118

stop, 144

stroke volume, 284

substrate, 214, 299; concentration, 214

surface air consumption (SAC) rate, 268

susceptibles, 200, 501

SVR (systemic vascular resistance), 284, 285

sympathetic nervous system, 284

system dynamics models, 15

system dynamics tools, 15; Berkeley Madonna, 15; STELLA, 15; Vensim, 15

systemic circulation, 283

systemic vascular resistance (SVR), 284, 285; regulatory factors, 285

systolic pressure, 284

tangent, 331, 334

terminal speed, 121

T-helper cells, 505

therapeutic range, 99

thrust, 150; average, 155

tightly coupled, 440

time delay, 155

torr, 265

total change, 61–62

total distance traveled, 54–60; left-hand sum, 59, 60; overestimate, 54; right-hand sum, 59, 60; underestimate, 54, 57

transient equilibrium, 257

transition rules, 392, 393

transmission constant, 201

tree traversal, 464

trigonometric functions, 331–334; amplitude, 333; cosine, 331, 333–334; sine, 331–334; tangent, 331, 334

truncate, 20

truncation error, 29

two-compartment model, 106–107

Types I, II, and III predator functional response, 233

Ulam, Stanislaus, 360

unconstrained decay, 82–84; half-life, 83

unconstrained growth, 73–86; analytical solution, 79–82; simulation program, 78–79

undamped vertical spring, 132–133

underestimate, 54, 57

underflow, 23

uniform distribution, 375

unstable, 92

updating rules, 408–411

upwelling, 275

V (volt), 289, 293

vagus nerve, 284

validation, 10

vasoconstriction, 285

vasodilation, 285

vasopressin-angiotensin system, 286

vectors, 238, 448–449; sum of, 448–449

velocity, 36–41, 114–116; average, 37–39

venous return, 285

Vensim, 15; diagram, 75

ventricles, 283

ventricular fibrillation, 288

verification, 10

vessel length, 286

v_{max}, 214

volt (V), 289, 293

voltage difference, 289

voltage, 289, 293

von Neumann neighborhood, 409

von Neumann, John, 359–360

water, density of, 118

weight, 117

weight in pendulum clock, 144

weight in SAW, 477

white blood cells, 259

y-intercept, 323